Designing our future

Sustainability Science Series

This book forms part of a series on sustainability science. The other titles in this series are:

Sustainability science: A multidisciplinary approach, edited by Hiroshi Komiyama, Kazuhiko Takeuchi, Hideaki Shiroyama and Takashi Mino, ISBN 978-82-808-1180-3

Climate change and global sustainability: A holistic approach, edited by Akimasa Sumi, Nobuo Mimura and Toshihiko Masui, ISBN 978-92-808-1181-0

Establishing a resource-circulating society in Asia: Challenges and opportunities, edited by Tohru Morioka, Keisuke Hanaki and Yuichi Moriguchi, ISBN 978-92-808-1182-7

Achieving global sustainability: Policy recommendations, edited by Takamitsu Sawa, Susumu Iai and Seiji Ikkatai, ISBN 978-92-808-1184-1

Designing our future: Local perspectives on bioproduction, ecosystems and humanity

Edited by Mitsuru Osaki, Ademola K. Braimoh and Ken'ichi Nakagami

TOKYO · NEW YORK · PARIS

© United Nations University, 2011

The views expressed in this publication are those of the authors and do not necessarily reflect the views of the United Nations University.

United Nations University Press
United Nations University, 53-70, Jingumae 5-chome,
Shibuya-ku, Tokyo 150-8925, Japan
Tel: +81-3-5467-1212   Fax: +81-3-3406-7345
E-mail: sales@unu.edu   general enquiries: press@unu.edu
http://www.unu.edu

United Nations University Office at the United Nations, New York
2 United Nations Plaza, Room DC2-2062, New York, NY 10017, USA
Tel: +1-212-963-6387   Fax: +1-212-371-9454
E-mail: unuony@unu.edu

United Nations University Press is the publishing division of the United Nations University.

Cover design by Mori Design Inc., Tokyo

Printed in Hong Kong

ISBN 978-92-808-1183-4

Library of Congress Cataloging-in-Publication Data

Designing our future : local perspectives on bioproduction, ecosystems and humanity / edited by Mitsuru Osaki, Ademola Braimoh, and Ken'ichi Nakagami.
    p. cm.
  Includes bibliographical references and index.
  ISBN 978-9280811834 (pbk.)
  1. Regional planning—Case studies. 2. Sustainable development—Case studies. 3. Human ecology—Case studies. I. Osaki, Mitsuru. II. Braimoh, Ademola K. III. Nakagami, Ken'ichi.
  HT391.D3823   2011
  307.1'2—dc22   2010042436

Contents

Plates ix

Figures x

Tables xiv

Contributors xvi

Preface xxiv

1 Introduction: From global to regional sustainability 1

   1-1 Designing our future: Society in harmony with nature, from local and regional perspectives 2
       Mitsuru Osaki

   1-2 The carrying capacity of the Earth 6
       Norihito Tambo

2 Sustainable land management 37

   2-1 Sustainable agriculture practices 38
       Masakazu Komatsuzaki and Hiroyuki Ohta

2-2 Soil quality and sustainable land use 50
 Ademola K. Braimoh

2-3 The role of biological control in sustainable agriculture 63
 Anthony R. Chittenden and Yutaka Saito

2-4 The role of biochar in land and ecosystem sustainability . . . 76
 Makoto Ogawa

3 How to make food, biological and water resources sustainable . . 87

3-1 Biomass as an energy resource 88
 Fumitaka Shiotsu, Taiichiro Hattori and Shigenori Morita

3-2 Worldwide cross-country pattern of ecosystem and
 agricultural productivities. 99
 Takashi S. Kohyama and Akihiko Ito

3-3 Sustainable forest management and evaluation of carbon
 sinks. 108
 Noriyuki Kobayashi

3-4 Ocean ecosystem conservation and seafood security 130
 *Masahide Kaeriyama, Michio J. Kishi, Sei-Ichi Saitoh and
 Yasunori Sakurai*

4 Regional initiatives for self-sustaining models 147

4-1 Biomass town development and opportunities for
 integrated biomass utilization 148
 *Ken'ichi Nakagami, Kazuyuki Doi, Yoshito Yuyama and
 Hidetsugu Morimoto*

4-2 Analysis of energy, food, fertilizer and feed: Self-sufficiency
 potentials 163
 Nobuyuki Tsuji and Toshiki Sato

4-3 Biogas plants in Hokkaido: Present situation and future
 prospects. 174
 Juzo Matsuda and Shiho Ishikawa

4-4 The bioenergy village in Germany: The Lighthouse Project
 for sustainable energy production in rural areas. 184
 Marianne Karpenstein-Machan and Peter Schmuck

4-5 Samsø: The Danish renewable energy island............. 194
    Ryoh Nakakubo and Søren Hermansen

4-6 Self-sustaining models in India: Biofuels, eco-cities,
    eco-villages and urban agriculture for a low-carbon future... 207
    B. Mohan Kumar

5 Self-sustaining local and regional societies.................. 219

5-1 Forest biomass for regional energy supply in Austria 220
    Yoshiki Yamagata, Florian Kraxner and Kentaro Aoki

5-2 Transition initiatives: A grassroots movement to prepare
    local communities for life after peak oil 234
    Noriyuki Tanaka

5-3 Rebuilding *satoyama* landscapes and human–nature
    relationships.. 244
    Kazuhiko Takeuchi

5-4 Local commons in a global context..................... 254
    Makoto Inoue

5-5 Risk and resource management 267
    Hiroyuki Matsuda

6 Bridging between sustainability and governance.............. 283

6-1 The concept of sustainability governance 284
    Nobuo Kurata

6-2 Problem-structuring for local sustainability governance:
    The case of Furano 288
    Yuka Motoda, Yasuhiko Kudo, Nobuyuki Tsuji, Hironori
    Kato and Hideaki Shiroyama

6-3 Decision-making in sustainability governance............ 309
    Seiichi Kagaya

7 How to sustain social, cultural and human well-being.......... 327

7-1 Ecology, sustainability science and "knowing" systems 328
    Osamu Saito and Richard Bawden

7-2 The preservation and creation of a regional culture for
    glocal sustainability 346
    Tatsuji Sawa

7-3 Sustainability and indigenous people: A case study of the
    Ainu people 360
    Koji Yamasaki

7-4 Sustainable rural/regional development by attracting
    value-added components into rural areas 375
    Kiyoto Kurokawa

7-5 Sustainable agricultural development in Asia and study
    partnerships 387
    Takumi Kondo and Hong Park

7-6 Nature therapy 407
    Yoshifumi Miyazaki, Bum-Jin Park and Juyoung Lee

7-7 Health risks and regional sustainability 413
    Tatsuo Omura

7-8 An integrated solution to a dual crisis: An assessment of
    Green New Deal policy 423
    Fumikazu Yoshida

7-9 Towards a sustainable world through positive feedback
    loops between critical issues 433
    Motoyoshi Ikeda

Index 439

Plates

2.1.1	Photo of *shizen-nou* field in Kanto, Japan.	40
4.4.1	Photo of storage for harvested energy crops; in the background the biogas plant in Juehnde with 2 fermentation units.	188
4.5.1	Map of Samsø.	196
4.5.2	Land-based wind turbines.	199
4.5.3	Offshore wind turbines.	200
5.1.1	Photo of a former barn used as additional storage facility for wood chips (left). Photo of the biomass boiler (centre). Photo of the indoor woodchip storage beside the boiler (right).	227
5.1.2	Photo of open-air storage place for chipped wood in front of the heat plant (left). Photo of low-quality timber storage before chipping (centre left). Photo of a heat boiler (centre right). Photo of a wide-bucket excavator used for loading the wood chips (right).	229
5.1.3	Photo of the CHP biomass plant (left). Photo of the town on a clear winter day (centre left). Photo of the town during inversion of weather conditions (centre right). Photo of a wide-bucket excavator under the solar panel roof of the biomass plant (right).	231
5.3.1	Restoration of paddy field in Machida City, Tokyo.	247
5.3.2	Rice terrace in Bali, Indonesia.	251
7.3.1	A spring bow.	366
7.4.1	The OTOP star logo on Thai products.	380

Figures

1.2.1	The global ecosystem and economic subsystem..........	7
1.2.2	The population of the world.........................	10
1.2.3	Population growth in the two regions in the world.......	12
1.2.4	The relationship between personal income and GDP growth rate..	12
1.2.5	The three areas of the Earth.........................	16
1.2.6	World grain production and population, 1950–2005.......	19
1.2.7	The amount of oil discovered and production...........	22
1.2.8	Global water demand, 1900–2000.....................	25
1.2.9	Water resources per capita..........................	26
1.2.10	A new water system in a large city....................	31
2.1.1	Effects of agro-ecosystem management and associated artificial practices on biological diversity and ecosystem services, and a comparison of innovation between sustainable farming with nature and sustainable farming with low-input practices...............................	41
2.1.2	Relationships between N_2O and N_2 production as a function of soil water content.........................	46
2.2.1	Inherent soil/land quality map for the Asia Pacific.......	53
2.2.2	The distribution of land classes in the Asia Pacific.......	54
2.2.3	Biome map for the Asia Pacific	55
2.2.4	The distribution of broad biome groupings in the Asia Pacific...	56

2.2.5	The extent of urban and agricultural land use on different land classes.	56
2.3.1	Three main categories of biological control using natural enemies.	66
2.3.2	Population dynamics of *S. nanjingensis* and *T. bambusae* on *P. pubescens* in a monoculture bamboo plantation.	70
2.3.3	Stable state of two plants and one common predator in a simulated system.	71
2.3.4	Hypothetical historical relationships in native moso bamboo forests	72
3.2.1	Latitudinal pattern of country-based net primary productivity (NPP) and residence time of plant carbon and of soil carbon, estimated by Sim-CYCLE	102
3.2.2	Latitudinal pattern of country-based annual production in 2005 of cereals, roots and tubers, and meat.	103
3.2.3	Latitudinal pattern of country-based human population density, annual population increase and per capita GDP	104
3.2.4	Correlation between human population density and annual production of cereals	105
3.3.1	Process and organization of J-VER	121
3.4.1	Temporal change in biomass of wild and hatchery populations of chum salmon in the North Pacific, 1925–2001.	131
3.4.2	Prediction of the global warming effect on chum salmon in the North Pacific Ocean	132
3.4.3	Scheme for adaptive management of sustainable fisheries of Pacific salmon based on the ecosystem approach	134
3.4.4	Schematic view of NEMURO.FISH flow chart	135
3.4.5	Summary of the features of NEMURO.	136
3.4.6	Onboard PC system and display of an SST image from the onboard GIS	138
3.4.7	Trajectories of Pacific saury fishing fleets showing a direct approach to fishing grounds from the mother port off eastern Hokkaido, Japan, for the period 1 September to 30 October 2006.	139
3.4.8	Overall site selection map, masked to depths in excess of 60 metres, for Japanese scallop aquaculture potential in Funka Bay, south-western Hokkaido.	140
3.4.9	Food web of the Shiretoko World Natural Heritage area.	141
4.1.1	Basic structure of cash flow.	150
4.1.2	Diagnostic Evaluation Model for Biomass Circulation.	154
4.1.3	Biomass utilization.	156
4.1.4	Consumption of fossil energy	156

xii FIGURES

4.1.5	Greenhouse gas emissions	157
4.1.6	Economic balance	157
4.1.7	Distribution of potential biomass abundance	159
4.1.8	Biomass utilization systems classified by introduction stage	160
4.2.1	Potential energy and material flows in Furano	164
4.2.2	The case where willow is not planted	168
4.2.3	The case where willow is planted	170
4.3.1	Number of biogas plants constructed by dairy farms in each year, 1999–2007	175
4.3.2	Number of biogas plants currently being operated by dairy farms, by year of construction	175
4.3.3	Degree of satisfaction with biogas plants	176
4.3.4	Achieving Green New Deal policies with biogas plants	181
4.4.1	The technical concept of the bioenergy village Juehnde	186
4.5.1	Production of renewable energy by type	195
4.5.2	The change in energy resources on Samsø	202
4.6.1	Potential demand for diesel and gasoline in India	208
5.1.1	Annual distribution of newly established biomass district heating plants in Austria, 1980–2008	223
5.1.2	Location and capacity of biomass district heating and CHP plants in Austria, 2008	224
5.1.3	Sector-wise distribution of wood demand for energy production in Austria	225
5.3.1	Biomass recycling potential in Saku City, Nagano Prefecture	249
5.4.1	Dynamic framework of cultural ecosystems	257
5.5.1	Schematic relationship between overfishing, maximum sustainable yield, maximum sustainable ecosystem services and no-take zone	269
5.5.2	The mean trophic level of fisheries landings and the total landings of Japanese fisheries	272
5.5.3	Allowable biological catch, total allowable catch and actual catch of Japanese sardine, 1997–2003	272
5.5.4	Total allowable biological catch for several major species	273
5.5.5	Seasonal fishing-ban areas of walleye pollock fishery in Shiretoko World Natural Heritage Site	275
5.5.6	Flow diagram for the Marine Management Plan of the Shiretoko World Natural Heritage Site	277
5.5.7	Fisheries yields for 11 major exploited taxa in the Shiretoko-daiichi, Utoro and Rausu Fisheries Cooperative Associations	278
6.2.1	Creating problem-structuring figures	292

6.2.2	Problem-structuring figure of member of Agricultural Committee: Before the interview	293
6.2.3	Problem-structuring figure of member of Agricultural Committee: After the interview	294
6.3.1	Planning process based on SEA	312
6.3.2	Practical application of the river environment improvement process	314
6.3.3	Map of the Aioi-Nakajima district	315
6.3.4	Differences in willingness to pay, depending on concerns about flood control	320
6.3.5	The partial utility of each attribute	322
6.3.6	The importance of each attribute for inhabitants with prior experience of flood drill activity	322
7.1.1	Major target and characteristics of sustainability science	333
7.1.2	Ecology and sustainability science from a systems perspective	336
7.1.3	Image of the learning process	338
7.1.4	A worldview matrix	338
7.1.5	Hard and soft systems approaches	339
7.1.6	Three dimensions of knowing systems	341
7.3.1	Variations in the population of captured deer and the cost of damage to agriculture and forests: Hokkaido, 1973–2004	363
7.4.1	Three basic principles of the OVOP movement	377
7.4.2	Roadside station and its additional functions	381
7.5.1	Agriculture of the Japanese islands and the East Asian corridor	396
7.7.1	A new framework for regional sustainability from the perspective of health risks	414
7.7.2	Causes of death in developed and developing countries, 1998	415
7.7.3	Infant mortality rates in East and Southeast Asia, 2004	416
7.7.4	Relationships between infectious diseases and regional sustainable development	417
7.7.5	Average concentrations of total coliforms in various water sources in the lower Mekong watershed	420
7.7.6	Estimated annual risk of infectious diarrhoea caused by *E. coli* in drinking water in the lower Mekong watershed	421
7.9.1	Interactions between critical issues	434
7.9.2	Throughout history, human beings have tried to solve one problem but often made others worse	436
7.9.3	Examples of the short-sightedness with which we have tried to solve one problem but caused other problems	436

Tables

2.1.1	Evaluation of farming practices adopted to increase carbon stocks.	43
2.2.1	Observed and ideal distribution of biomes by land class	58
2.3.1	World's major cassava producers by area and volume	67
2.3.2	Biological control programmes for two major cassava pests	68
4.1.1	The main sources of biomass in the database	151
4.1.2	Main condition setting for economic efficiency evaluation	151
4.1.3	Evaluation criteria and index	152
4.1.4	Transformation technologies	153
4.1.5	Evaluation standard for scenario assessment	158
4.1.6	Evaluation standard for present condition	158
4.2.1	Basic units, current values in Furano and definitions of vectors	166
4.6.1	Various categories of wasteland in India, 2003	210
4.6.2	Potential demand for biodiesel in India at different blending rates and related land requirements	210
4.6.3	Ethanol demand and availability for gasoline blending in India	212
5.1.1	Summary of system facility for micro-grid heating: Private farmer's initiative, province of Lower Austria (as of 2008)	226
5.1.2	Summary of system facility for medium-scale district heating: Farmers' association, province of Styria (as of 2008)	228

5.1.3	Summary of system facility for large-scale district heating with CHP: Province of Tyrol (as of 2008)	230
5.2.1	Designated Transition Towns (as of July 2009)	237
5.2.2	The 12 steps of Transition	238
6.2.1	Outline of the interviews: Sectors, actors and date of implementation	290
6.2.2	What should be done to achieve sustainability governance in Furano? Groups of issues identified from analysing information obtained from the interviews.	303
6.3.1	Several supporting technologies in systems analysis	314
6.3.2	Outline of the questionnaire survey conducted to ascertain the basic attitudes of the inhabitants and workshop members.	316
6.3.3	Results of the FSM analysis	317
6.3.4	Alternatives discussed in the workshop.	318
6.3.5	Evaluation of alternative projects based on the Choquet integral	319
6.3.6	Parameters in the model based on the contingent valuation method	320
6.3.7	Total willingness to pay	321
6.3.8	Attributes and their levels of conjoint analysis	321
6.3.9	Results of the subgroup discussion.	323
6.3.10	Consensus of the workshop.	324
7.1.1	Definitions and core questions of sustainability science	331
7.1.2	Three research priorities of the Sustainable Biosphere Initiative.	333
7.1.3	Reviewed results of the selected manuscripts.	334
7.1.4	Paradigm shift from technocentric to holocentric	341
7.7.1	Drinking water sources in the lower Mekong watershed	419

Contributors

Kentaro Aoki is a research scholar with the International Institute for Applied Systems Analysis (IIASA) in Laxenburg, Austria.

Richard Bawden is a fellow and founding director of the Systemic Development Institute in Australia, an emeritus professor of the University of Western Sydney, an adjunct professor of Michigan State University in the USA and a visiting professor of the Open University in the UK. For nearly 20 years, he was the dean of Agriculture and Rural Development at Hawkesbury Agricultural College/University of Western Sydney and thence, for eight years, a visiting distinguished university professor in residence at Michigan State University.

Ademola K. Braimoh, professor at the Center for Sustainability Science, is also the executive director of the Global Land Project, Sapporo Nodal Office, Hokkaido University. He recently edited the book *Land Use and Soil Resources* and a special feature on land-use and ecosystems for the journal *Sustainability Science*.

Anthony R. Chittenden is an assistant professor with the Center for Sustainability Science (CENSUS), Hokkaido University, Sapporo, Japan.

Kazuyuki Doi became chief engineer at the Naigai Engineering Co., Ltd (Japan) after graduating from Kyoto University, Japan. His research has included dialogue on the quality of water on paddy fields and zero-emission projects in Japanese rural areas. His present research focuses on the quantitative diagnosis of the effects of biomass utilization.

Taiichiro Hattori is a researcher in the National Agricultural Research Center for Kyushu Okinawa Region (KONARC). He specializes in crop physiology, and his main research topic is biomass production under unfavorable environmental conditions, especially under drought stress.

Søren Hermansen is director of the Samsø Energy Academy, which has made a large contribution toward implementation of a broad variety of local renewable energy projects on Samsø Island, Denmark.

Motoyoshi Ikeda has been working on oceanography, climate science and sustainability as a professor and professor emeritus at Hokkaido University. He has taken initiative in multi-disciplinary research and education towards the sustainable world.

Makoto Inoue is professor of Global Forest Environmental Studies at The University of Tokyo. He specializes in common-pool resource governance and sociology. He led the international research project and edited the book *People and Forest-Policy and Local Reality in Southeast Asia, the Russian Far East, and Japan* (Kluwer Academic Publishers, 2003), and contributed to *The State of the Environment in Asia 2006/2007* (UNU Press, 2009).

Shiho Ishikawa is an engineer at Hokuden Sogo Sekkei Ltd, Sapporo, Japan. She has been working on biomass energy and its economical analysis. She made a new utilization system of electricity from cogeneration in biogas plants.

Akihiko Ito is a researcher at the Center for Global Environmental Research, National Institute for Environment Studies, Tsukuba, Japan and a visiting associate professor at Nagoya University. He specializes in Terrestrial Ecosystem Models that include atmosphere-land-surface interactions and biogeochemical cycles. He has developed global-scale models such as SimCYCLE and VISIT.

Masahide Kaeriyama is a professor in the Faculty of Fisheries Sciences at Hokkaido University, and is also an affiliate professor at the University of Alaska Fairbanks. He is a member of the IUCN's Salmon Specialist Group, and PICES' AICE group. He specializes in ocean ecology and ichthyology of salmonids. He has written or edited 17 books.

Seiichi Kagaya is a professor in the Division of Engineering and Policy for Cold Regional Environment, Graduate School of Engineering, Hokkaido University, Japan. He specializes in regional science, strategic infrastructure planning and environmental planning and has a PhD in environmental science from Hokkaido University. He has worked as a visiting scholar at the University of Delaware, USA, Southbank University, London, UK and the University of Brasilia, Brazil.

Marianne Karpenstein-Machan is associate professor in the Interdisciplinary Centre of Sustainable Development at the University of Goettingen and senior lecturer in the Faculty Organic Agricultural Sciences, University of Kassel.

Hironori Kato is an associate professor in the Department of Civil Engineering, The University of Tokyo. His main research concerns are transportation planning and transportation policy. He has been publishing papers on public transit planning, project evaluation, and travel-demand modelling.

Michio J. Kishi is a professor in the Faculty of Fisheries Sciences and Graduate School of Environmental Science at Hokkaido University, and a principal scientist for the Environmental Biogeochemical Cycle Research Program at JAMSTEC (Japan Agency for Marine-Earth Science and Technology). He was awarded the JOS (Oceanographic Society of Japan) prize and the Uda prize (Japanese Society of Fisheries Oceanography, JSFO, prize) for his excellent work on marine ecosystem modelling, and has been engaged in JOS and JSFO activities for the past 20 years.

Noriyuki Kobayashi is professor of environmental law at Nihon University Law School and professor of forestry at the College of Bioresource Sciences, Nihon University. He is also a member of several Japanese national and municipal government committees involved in work on forest carbon sinks. He holds a PhD in agriculture from Hokkaido University. He was previously an expert reviewer of the IPCC Forth Assessment Report (involved in WGIII and Synthesis Reports).

Takashi S. Kohyama is professor in the Faculty of Environmental Earth Science and Center for Sustainability Science, Hokkaido University, Japan. His main field of study is forest community ecology, focusing on the maintenance of biodiversity. He served as a science committee member of IGBP from 2002 to 2007.

Masakazu Komatsuzaki is associate professor at the Center for Field Science Research and Education, College of Agriculture, Ibaraki University. Through research on sustainable soil management, organic farming and cover crop biology, he has written six books.

Takumi Kondo is associate professor in the Laboratory of Agricultural Development in the Research Faculty of Agriculture, Hokkaido University, Japan.

Florian Kraxner is deputy leader of the Forestry Program with IIASA in Austria, specializing in forest policy, socio-economics and bioenergy issues. As a visiting researcher at the National Institute for Environmental Studies (NIES) in Japan, he is working on integrated forest bioenergy and rural development concepts for eco-model cities.

Yasuhiko Kudo is a postdoctoral fellow at the Center for Sustainability Science (CENSUS), Hokkaido University, Japan. He holds a PhD in agricultural economics and specializes in agricultural and regional economics. He is currently studying sustainable community improvement, focusing on agriculture.

B. Mohan Kumar has been a faculty member of Kerala Agricultural University, Thrissur, India since 1975 and currently serves as professor and associate dean of the College of Forestry there. Trained as an agronomist and forester, he is a fellow at three national scientific academies in India, and has been a visiting scientist at different academic institutions in the USA,

UK and Japan. His current research interests include sustainable land-use systems with special reference to agroforestry and the effects of forest management practices on ecosystem processes, particularly nutrient cycling, soil fertility and vegetation dynamics. He is the editor of the *Journal of Tropical Agriculture* and an associate editor of *Agroforestry Systems*.

Nobuo Kurata is professor of applied ethics and philosophy in the Graduate School of Letters at Hokkaido University, Japan. He has published articles on environmental ethics and applied ethics. He has also contributed to activities of participatory technology assessment.

Kiyoto Kurokawa is professor of the Graduate School and Research Institute of Environment and Information Sciences, Yokohama National University, Japan and was a research fellow at the Research Institute of the Japan International Cooperation Agency (JICA) until the end of June 2010. He was previously director of the Small and Medium Enterprise Development Division of JICA. He has implemented various regional development projects and initiated collaboration between JICA and universities to enhance community development.

Juyoung Lee is a Japan Society for the Promotion of Science (JSPS) research fellow at the Center for Environment, Health and Field Sciences, Chiba University, Japan. He specializes in the development and management of urban green spaces promoting the health benefits of nature.

Hiroyuki Matsuda is professor of ecological risk management at Yokohama National University, visiting professor at Kochi University, visiting scientist at the Advanced Institute for Science and Technology, a Pew Marine Conservation Fellow (a role he has held since 2007 and is the first Japanese person to hold this position) and the representative director of the Society for Conservation of Fisheries Resources and Marine Environment. He has published several textbooks on ecological risk science in wildlife, fisheries, conservation and environmental impact assessments.

Juzo Matsuda is a professor emeritus in agricultural engineering at Hokkaido University, Sapporo, Japan, with many years experience as a researcher on the utilization and treatment of biomass waste, especially livestock waste and household solid waste. His work has been influential in the extension of biogas plants in Japan.

Yoshifumi Miyazaki is professor and vice director at the Center for Environment, Health and Field Sciences, Chiba University, Japan. Through research on nature therapy, he has written or edited nine books.

Hidetsugu Morimoto is a PhD candidate at Kyoto University. He received a Master of Arts degree from Mie University. His present research focuses on the creation of a recycling rural society in Japan using biomass resources.

Shigenori Morita is professor of crop ecology at the Graduate School of Agricultural and Life Sciences, The University of Tokyo. He was previously president of the Crop Science Society of Japan and founder of the Japanese Society for Root Research.

Yuka Motoda is professor in the Department of Political Studies at the Faculty of Law, Gakushuin University. Before that she earned her associate professorship at the Sustainability Governance Project, Creative Research Initiative "Sousei", Hokkaido University, Japan, and was involved in the project's various research and educational activities for the establishment of sustainability science. She is the author of *Development Aid as Intellectual Practice: Beyond the Rise and Fall of Agendas* (University of Tokyo Press, 2007, in Japanese).

Ken'ichi Nakagami is a professor of environmental policy, College of Policy Science, Ritsumeikan University. He has a PhD in environmental engineering and has developed expertise over more than 30 years in water resources and environmental management, sustainable region development programs using biomass towns, and strategic research on mitigation and adaptation for climate change. He has received academic awards from the Japan Association for Planning Administration (2009) and the Japan Society of Research and Information on Public and Co-operative Economy (2009).

Ryoh Nakakubo is a research fellow at Obihiro University of Agriculture and Veterinary Medicine. Before that he was a guest researcher at Aarhus University, Denmark. He specializes in the management and utilization of agricultural waste.

Makoto Ogawa is professor of environment technology, Osaka Institute of Technology and president of the Japan Biochar Association (JBA), Japan

Hiroyuki Ohta is a professor at the College of Agriculture, Ibaraki University, Japan. His research interests include soil microbial ecology, environmental toxicology and sustainable soil management.

Tatsuo Omura is professor in the Department of Civil and Environmental Engineering and a member of the Education and Research of Council at Tohoku University. Before that he was vice dean of the Graduate School of Engineering.

Mitsuru Osaki is a professor in the Faculty of Agriculture, Hokkaido University, Japan. He is a vice director of the Center for Sustainability Science (CENSUS), Hokkaido University, Japan. He trained as a plant physiologist and soil scientist. He has carried out many collaborative research and teaching projects on tropical land management and the rehabilitation of tropical forests. He is interested in sustainability from food, food production, bio-energy and *satoyama* perspectives.

Bum-Jin Park is an associate professor at the Department of Environment and Forest Resources, Chungnam National University, South Korea. His research interests focus on the relationship between forest environments and human health.

Hong Park is associate professor in the Laboratory of Agricultural Cooperative in the Research Faculty of Agriculture, Hokkaido University, Japan.

Osamu Saito is an assistant professor at the Waseda Institute for Advanced Study, Waseda University, Tokyo, Japan. His work focuses on ecosystem services management, with particular interest in the interlinkages between ecological, human and social systems. He has conducted socio-ecological studies on the ecosystem services provided by agricultural rural landscapes (*satoyama*) in both Japan and other Asian countries.

Sei-Ichi Saitoh is a professor in the Faculty of Fisheries Sciences and Center of Sustainability Science at Hokkaido University and executive adviser at SpaceFish LLP. He specializes in operational fisheries oceanography and satellite oceanography. He was a co-chair for the MONITOR Technical Committee of PICES (North Pacific Marine Science Organization).

Yutaka Saito is professor of animal ecology, Department of Ecology and Systematics, Research Institute of Agriculture, Hokkaido University. He was presented with the "Applied Entomology and Zoology" award by the Japanese Society of Applied Entomology and Zoology in 1995. He wrote *Plant Mites and Sociality: Diversity and Evolution* (Tokyo: Springer, 2010).

Yasunori Sakurai is a professor at the Graduate School of Fisheries Sciences, Hokkaido University, Hakodate, Hokkaido, Japan. His research focuses on reproductive biology, strategies and stock fluctuations of gadid fish (walleye pollock, Pacific cod, and Arctic cod), cephalopods (ommastrephid and loliginid squids) related to climate change, and the biology of marine mammals (Steller sea lion and seals). He has directed a number of national research projects and programmes focussing on ecosystem-based management for sustainable fisheries in Japan. He has been a member of the Cephalopod International Advisory Council (CIAC), the Ecosystem Study of Sub-Arctic Seas (ESSAS) and the PICES Programme. He has held the role of chair of the Japanese Society of Fisheries Oceanography (JSFO) since 2009.

Toshiki Sato is a postdoctoral fellow, Center for Sustainability Science (CENSUS), Hokkaido University, Japan. His areas of interest include the recycling of organic materials in agriculture, self-sufficiency of food production and regional energy use. He holds a PhD in applied biosciences from Hiroshima Prefectural University, Japan.

Tatsuji Sawa is a former researcher at Ritsumeikan Research Center for Sustainability Science, Ritsumeikan University, Japan. He specializes in modern Japanese art and culture. His book, *History of Japanese Noise Music*, will be published in 2010.

Peter Schmuck is associate professor in the Interdisciplinary Centre of Sustainable Development at the University Goettingen and senior lecturer at the Institute of Psychology, TU Berlin.

Fumitaka Shiotsu is a postdoctoral fellow at the Graduate School of Agricultural and Life Sciences,

The University of Tokyo. His research field is crop production ecology and he has been studying low-input sustainable agriculture of material crops for bio-ethanol.

Hideaki Shiroyama is a professor of public administration at the Graduate School of Law and Politics, and Graduate School of Public Policy, The University of Tokyo.

Kazuhiko Takeuchi is a professor at the Graduate School of Agricultural and Life Sciences at The University of Tokyo. He is the deputy executive director of the Integrated Research System for Sustainability Science (IR3S), which was launched in 2005 as a research alliance, comprising of 11 Japanese universities/institutes, for the establishment of the newly emerging discipline, sustainability science. He began his posts as vice-rector of the United Nations University on 1 July 2008 and as director of the newly established United Nations University Institute for Sustainability and Peace in January 2009, while keeping his two positions at The University of Tokyo. His research focuses on creating eco-friendly environments for the harmonious coexistence of man and nature, both on local and global scales. He is the editor-in-chief of the journal *Sustainability Science*, published by Springer.

Norihito Tambo is president of the Hokkaido Research Organization. His specialty is aquatic environmental engineering and urban water and wastewater systems.

Noriyuki Tanaka is professor at the Center for Sustainability Science (CENSUS), Hokkaido University, Japan. He holds a PhD in geochemistry. He is interested in sustainability education and manages all courses run by CENSUS. These courses are open for all Hokkaido University master's students.

Nobuyuki Tsuji is employed as an associate professor (fixed term) by the Sustainability Governance Project, Center for Sustainability Science, Hokkaido University, Japan. He has a PhD in mathematical ecology. He is currently focusing his studies on regional sustainability by mathematical analysis.

Yoshiki Yamagata works for the Center for Global Environmental Research at NIES, Japan. His current research topics include: climate change risk assessment, urban and regional carbon management and climate regime analysis. He is especially interested in land-use scenario analysis for climate change mitigation and adaptation.

Koji Yamasaki is assistant professor at the Center for Ainu and Indigenous Studies, Hokkaido University. He specializes in cultural anthropology and museum studies. His research, collaborated with the Ainu people, focuses on a modern meaning and use of museum materials.

Fumikazu Yoshida is a professor at the Graduate School of Economics, Hokkaido University, Japan. He has published the following books in English: *The Economics of Waste and Pollution Management in Japan* (Tokyo: Springer Verlag, 2002), *The Cyclical Economy of Japan*

(Sapporo: Hokkaido University Press, 2005) and *Sustainable Low-Carbon Society* (Sapporo: Hokkaido University Press, 2009, co-edited with M. Ikeda).

Yoshito Yuyama is team leader of the Research Team for Biomass Recycling System at the National Institute for Rural Engineering, National Agriculture and Food Research Organization, Japan. He received a PhD in agriculture from Kyoto University, Japan. His present research focuses on biomass resource utilization systems for rural development.

Preface

This book forms part of a series on sustainability science. Sustainability science is a newly emerging academic field that seeks to understand the dynamic linkages between global, social and human systems, and to provide a holistic perspective on the concerns and issues between and within these systems. It is a problem-oriented discipline encompassing visions and methods for examining and repairing these systems and linkages.

The Integrated Research System for Sustainability Science (IR3S) was launched in 2005 at The University of Tokyo with the aim of serving as a global research and educational platform for sustainability scientists. In 2006 IR3S expanded, becoming a university network including Kyoto University, Osaka University, Hokkaido University and Ibaraki University. In addition, Tohoku University, the National Institute for Environmental Studies, Toyo University, Chiba University, Waseda University, Ritsumeikan University and the United Nations University joined as associate members. Since the establishment of the IR3S network, member universities have launched sustainability science programmes at their institutions and collaborated on related research projects. The results of these projects have been published in prestigious research journals and presented at various academic, governmental and social meetings.

The *Sustainability Science* book series is based on the results of IR3S members' joint research activities over the past five years. The series provides directions on sustainability for society. These books are expected to be of interest to graduate students, educators teaching sustainability-related courses and those keen to start up similar programmes, active members of NGOs, government officials and people working in industry. We hope this series of books will provide readers with useful information on sustainability issues and present them with novel ways of thinking and solutions to the complex problems faced by people throughout the world.

Integrated Research System for Sustainability Science

1
Introduction: From global to regional sustainability

1-1
Designing our future: Society in harmony with nature, from local and regional perspectives

Mitsuru Osaki

This volume focuses primarily on the social conceptualization and materialization of a society in which nature and humans coexist. The basis of this ideal society is assumed to be the relationship between villages or towns and their natural environment. How these villages and towns can achieve local or regional independence in the face of globalization is examined. It requires food and energy independence so that an area can become independent both materially and in terms of its energy needs. Therefore, natural renewable energy and material circulation systems are necessary. Synergy can be achieved through collaboration among the various elements. In addition, public awareness of the importance of fostering local culture, traditions and ways of thinking is required so that areas can become independent indefinitely. Furthermore, cultural independence demands that public thinking be steered away from "plunder nature" towards "coexistence with nature". Instead of the globalization of the world economy, which has led to the standardization and simplification of human values, this book proposes the establishment of a society in which nature and humans coexist through the networking of diverse communities to promote and achieve local independence. It also describes the prospects for a society created through synergistic networking to assume local independence as its basic element.

A look at groundbreaking eras of human history from the perspective of the materials used for tools reveals stages characterized by stone, bronze and iron. It can be said that civilizations making good use of such tools prospered and were able either to overwhelm or to rule others in

---

Designing our future: Local perspectives on bioproduction, ecosystems and humanity,
Osaki, Braimoh and Nakagami (eds),
United Nations University Press, 2011, ISBN 978-92-808-1183-4

each period. This model served as the foundation of civilizations up until the nineteenth century. The evolution of tools from stone to bronze to iron facilitated environmental development (i.e. improvement of the lived environment) and resulted in a dramatic increase in food production and population growth, which in turn improved environments further. In this period, farming became easier – people cleared forests to increase areas of arable land and they developed water management, as exemplified by the introduction of irrigation systems. Environmental changes to improve living standards were considered part of the progress of civilization, and nobody thought that environmental development was tantamount to environmental destruction. This trend was especially clear in the West – the Western-style application of science and technology to control the natural environment was a perspective that advocated the pre-eminence of humans, as opposed to that of nature and the environment.

Up until the twentieth century, however, damage to the Earth was superficial to the extent that its surface was merely scratched and it was not systemically harmed; it would have been difficult to imagine such superficial damage extending to cover the whole globe. In the twentieth century, explosive environmental destruction was wrought owing to the combination of mineral and energy resources needed to underpin the coal-powered steam engine and the petroleum-fired internal combustion engine. It would not be far wide of the mark to summarize the twentieth century as an era of petroleum-fuelled civilization led by Western countries. In other words, the century was characterized by large-scale wars stemming from the scramble for fossil fuel resources, led by the United States – the country that wished to control these resources. Food has always been a restricting factor for human population growth in the past, but modern agriculture fuelled by fossil energy led to overproduction and gave rise to population growth as well as to trade wars over food products and to environmental degradation. Although food is the very basis for human survival, the trading of food caused the economic system to develop in an extremely distorted manner, to put it in stark terms.

As we move into the twenty-first century, an international consensus is developing that it would be extremely difficult to continue to promote a twentieth-century-type petroleum-based civilization. However, new models have yet to emerge. Because cheap fossil fuels served as the driving force for the petroleum-based civilization, three factors render the maintenance of this civilization difficult. First, it is becoming increasingly likely that the production of cheap fossil fuels will peak between 2010 and 2025 owing to rapidly rising excavation costs in addition to dwindling reserves (Rojey, 2009). Secondly, as a result of the impact of greenhouse gas emissions on the global environment, changes in temperature and the distribution of precipitation have caused a decline in the production of crops and other plants. Thirdly, soil and groundwater quality has been reduced by the use of large-scale machinery, monoculture and the application of large amounts of fertilizers and agricultural chemicals.

For the post-petroleum world, it is necessary to step up efforts to switch to natural renewable energy sources, to promote resource circulation and to increase reliance on resources and processes that are supplied by natural ecosystems. Given the deterioration in the global environment, these goals need to be reached within 50 years. Achieving them will require entirely new notions of values, economics and lifestyles. To produce the resources necessary for human survival, a regional resource production system must be established based on concepts such as food mileage, ecological footprints, and virtual water. It is vital to establish a self-supporting society based on regional self-sufficiency in food and natural renewable energy. Without this, it will be impossible to build a sustainable foundation for society. Unfortunately, the establishment of the mechanisms of sustainable development is still ongoing. In fact, in developing countries the situation is becoming worse because of food and energy crises.

This book summarizes past modes of sustainable development, and describes future ways to achieve sustainable development in rural areas and developing countries. For example, in Japan forest currently covers 75 per cent of land, but there have been three periods of severe deforestation: ancient predation (600–850 CE), early modern (1570–1670 CE), and the first half of the twentieth century, when forests were highly vulnerable to fire and the soil became dry and poor (Totman, 1998). As a result, the protective value of forests in capturing rainfall and moderating runoff, and hence the value of afforestation in harvested areas, were clearly recognized. Two clauses in the Yōrō Code of 718 expressly prohibited cultivation in mountain areas and advocated tree planting along river banks and dams to prevent erosion and water damage to cropland, and throughout the eighth century the government promoted tree planting. Thus, the recognized benefit of nature conservation from ancient times led to awareness of the concept of coexistence between humans and nature. In Japan and the East in general, there are many concepts concerning human relationships with nature that are difficult to translate into English. One example is *satoyama*, a Japanese phrase that applies to the border zone or area that encompasses both the foot of the mountain and the nearby arable flat land. Literally, *sato* means arable land, liveable land or home land, and *yama* means mountain, indicating that the *satoyama* concept is essential when considering how humans and nature might coexist. Thus, by drawing on various Asian concepts of how to live in harmony with nature, this volume is set to develop a new model for achieving regional sustainability.

REFERENCES

Rojey, Alexandre (2009) *Energy and Climate: How to Achieve a Successful Energy Transition*. Chichester: Wiley/SCI.

Totman, Conrad (1998) *The Green Archipelago: Forestry in Pre-industrial Japan*. Athens, OH: Ohio University Press.

1-2
The carrying capacity of the Earth

Norihito Tambo

1-2-1 Introduction

Globalization

The modern era is extremely unusual in the history of humankind.[1] It is an exceptional period lasting only around 300 years, in which the world's population is expected to explode from several hundred million to 10 billion, and its future is uncertain (because the modern era is driven by finite resources of fossil fuels). It seems that today's tragedy lies in our misunderstanding that the current situation is a permanent status quo that should last for years to come.

Rapid urbanization fuelled this change because metropolises are the easiest places to accumulate material wealth. In today's global society, especially in countries in the Organisation for Economic Co-operation and Development (OECD) and in the BRICs countries (Brazil, Russia, India and China), money tends to serve as the only barometer of activity; we have gone through the stages of computerization and penetration of the monetary economy, ultimately entering a period without a real economy, in which money alone is circulated by means of financial engineering and similar methods. Now, however, this economy without substance has begun to collapse, signalling, it is believed, the end of the modern era, and this phenomenon is spreading across the globe.

We are all probably familiar with the term *globalization*. The countries that colonized and underwent colonization over a period of some 200

Designing our future: Local perspectives on bioproduction, ecosystems and humanity,
Osaki, Braimoh and Nakagami (eds),
United Nations University Press, 2011, ISBN 978-92-808-1183-4

years that lasted until the mid-twentieth century have now become mature regions or substantially growing regions. It is a reality of present-day globalization that, although mature countries experience almost no population increase and no significant domestic economic growth, they nevertheless try to seize every opportunity to reap extra profits from the high economic growth rates of countries that are currently undergoing modernization. As globalization progresses, the world is in the midst of a process in which developing nations that are rapidly modernizing are trying to catch up with developed nations that have grown and become saturated in the modern era. In a nutshell, we are on a path to a situation in which all the countries of the world have grown or will grow to the point of saturation. Individual nations engage in self-promotion to try to benefit from other countries while continuing to be interdependent with them.

The energy that drives the global ecosystem and the economic subsystem

The Earth's main driving energy for activity comes from the sun (Figure 1.2.1). Our planet is a system powered mainly by solar energy, with the amount input to the Earth being 177,000 TW. A terawatt (TW) is 10^{12} watts, which is an enormous amount of energy. All the energy that reaches the Earth eventually radiates back into space as waste heat. The

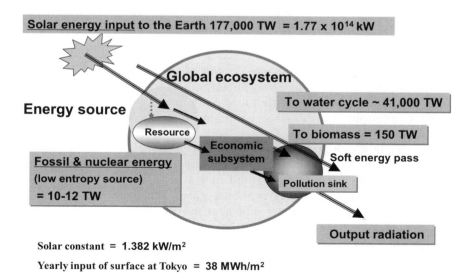

Figure 1.2.1 The global ecosystem and economic subsystem.
Source: the author.

Earth is therefore a neutral entity in terms of energy, and represents a closed system. In this set-up, materials circulate within the Earth's system but energy goes through the Earth. In addition to solar radiation, the gravitational force of the Earth–moon–sun system and geothermal energy may be relevant in some domains.

Although both the incoming (Q_{in}) and outgoing (Q_{out}) amounts of energy flowing through the earth are 177,000 TW, the quality of input energy differs significantly from that of output energy. The quality can be expressed as *entropy* (E), which is a key concept that is central to the second law of thermodynamics. One of the expressions of entropy is $E = Q/T$ (Q is the amount of energy, and T is the absolute temperature). In the Earth's system, the absolute temperature of incoming radiation is approximately 6,000 K, and that of re-radiation back to space is roughly 290 K. Therefore, the entropy of incoming solar radiation is $Q/6,000$ and that of re-radiated energy is $Q/290$ – that is, it leaves the Earth in the form of waste heat with a value 20 times as large. The high-quality solar energy (or solar energy with low entropy) that flows into the Earth helps the terrestrial system to maintain order – it would fall into a state of disorder without the solar energy input (the second law of thermodynamics). Instead, solar energy itself is consumed and discarded into space as low-quality (that is, high-entropy) waste heat.

Approximately one-third of the 177,000 TW of solar energy that reaches the Earth is reflected back into space. This degree of reflection is referred to as the albedo, and the albedo of the Earth is about 30 per cent on average. The value for ice, snow and clouds on the Earth's surface is almost 90 per cent.

Ultimately, all the energy that reaches the Earth's surface returns to space; however, in due course some of this energy is bounced back again by greenhouse gases in the atmosphere and returns to the planet's surface, thereby warming it a little more. Without greenhouse gases in the Earth's atmosphere, its surface temperature would be roughly 265 K (about −8°C) instead of the current 290 K (about 17°C). The most influential greenhouse materials are clouds and vapour, followed by carbon dioxide and methane, and then nitrogen dioxide. The Earth manages to maintain a temperate ecosystem and its status as a watery planet only because it has a blanket of greenhouse gases. The meteorological phenomena that accompany water and heat movement, such as those that cause wind, ocean currents and jet streams, consume most of the solar energy that reaches the Earth. All these phenomena involve vapour–liquid phase conversion and transportation.

Although humans act arrogantly, they do not have the ability to directly covert solar energy into organic substances like other animals. They can use only solar energy that has been trapped in organic compounds by plants. The energy of the biomass formed by plants is roughly 150 TW for the entire Earth, which is less than 0.1 per cent of the Earth's solar energy input (177,000 TW). Humans and animals have no option but to depend on this biomass for survival.

Most people in modern society are working for money to achieve a much higher living standard by engaging in the market economy in both urban industrial areas and green production areas. At the end of the twentieth century, the energy used for economic activities was about 10 TW, which is much less in absolute terms than the amount of energy that drives natural phenomena. However, with such limited energy consumption based on fossil fuels alone, global warming is resulting from carbon dioxide emissions. Experts believe that, if growth in the gross domestic product (GDP) continues on the basis of the current industrial structure and technological level, the rate of commercial energy consumption in urban industrial areas and their supporting green production areas will reach somewhere between 22 TW and 40 TW by 2050. If the energy consumption rate does increase, what kind of energy sources will be required to satisfy this ever-increasing demand? What will become of the Earth's heat balance at this point?

1-2-2 The evolution of civilization

Changes in the world's population

The world's population has increased rapidly in the modern era (Figure 1.2.2), and is now growing at its fastest rate ever. It is conceivable that humans – the crew of Spaceship Earth – are multiplying on a logistic curve that indicates a biocenotic increase (as seen in the multiplication of fruit flies in an enclosed space). We are at an inflection point where acceleration becomes the highest rate of multiplication. In retrospect, the success of modern civilization up until the twentieth century is supporting the behaviour of people today. This is the current situation of many developing nations. However, the situation is not globally uniform – success has already been localized in some areas (particularly in nations in a state of advanced modernization), and the looming prospect of saturation and stagnation has prompted many to maintain that a failure to change existing lifestyles may put the future in danger. People remain in two minds, accelerating with the right foot while braking with the left. Continuing along the current path may bring the world to a state of chaos.

The population is expected to grow to roughly 11 billion by around 2100, although some experts believe that this figure may be overstated and that it may exceed 10 billion but will then drop below that number.

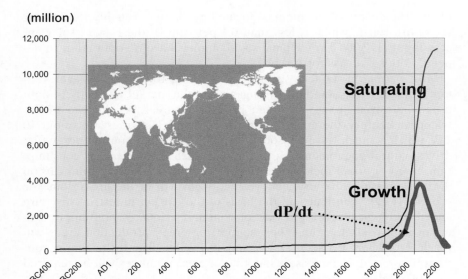

Figure 1.2.2 The population of the world.
Source: Tambo (2002).

The global population has never experienced a decrease in total numbers. In Japan, the population began declining in 2005; among other countries, only Italy and Russia are experiencing demographic decline. The population of Europe has reached saturation, and is seeking a state of equilibrium by damping oscillation. The growth of the modern era is gradually coming to an end (McEvedy and Jones, 1978; Tambo, 2002, 2009).

What, then, will become of the world's population in the future? It depends on what kind of lifestyle we choose and the future population density. Symbolically speaking, the answer lies in what we eat – will it be meat or grain? Modern livestock farming uses large amounts of grain for meat production. Therefore, grain for feeding people is greatly reduced. The total sustainable population will become much smaller if a lot of people shift their diet to meat from grain. It is thought that, if people lived only on grain-based foods, the global population could be sustainable at 14 billion (National Institute of Population and Social Security Research, 2004). Of course, in addition to the food supply, many social factors such as education, public health, social welfare, non-renewal resources, conservation of the natural environment, social regulation and self-governance also have both a qualitative and a quantitative influence in terms of the sustainable population.

It is said that we would need 3.2 earths if we all lived the lifestyle of Americans, but we have only one Earth. If we all ate and behaved like the Japanese, we would need two earths. Mathis Wackernagel and others in Canada made these calculations based on their concept of an ecological footprint (Wackernagel and Rees, 1996). Some ecological footprint estimates indicate that the world's population has already surpassed the level that the Earth can support by as much as 30 per cent, and that the carrying capacity of the Earth had already been exceeded by around 1985 (WWF, 2006).

When global modernization began in the middle of the nineteenth century, Japan's population was 30 million and that of the United States was 25 million. To ensure the sustainable survival of human beings, the synergistic effects of a declining population and decreasing consumption need to be leveraged. A significant population increase in developing countries, particularly China, the Indian subcontinent, Southeast Asia, the Near East and the Middle East and Africa, will serve as a major destabilizing factor for the future of the Earth. Referring to this may be taboo; however, this issue must be faced sooner rather than later.

World regions with population increases and those with saturated populations

Figure 1.2.3 outlines patterns in the world's population increases. The horizontal axis shows time, and the vertical axis indicates population. The zone at the bottom shows developed countries – the world powers that had colonies for decades from the eighteenth century to the twentieth century. The populations of these countries now show hardly any increase – they have become mature and saturated. The nations currently experiencing population growth are those that were once colonies and states alike. In these countries, GDP is also growing at an annualized rate of several percentage points or more. In so-called developed nations (the G7 countries in particular), not only has population growth stagnated, but GDP growth too has been quite limited. In the developed nations, labour costs are high, and people demand high-quality and expensive food, other goods and services, making it difficult for these countries to achieve effective growth on the basis of their own countries' resources and domestic markets. In an effort to compensate for this and to make money in other ways, these countries use developing nations as a stepping stone for their own growth. The development of efficient new technologies is the key to gaining priority in the globalized competition of developed countries within developing regions.

Figure 1.2.4 indicates the pattern of the relationship between personal income and GDP growth rates. The horizontal axis indicates annual GDP

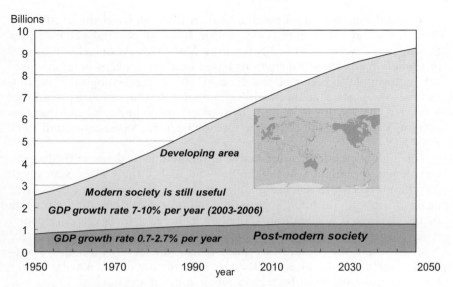

Figure 1.2.3 Population growth in the two regions in the world.
Source: Based on Population Division of the Department of Economic and Social Affairs of the United Nations Secretariat, *World Population Prospects: The 2006 Revision*, and *World Urbanization Prospects: The 2005 Revision*, <http://esa.un.org/unpp>.

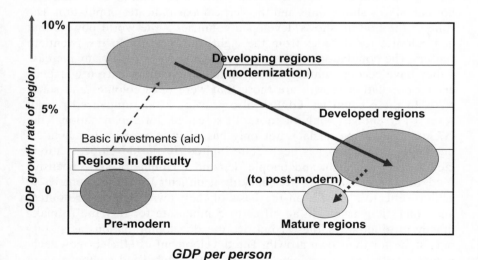

Figure 1.2.4 The relationship between personal income and GDP growth rate.
Source: the author.

12

per capita, and the vertical axis indicates the annual growth rate of total GDP. The graph's lower-left corner shows pre-modern areas – poor, low-income regions (annual GDP of about USD 100 per capita) with little prospect of growth. Such regions are incapable of providing even elementary education to all residents and they lack electricity, water and road infrastructures. They are trapped in this situation for the time being.

Once an elementary education system is put in place, however, road construction proceeds, electricity and water services are started, and countries start moving towards modernization. In due course, these countries will need to overcome much harder barriers to establish good governance, a social fabric, a public health system, etc., even if there is some outside material aid. When annual GDP per capita increases to roughly USD 1,000–2,000 and a rudimentary social infrastructure has been developed, the growth rate then increases rapidly. This is the beginning of modernization. Currently, per capita GDP in China is approximately USD 2,000 (and is over USD 3,000 in coastal areas), but it was less than USD 1,000 until recently. The growth rate rose rapidly to 10 per cent. This situation corresponds to that of the developing/high-growth regions at the top of the polygon in Figure 1.2.4. In the stage of growth when the total GDP of the developing nations becomes relatively significant to the world economy, their low currency exchange rates become barriers that need to be overcome to achieve both international and domestic social and economic equity and development. Under the occupation regime in Japan, the rate of the Japanese yen to the US dollar was compulsorily fixed at 360:1 until the early 1970s. Now it is 90:1 under the full adoption of a floating exchange rate system and Japan enjoys high living standards, as well as being a world leader with a stable and peaceful social system. Modern-day China is chasing the developed nations with a slightly biased currency exchange rate set by the Chinese government, even in a floating exchange rate system.

As people work hard, their personal income gradually increases. Japan's per capita nominal GDP in 2006 was roughly USD 35,000, making it a high-income country with much more than 10 times the per capita GDP of the 1970s, when the nation's greatest growth was recorded. The corresponding figure for Switzerland is about USD 50,000. When per capita GDP becomes this high, it tends to grow at an annualized rate of up to 2 per cent. The figure for Japan is also expected to fluctuate around this level. In the United States – a country with a history of immigration – the pattern of population increase has been slightly different from that of other developed nations. Despite the United States' ability to manipulate the dollar – a key currency – and its status as the only nation capable of producing growth on its own for many years, it seems to be approaching a condition similar to that of other advanced nations owing to the subprime mortgage crisis.

Developed regions, where modernization is mostly complete (particularly the G7 nations), cannot allow themselves to continue declining as they are; they therefore try to avoid significant decreases in their total GDP while recognizing their inability to achieve high personal growth rates. This is what Japan is experiencing right now.

In mature regions, there is a graduation from modern society to postmodern society. Mature countries will experience gradual reductions in total GDP without a significant fall in per capita GDP, and so will stay in the lower-right corner of the polygon in Figure 1.2.4. At this point, people begin efforts to create harmonious and sustainable societies for the future.

Increases in water and energy consumption and economic growth

During the twentieth century – the core of the modern era – the world's population quadrupled from 1.6 to 6 billion. In the same period, water consumption increased approximately tenfold, which meant that per capita water consumption increased by roughly a factor of 2.5.[2] Meanwhile, GDP in market prices expanded by a factor of 17 and per capita GDP more than quadrupled. These figures are spectacular because they represent the average for the entire world, including Africa, the United States and Japan. As such, the characteristics of the modern era are very simple – people have consumed enormous amounts of resources to build material wealth, and their incomes have risen in proportion to their consumption of natural resources.

Industrialization grew rapidly in this period, multiplying energy consumption by a factor of 11. This means that per capita energy consumption nearly tripled. Therefore, given the fourfold per capita GDP increase, the energy efficiency required to raise GDP went up by a factor of 1.33. This phenomenon can be taken as technological progress because levels of efficiency measured by monetary standards improved. However, because per capita lifetime energy consumption rose threefold, the lifetime physical efficiency for each person declined by one-third. Oil reserves will be depleted in the not-so-distant future as well as rare metals and phosphorus, which are indispensable for modern civilization.

How do people today see the resulting progress in terms of happiness for mankind? It seems that a small number of people aged 70 or older, who were directly involved in these changes, are really feeling the benefits of this rapid economic growth. In China today, a somewhat reckless confidence among Chinese people is evident. This attitude is understandable to the author but it is to be regretted that it has led to the global climate change issue.

The tenfold growth in water consumption during the twentieth century stemmed from huge increases in the area of irrigated farmland. More than 35 per cent of the world's food supply comes from this type of farmland. As a result of these practices, groundwater began to be depleted. For example, the Ogallala Aquifer in the breadbasket of the Great Plains, which cover much of the central United States, has now discharged half of its reserves. Thus, half the water that had been stored over a timeframe of more than 300 years was consumed in just 30 years. In China, the water in the Yellow River sometimes does not run from the riverhead through to the river mouth for more than 150 days a year (known as discontinuous flow) owing to water abstraction for irrigation along the way. In the former Soviet Union and its republics such as Kazakhstan and Uzbekistan, the water in the Aral Sea, once the world's fourth-largest freshwater lake, is close to disappearing. This situation has arisen because huge amounts of water have been used to irrigate large-scale cotton fields in the basins of the Amu Darya and Syr Darya rivers, which flow from the Himalayas. As a result, the amount of water in the Aral Sea, located in the area furthest downstream, drastically decreased to just a third in terms of the lake's area and to a mere tenth in terms of water volume. In other areas, for example China, there is serious widespread water contamination as a result of the reckless use of water as a medium of transport and/or recreation.

If sufficient resources are available, people can evaluate progress only in terms of the value gained. However, now that the Earth's raw materials and space have become insufficient, constraints on these commodities have to be considered for achieving sustainable world. This represents the arrival of the age of the global environment, with requirements that clearly distinguish the modern era from the period that now follows it.

1-2-3 Anthropogenic activities and urbanization

The United Nations defines a city as a settlement with a population of 100,000 or more. In 1950, 30 per cent of the world's population (over 2.5 billion people) lived in cities. Now, in the twenty-first century, urban dwelling accounts for roughly half the world's population of 6.6 billion (see UNFPA, 2007), and is expected to make up 70 per cent in 2050. People move to cities because they can acquire more material wealth and information there, and even the poor can survive in these areas. Suburban slums are a typical problem facing major cities in developing nations.

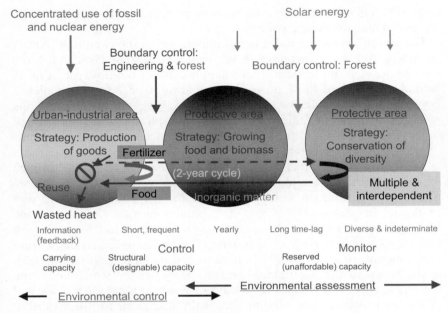

Figure 1.2.5 The three areas of the Earth.
Sources: Tambo (1986, 2002).

The Earth's space can be divided into three regimes, based on use characteristics (Figure 1.2.5). In this way, various social and environmental conditions can be easily understood. As can be seen in Figure 1.2.5, each regime has its own individual strategic goal, choice of energy as its driving force, degree of complexity in system configuration, capability for control and monitoring, and method of evaluation, etc.

One of the three divisions is the urban-industrial area – the space in which human activities are most highly concentrated and whose goal is to maximize the acquisition of material wealth efficiently. It is characterized by the prevalence of industries involved in the production and distribution of material wealth as well as those dealing in information management. The area has little biological production functionality, and is driven by the intensive use of commercially available energy such as fossil and atomic power resources. For people to survive in this area, almost all food for human consumption must be procured from elsewhere.

Another area of human activity is that of biomass production and food production, in which food and biomass are supplied for people working in the urban-industrial area. The strategic goal of this region is to maximize the production of biomass. People attempt to reap as much wheat grain as possible from each unit they sow.

Before chemical fertilizers became popular, human excrement in cities was transported and used on farmland as fertilizer, and produce was harvested before being delivered back to cities, forming a two-year cycle of organic waste, phosphorous and nitrogen. In Japan today, however, this pattern has been disrupted owing to mass food imports from abroad and the mass application of chemical fertilizers.

Another major component of the green production area is plantation forests. Trees are replaced through repeated felling and planting every 30 to 50 years at the earliest in plantation forests, and in a cycle of several decades to 100 years in mature forests. Such cultivation areas are different from farmland, where sowing and harvesting are repeated every year, but are similar in that people try to harvest and reproduce as much timber as possible from each seedling they plant. The goal of plantation forests is to artificially reproduce organic matter but, because it takes more than 30 years to harvest trees, the forests provide small, short-lived animals, such as raccoon dogs, foxes, rabbits, squirrels and birds, with artificial yet relatively stable spaces where they can continue to live for several generations. In Japan – a country with a high population density – plantation forests play roles that conform to those of the environmental conservation area (to be explained next) as *satoyama* (village-vicinity mountains). Although solar energy is the main driving energy for this area, mechanical power and additional fertilizer application have also significantly increased the productivity of biomass, giving rise to the Green Revolution in the modern era.

The other characteristic area is that of ecosystem conservation. Unlike the other two areas, human involvement is kept to the minimum necessary here, and the work of the system is left to material circulation by solar energy and the forces of nature. Although relationships with human beings are specified in the Convention on Biological Diversity, the goal of this area is to help as many biological species as possible to make the most of solar energy and water in order to form and live in this diverse ecosystem. In other words, the area aims to protect the very foundation of human survival, because humans are also an animal species and cannot survive without the blessings of nature. However, the Convention itself seems to be political in nature, because it relates to the exercising of regional sovereignty.

The penetration of human activities from the biomass production area into the conservation area appears to pose the most serious challenge to the next generation, along with population increase and the expansion of the biomass-related industry in particular. The possible disruption of nature in the tropical zone and the southern hemisphere, already evident in the developed northern hemisphere, will be the Earth's next great crisis. From the viewpoint of control, environmental impact assessment is primarily conducted in the ecosystem conservation area. The accumulation of a scientific database and in-depth recording of various phenomena, including constant monitoring/observation, are indispensable because the phenomena that occur are diverse and complicated, and it takes a long time for some responses to variations to emerge.

1-2-4 Food production and agriculture

Farmland areas and the farming population of regions

Feeding the people who earn their living in urbanized areas is undertaken by large-scale farmers who have formed partnerships with these areas.[3]

The United States has the world's highest level of agricultural productivity. At the beginning of the twentieth century, 1 American farmer could feed approximately 7 non-farmers. Today, however, it is assumed that 1 farmer can feed more than 100 non-farmers. Many years ago, Japanese farmers consumed half the food they produced and sold the remaining half. This meant that they could feed only two families, and so the farming population had to support at least 50 per cent of the total population. The ratio of farmers to the total population in China and the developing nations is close to this figure. However, under the World Trade Organization system, these countries cannot thrive economically if they try to go head to head with their counterparts in large agricultural nations such as the United States, Australia and South Africa. As a result of this inability to compete, local agriculture has collapsed.

In the United States, each farm household has more than 100 hectares of land on average; however, the farming population accounts for less than 2 per cent of the total population. Large-scale farmers in other regions, such as Oceania and South America, also feed the world's urban dwellers. In most Asian countries, the farming population represents more than 50 per cent of the total population in each country, but their farmland areas are very small. China cannot produce more than the minimum required to support its own population as urbanization progresses. This is why large-scale farming areas have linked up with urbanized industrial areas in the modern era.

Over 6 billion people inhabit the Earth. These individuals and the livestock that exist solely for the needs of human beings make up over 70 per cent of the total animal mass on the land. People produce large quantities of beef and pork to meet consumer demand, while talking about the importance of biodiversity in the environment. The extinction of

AD	Grain production (million tons)		World population (million)	(tons per person)
1950		691	2519	0.274
1955	Green Revolution	790	2756	0.265
1960		889	3021	0.294
1965		988	3335	0.296
1970		1246	3692	0.337
1975		1417	4068	0.348
1980		1588	4435	0.358
1985	Serving growth	1745	4831	0.361
1990		1902	5264	0.361
1995		1902	5674	0.335
2000		2100	6071	0.346
2005		2264	6454	0.351

Figure 1.2.6 World grain production and population, 1950–2005. *Source*: the author.

some species has led to a fuss about the risk of polar bears dying out, but the disappearance of some species is a result of the human population growing unchecked. People advocate symbiosis with nature and the conservation of biodiversity without considering the excessive multiplication of their own kind.

Figure 1.2.6 shows the relationship between world grain production and population. Around 1950, per capita daily production was just 750 grams. In the 1980s, during the so-called Green Revolution, production increased to between 850 and 900 grams per capita through the worldwide promotion of large-scale farm irrigation. This involved the use of deep-well pumps to pump huge volumes of groundwater and the construction of a number of dams. This development was also made possible through the use of large quantities of chemical fertilizers created by fixing atmospheric nitrogen produced as a result of fossil fuel consumption. It also became feasible to perform several different tasks on a large scale using oil to drive tractors. The Green Revolution boosted the supply of grain to the rapidly increasing global population by 10 per cent per person.

Beginning in the 1980s, the world population continued to rise, and reached 6.5 billion in 2005. Over the period 1980–2005, grain production increased by a factor of nearly 1.5, and agricultural production also increased to keep up with the growth in population. However, per capita daily grain production has not risen at all, and hovers around 850–900 grams. This may be because of market controls, but it is also possible that the Green Revolution is over. In other words, population growth has caught up with increased food production, and the two are now expanding in parallel.

It is believed that the worldwide ratio of food stocks in the 1970s was only 15 per cent of food production in the off-crop season of each year. As a result of the Green Revolution, this ratio had grown to over 30 per cent by the latter half of the 1980s, prompting people to believe that they could buy grain from anywhere in the world as long as they had money. However, the ratio has now reverted to 15 per cent. In light of this decrease, it could be said that human beings may fall into the Malthusian trap, as explained in *An Essay on the Principle of Population* by Thomas Robert Malthus – a trap in which food production (following an arithmetic increase) is eventually outstripped by population growth (following an exponential increase), thus causing food shortages. Grain serves as the foundation for the conversion of solar energy into food, which is necessary for all human beings to survive. In Japan, the food self-sufficiency ratio is 27 per cent at the grain level, which is extremely low compared with other countries in the world; in terms of calories, it is 39 per cent because Japanese people tend to consume high-energy food. When measured by cost, the ratio is 68 per cent because the items the Japanese eat tend to be expensive.

Changes in fish yields

It seems that Japanese people consume a large percentage of the fish in the world.[4] Around 1950, the global fish catch was only 20 million tons per annum; however, over time the figure has increased in a linear fashion. With advances in fish detection technology, the increased size and performance of fishing vessels and lower petroleum fuel prices, fishermen began to travel globally to catch fish in seas worldwide. Even so, the annual oceanic fish catch has never breached the 100 million ton mark. Fish yields peaked prior to 2000, before starting to decrease. This means that there is no longer a surplus of fish in the sea, and we should refrain from catching them.

Aquaculture is now rapidly expanding to fill the role once played by fishing. As of 2000, aquaculture produced more than 50 per cent of the amount of natural fish catches. Cultivation targets include eels, salmon, prawns and sea bream, and Kinki University in Japan has also succeeded in farming tuna.

Aquaculture is an industry that uses cheap fish and meat as feed in order to produce large, expensive fish. In terms of the hierarchy in the ecological food chain, farmed fish are apex predators. Solar energy is fixed by plants on land and by phytoplankton in the sea. Phytoplankton are eaten by zooplankton, which are in turn consumed by small fish. These small fish are then eaten by medium-sized fish, which subsequently become meals for large fish. Ultimately, the large fish are eaten by human beings. This is how the food chain works. Solar energy availability drops by one digit with every step up the food chain (meaning that it becomes one-tenth of its previous level). In aquaculture, which produces apex predators, biomass availability is extremely low. Grass consumption by cows causes solar energy availability to decrease by one digit, but the consumption of marine resources causes it to drop by a much larger factor. For example, if people can make 100 times more money by having tuna consume sardines, they prefer to farm tuna, even if solar energy availability is further reduced by one digit as a result. In short, this is an economic activity aimed at making as much money as possible. It is solely based on economic considerations, and is entirely unrelated to biological energy flows.

1-2-5 Energy issues

Relationships between human beings and fossil energy

The mass consumption of fossil energy contributed the most to the intensive anthropogenic activities of the modern era.[5] The world population increased at a rate unprecedented in all of human history, beginning with the use of coal at the end of the seventeenth century, followed by the mass consumption of oil in the twentieth century. This kind of population growth will culminate in the twenty-second century. Presently, at the beginning of the twenty-first century, the population of the world is rising at a rate that appears to be the fastest in history – an increase that shows an almost one-to-one correspondence with the rate of energy consumption. Lying beyond this population increase curve in the modern era is a major turning point: if oil and other non-renewable energy resources run out, will it be possible to continue to develop anthropogenic activities using different energy sources, or will such activities stagnate and enter a period of rapid decline?

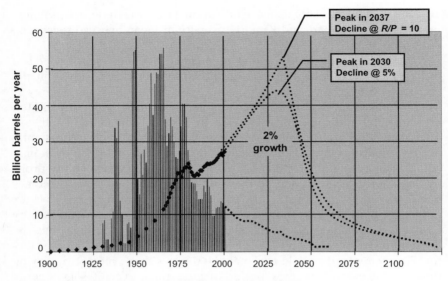

Figure 1.2.7 The amount of oil discovered and production.
Source: Based on Energy Information Administration (2000), slide 18.
Notes: The bars indicate the amount of oil discovered each year, with the lower dashed line indicating the projected reserve curve. The dotted lines indicate changes in production under different growth and decline scenarios. R/P = ratio of reserves to annual production.

When will the oil run out? Figure 1.2.7 shows that oil production has already outstripped the amount of oil discovered. In the future, the area under the reserve curve, showing the amount of oil that it is anticipated will be discovered, will match oil production. It is believed that under these conditions oil will become scarce around 2050 to 2075. In the long history of humankind, our dependence on fossil fuels, including coal, can be sustained for only for 300 years or so. We need to realize that we are living in an unusual period of human history in which the population has been able to increase owing to the availability of fossil fuels. It has become clear that non-renewable resources will run short, and that the pattern of economic growth seen throughout the twentieth century will no longer be feasible.

Global warming

According to long-term observations by Charles David Keeling et al. conducted at Hawaii's Mauna Loa Volcano beginning in 1957, atmospheric carbon dioxide concentration increased from 310 ppmv (parts per million by volume) in the 1950s to 380 ppmv in the 2000s. Mainstream opinion worldwide is that large amounts of carbon dioxide emissions from human activities have caused global warming. The Intergovernmental Panel on Climate Change (IPCC) has played a central role in such observations and arguments.[6]

The IPCC, which was established in November 1988, issued its Third Assessment Report in 2001 (IPCC, 2001) and its Fourth Assessment Report (AR4) in 2007 (IPCC, 2007), thereby catapulting the relationship between greenhouse gases, such as carbon dioxide, and global warming to the status of a major intergovernmental issue. The IPCC concluded that a twofold increase in carbon dioxide concentration would cause the temperature of the Earth's surface to rise by 1.4–4.5°C. It is surprising that consumption of only 10–12 TW of commercially provided energy (fossil and nuclear), which is a very small amount set against solar input of 170,000 TW, is generating such a notable climate abnormality.

Short-term equilibrium flow-type energy societies and accumulated resource (stock) utilization energy societies

Modern Western-type civilization has been established using a combination of solar, fossil and nuclear energy.[7] If people lived solely on biomass, wind power and solar energy as they did in the Edo period (1600–1867) when fossil and nuclear energy were unavailable – if they *lived green*, to use a current expression – the exclusive use of solar energy input would mean that a short-term flow-type energy balance, based on a roughly 30-year cycle of biomass generation and incineration, would be needed. The heat balance (probably a maximum of 150 TW) in this scenario would keep short-term heat input and output in perfect equilibrium. It is clear that modern Western-type society has switched from a flow-type society to a stock-utilization society in which people consume fossil energy and nuclear energy. This change gave rise to significant population increases and the expansion of civilization. Essentially, the process is akin to eating away at the Earth's long-term resource savings over a period of roughly 300 years.

Let us consider nuclear power generation, which has low carbon dioxide emissions (Yoda, 1998). The promotion of power generation that utilizes nuclear fission is under consideration worldwide because of its status as a centralized energy system that provides the most immediate replacement for currently used fossil energy (oil and natural gas). For the time being, however, large-scale light-water reactors that burn uranium-235, plutonium-239 and plutonium-241 will serve as the mainstay. Uranium reserves as an energy resource (exploitable reserves vary depending on the amount of funding available) totalled roughly 4.74 million tons as of 2005, and current world consumption stands at 54,500 tons per year, suggesting that, at current levels of usage, uranium resources would last roughly 85 years. In Japan, the use of mixed oxide (or MOX) fuel manufactured by reprocessing plutonium-239, which is produced from the nuclear fission of uranium-235, is permitted in reactors in amounts equivalent to 25 per cent of conventional uranium-235 fuel assemblies. Under this provision, the life of uranium fuel stocks can be extended by roughly 20 per cent, making it a resource that will be available for about 100 years. Even so, the world's uranium will be depleted in the not-so-distant future. If the demand for nuclear power rapidly increases in developing nations, there is a possibility that the timing of uranium depletion at current levels of usage will coincide with that of oil and natural gas.

However, it is feasible to secure a resource quantity that will last 1,000 to 2,000 years by burning inactive uranium-238, which accounts for 99.3 per cent of natural uranium and cannot be used in current light-water reactors, by striking fast neutrons against uranium in fast-breeder reactors. The coolant for fast-breeder reactors is metallic sodium – a liquid metal that is difficult to handle because it explodes on contact with water (which is an increased risk factor).

People will be left with no choice but to use fast-breeder reactors if all other efforts fail – that is, if measures for energy conservation are insufficient, if the development of renewable energy, distributed compound energy, cogeneration systems and so forth do not suffice, and if various technological innovations fail. Like it or not, people will be forced to choose between the scenario of giving up on using fast-breeder reactors and allowing themselves to perish, or adopting them despite the inherent risks and using advanced technology and the utmost care. In some distant future, a technology to use atomic fusion will, it is hoped, be realized.

1-2-6 Water controls the survival of humans

The presence of water on the Earth and demand for it

Water, along with energy issues, will be the most important consideration in achieving a sustainable world tomorrow.[8] Water covers two-thirds of the Earth's surface. Seawater accounts for 97.3 per cent of this, but humans and terrestrial fauna and flora directly depend on freshwater. Unfortunately, the vast majority of freshwater on the planet is contained in glaciers. Hence, we are directly dependent on flowing water resources, and even the combination of river water, water in lakes and marshes, groundwater and so forth makes up less than 0.01 per cent of the Earth's

THE CARRYING CAPACITY OF THE EARTH   25

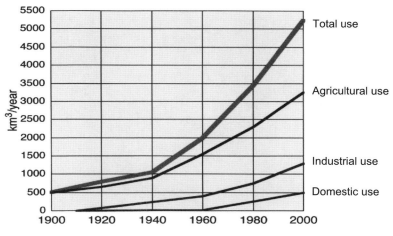

Figure 1.2.8 Global water demand, 1900–2000.
Source: Tolba et al. (1992).

total. Freshwater exists in the form of vapour in the air; however, this stays in the atmosphere for an average of only about 10 days. In other words, it is a high-speed circulation resource that returns to the Earth's surface as precipitation, with an average 10-day cycle. In addition, water is constantly generated from the condensation of vapour (water evaporated by solar energy) and returns to the Earth's surface after being recycled into freshwater that is almost pure. On the Earth, this process represents a gigantic flux amounting to 111×10^3 km^3 per year.

Figure 1.2.8 indicates that total water demand in the early twentieth century was a mere 500 km^3 per annum, but that this total had reached 5,500 km^3 by the end of the century. Agriculture accounts for the largest proportion of global water consumption, and the amount of water used for industrial purposes is increasing, particularly in developing nations. In Japan, however, because the reuse of wastewater is actively promoted (as epitomized by the iron and steel industry, in which 97 per cent of used water is reused), the amount of industrial water intake from natural rivers has hardly risen at all. On the other hand, the per capita basic unit (the standard amount of raw materials and labour required to manufacture a given quantity of a product) for municipal water, particularly that for domestic use, has seen a gradual increase. However, owing in part to the fact that population growth has peaked, and partly because of enhanced awareness of the need for environmental conservation, the expansion of scale has stopped and the focus of policy planning/implementation has seen a shift to areas such as water quality improvement and enhanced safety in the event of disasters. The biggest problem that the world's

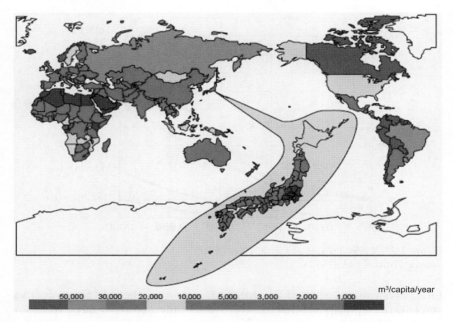

Figure 1.2.9 Water resources per capita (please see page 471 for a colour version of this figure).
Source: Ministry of Land, Infrastructure, Transport and Tourism, Water Resources Department (2003).

water system will face in the future will be to expand farmland and secure the amount of agricultural water necessary to accommodate rapid population growth and the expansion of feed-grain production to cope with the change of diet from grain to meat in developing nations.

Figure 1.2.9 shows per capita annual water availability in various parts of the world. Water stress is considered very strong when only 3,000 litres of water are available per capita per day – equivalent to less than 10 bathtubs. With roughly 10 bathtubs of water per person per day, it is possible to cultivate crops, maintain industry and lead daily life. The regions where life is the most difficult are where per capita water availability is a mere 1,000 m^3 or less per annum. These areas extend from the Arabian Peninsula to Egypt and from the northern part of the Sahara Desert up to Morocco. It is easy to understand these regions' water shortages given their lack of precipitation.

Surprisingly, another area that has one of the world's worst per capita water availability figures is Japan's Kanto (Tokyo) region. Despite annual precipitation of 1,600 mm, it is one of the world's most water-deficient areas, with a per capita annual water availability of 1,000 m^3 or less. This low availability is due to the area's enormous production output, with a population of roughly 30 million people achieving a GDP close to that of Germany and higher than that of France. It is natural that such large-scale activity should cause water shortages.

Endangerment of the world's many water systems

More than 35 per cent of grain production in the modern era is a result of irrigation farming, which gave rise to the Green Revolution. In irrigation farming, river water and groundwater are diverted to fields and used in sprinkler systems or paddy fields to assist crop growth. Concerns related to the irrigation of fields in arid or semi-arid areas include topsoil erosion and devastation owing to salt damage. Many irrigated fields are located in arid zones, and river water and groundwater in such regions are generally hard, with a relatively high saline concentration. Irrigation-related water sprinkling results in the evaporation of most of the water applied, causing the salt it contains to remain on the Earth's surface (with the maximum concentration being approximately 40–50 cm from the ground surface). In the worst-case scenario, hard chloride layers may be formed or the monovalent sodium ions in irrigation water may break the soil's aggregate structure, resulting in the dispersal/loss of soil particles. The decline of the Fertile Crescent in Mesopotamia appears to have been caused by salt accumulation in fields. Only in paddy-field farming in monsoonal regions such as East Asia and Southeast Asia is irrigation-based agriculture permanently free of salt damage.

Many large coastal cities around the world have suffered serious damage owing to ground subsidence and saline wedge intrusion into subsurface aquifers that are the result of the pumping of large quantities of groundwater. Although avoiding excessive pumping of water is the primary way to prevent such damage, recent sea-level rises as a result of global warming have posed new problems to modern urban societies, many of which are located in coastal areas. The progress of global warming causes continuous sea-level rises because even deep seawater temperatures gradually increase, resulting in the thermal expansion of water in all sea layers, and continental glacier thaw flows into the sea from places such as Greenland and the Himalayas.

It is said that, if temperatures rise by 3°C or more by the year 3000, the sea level will rise by approximately 3 metres. The IPCC AR4 simulations predicted that, with twice the current carbon dioxide concentration, that is, 750 ppmv, the sea level would rise 20 cm by the year 2100 and then roughly 10 cm every subsequent 100 years to reach approximately 50 cm by 2400.

Since the beginning of the twenty-first century, flood damage has also occurred frequently worldwide. With the progress of urbanization, people have come to live near the sea, and the wealthy have moved to waterfront areas to enjoy the presence of the water. This is true of the United States, Australia, Japan and European lowland areas, including the United Kingdom. In waterfronts with high population density and increased economic activities, flooding causes tremendous damage. The conditions of riverside slums in large cities also aggravate this type of damage, as illustrated by the example of New Orleans.

Local floods stemming from sudden heavy rains, tornadoes, heat waves and other extreme weather conditions have been frequently observed in Europe, the United States and even Japan. Whereas Japan has long suffered heavy flooding arising from typhoons during the monsoon season, modern European civilization arose in a mild climate that continued since its establishment. Until around 1980, participants at international conferences did not talk about the same things when it came to water issues: Europeans focused primarily on water resource conservation and pollution control in river basins, whereas the Japanese participants were more concerned about controlling floods and droughts. The 3rd World Water Forum, held in Kyoto in 2003, was the first international conference to address flood measures as a major agenda item. This new focus seems to be attributable to the repeated heavy flooding that occurred in central Europe in the preceding year, and conference attendees discussed the relationship of flooding to global warming.

Water system integration

The majority of freshwater circulating on the Earth is used for agriculture, with its primary goal being to produce food. Water necessary to conserve biodiversity and the natural environment is referred to as ecosystem conservation water. If the amount used for irrigation is further increased to expand agricultural production, water availability for other purposes such as ecosystem conservation and municipal/industrial use will decrease. As the introduction of reuse and recirculation systems to secure municipal/industrial water will also involve further energy consumption, comprehensive watershed management must be implemented, with energy use minimized as much as possible. The basic philosophy to adopt in achieving this is to use a quality of water appropriate to the purpose at hand. In the modern era, large-scale, centralized systems have been applied: major clean-water sources are used for both urban water supply and sewerage systems; water is conveyed over long distances and drinking water is provided for all purposes; all used water and wastes are then discharged as sewage in downstream basins. Except for the modern technological methods of water and wastewater treatment, there is no major foundational difference between the water supply and sewerage systems used in Rome roughly 2,000 years ago and those of today. In the twenty-second century, the available water resources will be exhausted unless we implement a totally new paradigm for water use; humans may not be able to ensure a sustainable future without finding new ways to coexist with nature and graduating from the modern era.

Originating in the era before the energy revolution, independent water supply and drainage systems distributed regionally have been used for more than 1,000 years in Europe, Southeast Asia and semi-arid areas. Such systems, which utilize local water resources and are aimed at local agriculture and the circulation of organic matter with the maximum use of human strength, animal power and natural energy, will attract renewed attention as a means to overcome environmental constraints in combination with modern technology.

With creative unintegrated water systems, the number of upstream dams may be reduced, thus enhancing the possibility of coexistence among river channel ecosystems. This may also help to moderate conflicts between people in upstream and downstream areas. However, it is not possible to state categorically that no dams will be necessary as long as modern centralized water systems are used. As for agricultural water, unintegrated irrigation will become a topic of discussion, and the possibility of employing local circulation-type irrigation will increase for paddy fields. These techniques will also become a trump card as measures for addressing water issues to increase the self-sufficiency ratio.

The human body can be used as an analogy for the developed forms of post-modern cities with high population density (i.e. compact cities) – at least in terms of water use, water treatment, methods of distribution and ways of using electric and thermal energy, as well as information manipulation. All water treatment processes can be simply redesigned to be similar to living organisms, using the functional separation membrane technology, which has seen significant development since the latter half of the twentieth century based on biology and physiology being developed in modern polymer science and the like. With this technology, almost all mass flux can be separated in a compact manner with high precision. Even seawater can be desalinated. On the other hand, energy consumption will see a slight increase. Related measures will involve discussions aimed at designing the sustainable distribution of water, materials and energy for entire watershed systems with comprehensive municipal water networks, including transportation, the reduction of energy used in long-distance transportation and enhanced safety in the event of disasters.

All water systems should be based on the assumption that individuals who play a part in them will take responsibility by themselves, and should be so designed as to be operated in a manner that will maximize levels of self-sustainability within the main hydrological cycle. Therefore, water use and disposal (referred to as water metabolism) in cities and regions will be based on the principle of the polluter (user) pays.

Animals and humans maintain an appropriate metabolism of water, consuming and discharging the minimum necessary amounts of water according to their biological requirements. The base daily water balance for an adult is 2 litres, but the water ingested by the body is recycled in the urinary system, which includes the kidneys and the bladder. An average male weighing 60 kg has 36 litres of bodily fluids (16 litres of which are extracellular), and these are filtered through the kidneys an average of four or five times a day (150–200 litres). Assuming a man drinks 2 litres of water per day, water remains in the body for 18 days (or 8 days if extracellular fluids are considered to circulate primarily in the body). Therefore, this water is reused 80 to 90 times (or 30 to 40 times, if extracellular fluids are considered) before being discharged in the form of urine and sweat. Waste is returned to nature in minimum quantities as human excrement (approximately 1.5 litres). Animals live in ecosystems based on this physiological cycle featuring the processes of water recycling and waste condensation. In a world with a population of 10 billion people, it will become necessary to introduce the mechanism seen in the metabolism of individual organisms (consisting of circulation within the body and condensation and discharge from it) in order to establish systems for social infrastructure and to achieve coexistence between cities and villages as well as with the natural environment.

The basic structure of water districts

Figure 1.2.10 shows a schematic diagram of a water district first proposed by Norihito Tambo in 1976 (Tambo, 1976, 2002). This diagram represents a proposal for the basic structure of cities with a high population density in drainage basins, and consists of an independent water metabolism with minimum energy consumption and environmental loads that resembles the metabolism of individual organisms. Such a set-up creates harmonious relationships with the natural environment, downstream basin areas and agriculture.

In the 1992 Dublin Principles, the United Nations expressed its hope to provide 50 litres of clean water per capita per day to all humans – a plan that has not yet materialized. In many developed nations, as much as 300–400 litres of drinking water are consumed per person per day because it is used for purposes other than drinking, such as washing, bathing, showering, flushing toilets and watering plants. There is not much difference between the amounts of drinking water required by developed and developing nations. Moreover, downstream cities and river

THE CARRYING CAPACITY OF THE EARTH 31

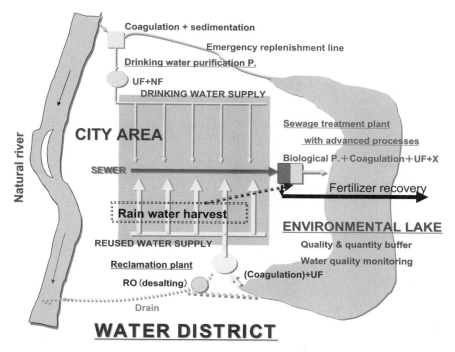

Figure 1.2.10 A new water system in a large city (please see page 472 for a colour version of this figure).
Sources: Tambo (1976, 2002, 2010).

ecosystems alike demand this kind of high-quality water; if we seek to coexist with nature, we cannot justify allowing only humans in upstream areas to use high-quality water for all purposes to satisfy the maximum water demand.

As a rough concept, cities should leave as much clean water as possible intact in river channels for people living in downstream areas as well as natural ecosystems, and collect, from natural clean water areas, only the minimum amount of drinkable water (or high-quality water) necessary for human health and survival – 50 litres per person per day as specified in the Dublin Principles. Urban dwellers will use/recycle the rainwater and used water (referred to as sewage) available in their areas as municipal water for all purposes other than drinking and so forth. They will then recycle used drainage (referred to as sewage) for systems intended to supply the pipeline system or the groundwater from wells and water not intended for drinking (referred to as non-potable water) after processing it biologically and physico-chemically to appropriate levels and storing it in nearby environmental lakes (or groundwater basins).

With the new advanced water treatment technology such as membrane separation, it is possible to store large amounts of recycled water in these reservoirs and use it repeatedly for purposes other than drinking, as organisms do. If clean water in environmental lakes supports healthy fish (determined through bioassay) and the mud in the lake bed remains clean (through integral environmental management), and if constant monitoring can guarantee water treatment results, it would be wasteful for it to be discharged. It should be returned to aquifers and pumped up for purposes other than drinking. Environmental lakes can accommodate fluctuations in water consumption and discharge between day and night, and are also capable of harvesting rainwater in urban areas, adjusting water quantity and quality, and enabling accurate and continuous monitoring in urban areas. Routine technical management of environmental lakes will allow advanced quality/quantity management of drainage in urban areas without the need for large-scale monitoring to cover entire downstream basins in large natural systems, which is difficult both economically and technologically.

Only concentrated domestic wastewater derived from separated wastewater systems will be collected for sludge recovery treatment; thick sludge will be separated and most fertilizer components, such as finite phosphorus and nitrogen, will be recovered for return to farmland. The collected materials will be either composted or anaerobically fermented. In the latter process, methane gas will be removed for energy recovery, and the residue will be returned to fields and forests as organic soil. World reserves of phosphate ore (a non-renewable resource indispensable for agriculture) are estimated at just 18 billion tons; if we continue with the current metabolism of modern water supply and sewerage systems and use up these reserves, we will no longer be able to produce food. Every year, 150 million tons of phosphate ore are used, and demand is expected to increase owing to population growth; thus reserves will run out in roughly 100 years. It is therefore clear that we cannot afford to simply dispose of phosphorus. As much as 80 per cent of this valuable material is contained in wastewater from toilets, and 9 per cent is in wastewater from kitchens, meaning that nearly 90 per cent of phosphorus is found in domestic wastewater. Therefore, as long as kitchen and toilet wastewater are recovered, there will be a continuous supply of phosphorus for 1,000 years.

If cities and villages introduce the water metabolism structure proposed in water districts, domestic conflicts between upstream and downstream areas as well as international conflicts over river channels will be resolved. In addition, it will become possible for farmers and urban dwellers to live in harmony and for clean water to be left in river channels to ensure coexistence with nature and populations downstream.

Coastal water contamination phenomena such as red tides will hardly ever occur in systems like this. A comprehensive local water metabolism system aimed at integrated basin management will be able to serve as the foundation of a new civilization that is viable for the twenty-second century simply by establishing local routine systems that link quality-based utilization with quality monitoring, as described above.

In the area of irrigation, water-deficient regions will come to employ drip irrigation and other upland cropping methods out of consideration for water usage that will enable the maintenance of green production areas with the minimum amount of consumption. In paddy fields, water will be circulated for reuse between upstream and downstream areas of fields divided into blocks. This strategy will enable irrigation with minimal water consumption, thereby reducing fertilizer runoff and minimizing the environmental burden on clean rivers.

1-2-7 Ways to determine carrying capacity

Sueishi et al. (1972) published a *Report on Carrying Capacity Quantification Research*, which contained research results that created an initial path for investigative activities on carrying capacity in Japan. The United States too coined the term 'environmental capacity', and discussions on this had attracted keen attention among natural environmentalists. The authors of the report divided carrying capacity into four types – Type I: Nature's Assimilation Capacity; Type II: Pollution Control Capacity; Type III: Capacity for Regional Activity; and Type IV: Duration (Future time) Activity Capacity – and quantified a controllable structural capacity that considered future time through a linkage between regions.

Jay Wright Forrester published his *Urban Dynamics* in 1970. His book established urban structure models that took numerous factors into consideration and analysed the dynamic behaviour of urban areas, which constitute the central structure of the modern era, thereby setting out a course for self-determination. In 1972, the Club of Rome released a report entitled *The Limits to Growth* as a global version of similar dynamic models (Meadows et al., 1972), thereby spreading an understanding of our finite planet worldwide. In both publications, the sectors comprising the dynamic models are actually modern in structure, and policies and impact are evaluated via operation parameters.

The argument outlined here is based on the earlier argument of Sueishi et al., and has been developed with the era of global environmental constraints in the twenty-first century in mind. It is a proposal for the structure of a survival system that humans can employ in the future – a system that will respond to resource and environmental constraints and that will feature a social infrastructure system different from that of the modern era. This reasoning is a manifestation of a desire to establish dynamics that can enable continued human survival in the post-modern era and beyond the twenty-second century.

The ecological footprint concept as we know it today represents an excellent approach to the evaluation of static carrying capacity in the modern era. However, if new technical/civilization proposals are incorporated that change the structures of the three characteristic areas – urban-industrial areas, biomass production areas and environmental conservation areas – to suit regional conditions, the concept will also serve as an effective planning tool to view the dynamic future differently. Even residents who have no special training will be able to use the tool to easily understand the situation. Of course, the dynamic models proposed by Forrester and the Club of Rome can also provide more accurate information for making decisions on the steps society should take through sensitivity analysis of various proposal factors. This will require the consideration of new technical and philosophical proposals for structures designed to sustain activities with significantly different population densities and levels of energy/material consumption in such characteristic regions as the mature regions of modern civilization, developing areas, tropical rain forests, arid zones, cold regions, monsoonal areas and so forth. This will be the next challenge to be addressed by Earth simulators. It would be a waste of time to continue discussing climate change only on the basis of the theory of carbon dioxide emissions and control only on the basis of modern social structure.

Whatever the case may be, it is not possible to pinpoint a simple value for the carrying capacity of the Earth. It is futile to talk about coexistence with the natural environment or sustainability to suit the present convenience of humans alone as a mere extension of modern civilization. The key to human survival lies in the evolution of human ethics and daily behaviour in terms of the meaning of sustainable development. Now is the time for carrying capacity as a form of environmental structural capacity to be discussed by proposing specific structures for areas and regions and by clarifying driving forces and operational methods.

At the same time, the prospect of an imminent end to the modern era of growth– a period in which people worshipped at the altar of progress – has become clear. A shift in values will take place by around 2050, and the modern era will be replaced by a new era in which the world shares a new set of universal values (not simple monetary values). It is not clear what these values will be. Without such an alteration in our thinking, however, we will continue to consume large amounts of energy and resources to achieve growth; in our continuing scramble to compete, we will succeed in ruining ourselves.

Even today, in addition to the evaluation of material growth (GDP), the United Nations' Human Development Index (HDI) has been used since 1990 to scale human well-being. The HDI combines "life expectancy at birth", "knowledge and education (the adult literacy rate)" and "GDP per capita" of a nation. However, using only these kinds of indices makes it difficult to capture an alteration in the value system from the modern to the post-modern that expresses an essential change in the human way of life as well as the alteration in our ecological background.

Development, in contrast to simple growth, connotes a change in values or content. Yet, development is still largely considered to be the same as progress. The current definition of progress will be questioned in the future.

Notes

1. This subsection is based on Tambo (2009) and Tambo et al. (2002).
2. This subsection is based on Tambo (2009).
3. This subsection is based on Tambo (2009).
4. This subsection is based on Tambo (2009).
5. This subsection is based on Tambo (2009).
6. This subsection is based on Council for Science and Technology Policy (2002) and Koike (2006).
7. This subsection is based on Tambo (2009).
8. This subsection is based on Van der Leeden et al. (1990).

REFERENCES

Council for Science and Technology Policy (2002) "The Battle Front of the Global Warming Study", Cabinet Office, Government of Japan.

Energy Information Administration (2000) "Long Term World Oil Supply", summary of EIA presentation, 28 July, <http://www.eia.doe.gov/pub/oil_gas/petroleum/presentations/2000/long_term_supply/sld001.htm> (accessed 1 March 2010).

Forrester, Jay Wright (1970) *Urban Dynamics*. Cambridge, MA: MIT Press.

IPCC [Intergovernmental Panel on Climate Change] (2001) *Climate Change 2001: Synthesis Report. A Contribution of Working Groups I, II, and III to the Third Assessment Report of the Intergovernmental Panel on Climate Change* [R. T. Watson and the Core Writing Team (eds)]. Cambridge: Cambridge University Press.

IPCC [Intergovernmental Panel on Climate Change] (2007) *Climate Change 2007: Synthesis Report. Contribution of Working Groups I, II and III to the Fourth Assessment Report of the Intergovernmental Panel on Climate Change* [Core Writing Team, R. K. Pachauri and A. Reisinger (eds)]. Geneva, Switzerland: IPCC.

Koike, I. (2006) *How Much Do We Know about Global Warming?* Tokyo: Maruzen.
McEvedy, Colin and Richard Jones (1978) *Atlas of World Population History*. Harmondsworth: Penguin Reference Books.
Meadows, Donella H., Dennis L. Meadows, Jorgen Randers and William W. Behrens III (1972) *The Limits to Growth*. New York: Universe Books.
Ministry of Land, Infrastructure, Transport and Tourism, Water Resources Department (2003) "Per Capita Annual Water Availability in Various Parts of the World".
National Institute of Population and Social Security Research (2004) *A Trend of the Population – Japan and the World 2004*.
Sueishi, T., Y. Nambu and N. Tambo (1972) *Report on Carrying Capacity Quantification Research*. Institute for Cultural and Environmental Studies.
Tambo, N. (1976) "Structure and Capacity of Urban/Regional Metabolic Systems of Water", Japan Water Works Association journal, vol. 497.
Tambo, N. (1986) "Resources Circulation and Environmental Safeguard", keynote speech at the 1st Environmental Engineering Symposium held by the Science Council of Japan.
Tambo, N. (2002) "A New Metabolic System", *Water 21*, December, International Water Association.
Tambo, N. (2009) "Japan and Hokkaido in the 21st Century – Aim at the Sustainable Society", Hokkaido Area Management and Support (HAMANAS) Foundation.
Tambo, N. (2010) "A Person and Water [*Hito to Mizu*]". HAMANAS Foundation, Hokkaido.
Tambo, N., et al. (2002) "Preparation of Social Infrastructure under Population Declining Stage of the Nation", Japan Society of Civil Engineers.
Tolba, M. K., et al. (1992) *The World Environment 1972–1992: Two Decades of Challenge*. United Nations Environment Programme (UNEP). London: Chapman & Hall.
UNFPA [United Nations Population Fund] (2007) *State of World Population 2007: Unleashing the Potential of Urban Growth*. New York: United Nations Population Fund.
Van der Leeden, Frits, Fred L. Trois, and David Keith Todd (1990) *The Water Encyclopedia*, 2nd edn. Chelsea, MI: Lewis Publishers.
Wackernagel, Mathis, and William E. Rees (1996) *Our Ecological Footprint: Reducing Human Impact on the Earth*. Gabriola Island, BC: New Society Press.
WWF (2006) *Living Planet Report 2006*. Gland, Switzerland: WWF International.
Yoda, N. (1998) "A Design of the Next Generation Energy", *Energy Forum*, Central Research Institute of Electric Power Industry.

2
Sustainable land management

2-1
Sustainable agriculture practices

Masakazu Komatsuzaki and Hiroyuki Ohta

2-1-1 Agro-ecosystem management and ecosystem service

During the latter half of the twentieth century, though intensive agriculture increased crop yields and successfully met the growing demand for food, it degraded the natural resources upon which agriculture depends – soil, water resources and natural genetic diversity (Pimentel et al., 1995; Gliessman, 2006). Today, conventional agriculture is built around two related goals: maximization of production and maximization of profit. In pursuit of these economic goals, a host of practices have been developed without regard for their unintended long-term consequences and without consideration of the ecological dynamics of agro-ecosystems. The Millennium Ecosystem Assessment (2005) revealed that the overuse and mismanagement of pesticides poison water and soil, and nitrogen (N) and phosphorus (P) inputs and livestock wastes have become major pollutants of surface water, aquifers, and coastal wetlands and estuaries.

The world's human population is expected to grow from a little over 6 billion today to over 8 billion by 2030, an increase of about a third, with another 2 to 4 billion added in the subsequent 50 years (Cohen, 2003). Tilman et al. (2001) predict that feeding a population of 9 billion using conventional methods would mean converting another 1 billion hectares (ha) of natural habitat to agriculture, primarily in the developing world, together with a doubling or tripling of N and P inputs, a twofold increase in water consumption and a threefold increase in pesticide usage.

Designing our future: Local perspectives on bioproduction, ecosystems and humanity,
Osaki, Braimoh and Nakagami (eds),
United Nations University Press, 2011, ISBN 978-92-808-1183-4

As environmental degradation as a result of farming is becoming a serious problem at both local and global scales, many conventional farmers are choosing to make the transition to practices that are more environmentally sound and could potentially contribute to long-term sustainability of agriculture. Sustainable agriculture would ideally produce good crop yields with reduced impact on ecological factors such as soil fertility and utilize minimal input for production (Pimentel et al., 1997).

With the growing interest in reducing excessive chemical inputs in farming, the importance of low-input sustainable agriculture in reducing the usage of energy and chemical inputs has been recognized (Poincelot, 1986). For example, the policies of the Japanese government to develop more environmentally friendly farming practices and growing awareness regarding the importance of a reduction in chemical materials have led to a widespread interest in conservation farming. According to recent statistical data, 167,995 farms in Japan are engaged in conservation farming, accounting for 21.5 per cent of the total cropping area in the country (Sustainable Agriculture Office, 2008). Conservation management increases the efficiency of conventional practices in order to reduce or eliminate the usage and consumption of costly, scarce or environmentally damaging inputs such as synthetic pesticides and fertilizers. Although these kinds of efforts have helped to reduce the negative impacts of conventional agriculture, they have not eliminated its dependence on external human inputs and their damage to local environments.

To develop sustainable farming, conventional farming should replace conventional inputs and practices with alternative practices. Organic farming systems are one of the alternatives to conventional agriculture. Instead of synthetic inputs, organic farming uses cover crops, compost and animal manure to increase soil fertility, and this would ideally produce good crop yields with minimal impact on ecological factors (Pimentel et al., 1997; Mäder et al., 2002). Producers, sellers and consumers of organic food regularly mention the "natural" aspect to characterize organic farming or organic food, in contrast to the unnaturalness of conventional farming (Verhoog et al., 2003).

However, large-scale organic farming can cause environmental damage and use massive amounts of energy. For example, intensive organic vegetable production has caused nitrate leaching from soil (Maeda et al., 2003) and commercial organic markets have used huge amounts of energy and destroyed local production (Gliessman, 2006). In this regard, farming practices for sustainable agriculture to promote coexistence between humans and nature should focus not only on replacing the chemicals used in farming but also on redesigning the agro-ecosystem to maximize the ecological, economic and social synergies among them and minimize the conflicts.

Plate 2.1.1 Photo of *shizen-nou* field in Kanto, Japan. In the *shizen-nou* farming system, farmers do not eliminate weeds completely, do not add any chemicals and use small amounts of organic matter as nutrients to grow vegetables (Photo A) and sweetcorn (Photo B) naturally.

The concepts of agro-ecology, eco-agriculture and *shizen-nou* (see Plate 2.1.1) are intended to manage farming by providing conditions with ecological qualities that are beneficial to wild biodiversity and ecosystem services, and by developing a network of diverse communities to promote and achieve local independence. Although each approach is different in terms of the composition of the food system, landscape management and farming practices, they have similar farming practices that coexist with nature; however, these farming practices show a sharp contrast with low-input sustainable agriculture or large-scale organic farming.

Figure 2.1.1 illustrates the difference between low-input sustainable farming and sustainable farming practices that coexist with nature as a result of agro-ecosystem management. There are many agricultural practices and designs that have the potential to enhance functional biodiversity, and others that negatively affect it. The idea is to apply sustainable management practices in order to enhance or regenerate the kind of biodiversity that can supplement the sustainability of agro-ecosystems by providing ecological services such as biological pest control, but also nutrient cycling, and water and soil conservation. The goals of farming systems are also different between low-input sustainable agriculture and sustainable farming practices that coexist with nature. Whereas the low-input system focuses on the efficiency of inputs, sustainable farming practices focus on developing an agro-ecosystem that provides positive feedback to natural ecosystems (e.g. increasing ecosystem services) (Reijntjes et al., 1992; Altieri, 1999). Sustainable farming in a society where humans and nature coexist should encourage practices that increase the abundance and diversity of above- and below-ground organisms, which provide key ecological services to agro-ecosystems. This

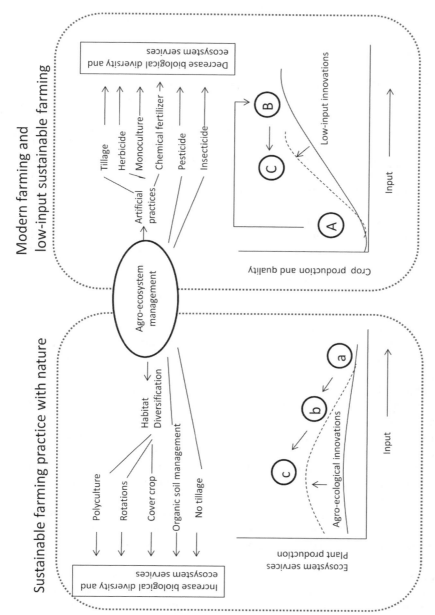

Figure 2.1.1 Effects of agro-ecosystem management and associated artificial practices on biological diversity and ecosystem services, and a comparison of innovation between sustainable farming with nature and sustainable farming with low-input practices.

farming also needs to diversify to adapt to climate change, including rising temperatures and more extreme weather events.

2-1-2 Agro-ecosystem and farming practices

Carbon and nitrogen dynamics in agro-ecosystems

The maintenance and improvement of soil quality in croplands are critical to sustaining agricultural productivity and environmental quality for future generations. Fertile soil provides essential nutrients for crop plant growth, supports a diverse and active biotic community, exhibits a typical soil structure and allows for undisturbed decomposition. In general, an increase in soil organic matter (SOM) increases crop yield response and conserves water, thus improving soil quality. SOM, which includes a vast array of carbon (C) compounds originally created by plants, microbes and other organisms, helps to maintain soil fertility and plays a variety of roles in the nutrient, water and biological cycles (Tiessen et al., 1994; Reeves, 1997). SOM is also critical for its function to support crop growth naturally, and provides a place for water, air and biological ecosystems to exist in the soil. Proper soil management also has great potential to contribute to C sequestration by transferring atmospheric CO_2 into long-lived pools (or "sinks") and storing it securely so that it is not immediately re-emitted (Lal, 2004). Soil management practices that improve soil quality through enhancing SOM and fertility will become more widespread, because soil management also determines the level of food production and largely the state of the global environment. Moreover, the current pressure on land resources of the world is enormous (Komatsuzaki and Ohta, 2007).

Table 2.1.1 shows an evaluation of farming practices to increase C stocks and their positive and negative impacts on agro-ecosystems. Practices directed towards effective management of soil C are available and identified, and many of these are feasible and relatively inexpensive to implement. There exist agronomic practices that increase the return of biomass C to the soil, and improve tillage and residue management. Increased N fertilization has made a large contribution to the growth in productivity, but further increases will lead to greater emissions of nitrous oxide (N_2O).

The strategy of soil management for sustainable agro-ecosystems should be compatible with increasing SOM to improve soil quality for sustaining food productivity and to control soil residual nutrients that aggravate environmental problems. Cover cropping is a unique technique for improving the N cycle in cultivated soil because it scavenges the

Table 2.1.1 Evaluation of farming practices adopted to increase carbon stocks

Treatment	Effect on OM input (changes to primary production and/or supplied to the soil)	Effect on OM output (rate of mineralization)	Other positive effects	Negative secondary environmental effects	Additional carbon stock (t-C ha^{-1} y^{-1})	References
No-till	Slightly low production, slightly low level of OM conversion into humus	Low rate (increased protection of OM owing to improved soil aggregate)	Erosion control, reduced fuel consumption, enhanced soil biological diversity	Slightly low production, use of pesticides, emission of N_2O to be confirmed	0.07–0.33	Robertson et al., 2000 Smith et al., 2008 Komatsuzaki et al., 2008
Crop rotation	Increased OM input	Increased soil respiration	Break in insect and pest cycle	None	0.05–0.25	Lal, 2004 Smith et al., 2008
Cover crop	Annual production and increased OM returned (crop not harvested)	Increased soil respiration	Scavenging residual nutrient erosion control reduces fertilizer consumption, enhancing soil biological diversity	Emission of N_2O to be confirmed	0.15–0.25	Lal, 2004 Smith et al., 2008 Komatsuzaki et al., 2008
Manure application	Increased OM input Increased production through additional nutrients	Increased soil respiration	Improved soil productivity	If excessive inputs occur, N leaching and N_2O emission	0.05–0.75	Robertson et al., 2000 Lal, 2004 Smith et al., 2008

Note: OM = organic matter

soil residual N and turns residuals into nutrients for subsequent crops (Komatsuzaki and Ohta, 2007). Komatsuzaki et al. (2008) reported that rye cover crops accumulated soil N as the residual N level in the soil increased, and the distribution of soil inorganic N showed that inorganic N concentration at a depth of 60–90 cm was significantly low for rye compared with hairy vetch and fallow at cover crop growth termination; this reduction in soil inorganic N was observed to occur year-round. In addition, no-till with a rye cover crop showed the highest increase in the ratio of soil C storage, although winter fallow showed a decrease in soil C storage during a five-year experiment.

No-till with a cover crop, however, showed higher N_2O emissions compared with conventional treatment. Komatsuzaki et al. (2008) showed the capacity for soil C storage that was offset by annual N_2O emissions for each farming system. Conventional tillage had a negative impact on mitigating the global warming potential because it showed an increase of 0.253 ton CO_2 equivalent per hectare (equiv. ha^{-1}) of greenhouse gas emissions; on the other hand, no-till with rye cover cropping had a positive impact on mitigating the global warming potential, with a reduction of 0.486 ton CO_2 equiv. ha^{-1}.

Soil biological function and farming practices

Russell (1957: 137) observed: "Much of our agricultural effort goes toward sustaining the large and varied population of living things in the soil: we get only the by-products of their activity." Russell's great insight was based on the calculation of dissipated calories per acre per annum for plots of Broadbalk wheat at Rothamsted research centre in the United Kingdom. Today we understand that "living things in the soil" represents soil microbial biomass, which comprises only about 1–4 per cent of total SOM (C base, e.g. Jenkinson and Powlson, 1976), but it is an important reservoir of essential plant nutrients such as N, P and sulphate (S). Later, Jenkinson and Ladd (1980) estimated that the microbial biomass in the plough layer of the unfertilized plot of Broadbalk wheat contained 95 kg N ha^{-1} and the N flux from the biomass was 38 kg N ha^{-1} y^{-1}, close to the N amount taken up by wheat (24 kg N ha^{-1} y^{-1}). Therefore, it is clear that, in low-input soils, a large proportion of the plant-available N, P and S is derived from the mineralization of nutrients immobilized within the cells of the microbial biomass (Marumoto et al., 1982; Brookes, 2001).

Because soil microbes decompose plant residues and utilize nutrients preferentially from them rather than from the soil nutrient pool (Ocio et al., 1991), the C/N ratio and percentage of readily decomposable and resistant tissues in plant residues are critical factors in determining rates of microbial activity and ultimately the availability of nutrients to crops. The major plant matter components such as cellulose, hemicellulose and lignin require a diverse array of hydrolytic enzymes for degradation, which probably explains the great diversity of hydrolytic microbes in soils. These plant materials also contain very little N and thus have a major impact on the N cycle. In general, microbial degradation of organic matter with C/N ratios >20 results in less mineralization but more assimilation (or immobilization), and the balance between the two processes is dependent on the N content of microbial biomass. Therefore, the C/N ratio of plant residues determines the balance between mineralization and assimilation and ultimately the availability of N to crops.

Soil water content and the aggregate microstructure, which are related to agricultural water management and the tillage system, also act as strong selective forces that determine the soil's biological functions. Generally, soil water content is regarded as a master variable for the ability of bacteria to migrate from one microhabitat to another and as a transport mechanism for solutes and gases. Further, soil texture and aggregate size contribute to substantial physical heterogeneity and affect water distribution among soil particles. This structural heterogeneity results in the coexistence at the millimetre or submillimetre level of oxic and anoxic conditions and thus facilitates both nitrification (oxic process) and denitrification (anoxic process). Further, these processes are more complex when the physiological versatility of ammonia oxidizers and denitrifiers is considered.

In actual soil environments, two important by-products and intermediates of these processes – NOx (= NO + NO_2) and N_2O – are both produced and consumed (Figure 2.1.2; see also Fenchel et al., 1998). Ammonia oxidizers tend to produce more NOx under fully oxic conditions but N_2O under oxygen-limited conditions. On the other hand, denitrifiers produce N_2O and further N_2 under fully reduced conditions but NO under less reduced conditions. Agricultural production on well-drained soils and with high rates of N fertilizer usage will allow the soil condition to be aerobic and favour large fluxes of N_2O and NO. This seems to be particularly true in tropical agricultural systems (Bouwman, 1998).

Tillage practices greatly influence the soil aggregate structure and thus determine soil microbial activity. Conventional tillage by ploughing and disking to prepare the land can continuously reduce SOM content and increase CO_2 emissions because mechanical practices that disintegrate soil macro-aggregates expose the protected organic matter for microbial degradation and increase air permeability, thereby accelerating microbial oxidation (Beare et al., 1994). No-till accumulates a mulch of crop residue on the soil surface, which can result in higher soil moisture content. It is possible that wetter soil conditions increase the microbial production

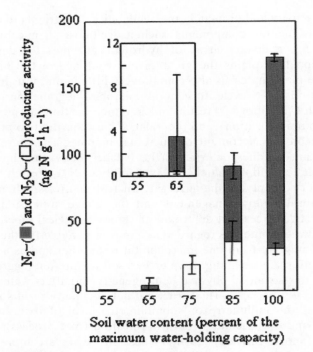

Figure 2.1.2 Relationships between N_2O and N_2 production as a function of soil water content.
Source: unpublished results, M. Umezu, Y. Sato, M. Komatsuzaki and H. Ohta.
Notes: Soil samples were taken from an upland rice field treated with NH_4^+-N (100 kg N ha^{-1} y^{-1}) and rye cover crop. Samples were incubated in serum vials, and N_2O in the headspace gas was analysed by electron capture detector-equipped gas chromatography. N_2-producing (total denitrification) activity was determined in the presence of acetylene (10 KPa) as an inhibitor of N_2O reductases.

of N_2O. This is the essence of the trade-off between soil C sequestration and the increased emission of N_2O in no-till practices.

2-1-3 Conclusions

Agriculture dominates land and water usage like no other human enterprise, with landscapes providing critical products for human sustenance. Farmers, consumers, researchers and policymakers in many parts of the world have begun to develop and promote sustainable agriculture. However, sustainable farming needs to be adopted on a much larger scale to achieve the Millennium Development Goals for hunger, poverty and environmental sustainability in developing countries, and to sustain ecosystems in the rural areas of industrialized countries. Recent intensive research has revealed that sustainable farming practices can help to mitigate global warming, conserve biodiversity and maintain soil fertility and productivity. However, these farming practices often do not return enough to farmers directly. Therefore, political and social incentives will be required based on the common understanding that soil and agro-ecosystems are essential to developing a society in this century in which nature and humans can coexist.

REFERENCES

Altieri, Mibuel A. (1999) "The Ecological Role of Biodiversity in Agroecosystems", *Agriculture Ecosystem & Environment* 74: 19–31.

Beare, Mike H., M. L. Cabrera, P. F. Hendrix and D. C. Coleman (1994) "Aggregate-protected and Unprotected Organic Matter Pools in Conventional and No-tillage Soils", *Soil Science Society of America Journal* 58: 787–795.

Bouwman, A. F. (1998) "Environmental Science: Nitrogen Oxides and Tropical Agriculture", *Nature* 392: 866–867.

Brookes, Philip (2001) "The Soil Microbial Biomass: Concept, Measurement and Applications in Soil Ecosystem Research", *Microbes and Environments* 16(3): 131–140.

Cohen, Joel E. (2003) "Human Population: The Next Half Century", *Science* 302: 1172–1175.

Fenchel, Tom, G. M. King and T. H. Blackburn (1998) *Bacterial Biogeochemistry: The Ecophysiology of Mineral Cycling*. London: Academic Press.

Gliessman, Stephen R. (2006) *Agroecology: The Ecology of Sustainable Food Systems*. Boca Raton, FL: CRC Press.

Jenkinson, David S. and J. N. Ladd (1980) "Microbial Biomass in Soil: Measurement and Turnover", in E.A. Paul and J. N. Ladd (eds), *Soil Biochemistry* 5. New York: Marcel Dekker.

Jenkinson, David S. and D. S. Powlson (1976) "The Effects of Biocidal Treatments on Metabolism in Soil. V. A Method for Measuring Soil Biomass", *Soil Biology and Biochemistry* 8: 209–213.

Komatsuzaki, Masakazu and H. Ohta (2007) "Soil Management Practices for Sustainable Agro-ecosystems", *Sustainability Science* 2: 103–120.

Komatsuzaki, Masakazu, Y. Mu, R. Zhaorilgetu, H. Ohta, M. Araki, S. Hirata and S. Miura (2008) "Cover Crop and No Tillage Practices Enhance the Ecological Significance of Soil Biodiversity and Carbon Sequestration", in *Proceedings of the International Conference on Sustainable Agriculture for Food, Energy, and Industry 2008*. Sapporo: International Conference on Sustainable Agriculture, pp. 74–78.

Lal, Rattan (2004) "Soil Carbon Sequestration Impacts on Global Climate Change and Food Security", *Science* 304: 1623–1627.
Maeda, Morihiro, B. Zhao, Y. Ozaki and T. Yoneyama (2003) "Nitrate Leaching in an Andisol Treated with Different Types of Fertilisers", *Environmental Pollution* 121(3): 477–487.
Mäder, Paul, A. Fließbach, D. Doubois, L. Gunst, P. Fried and U. Niggli (2002) "Soil Fertility and Biodiversity in Organic Farming", *Science* 296: 1694–1697.
Marumoto, Takuya, J. P. E. Anderson and K. H. Domsch (1982) "Mineralization of Nutrients from Soil Microbial Biomass", *Soil Biology and Biochemistry* 14: 469–475.
Millennium Ecosystem Assessment (2005) *Ecosystems and Human Well-being: Synthesis*. Washington DC: World Resource Institute.
Ocio, J. A., J. Martinez and P. C. Brookes (1991) "Contribution of Straw-derived N to Total Microbial Biomass N Following Incorporation of Cereal Straw to Soil", *Soil Biology and Biochemistry* 23: 655–659.
Pimentel, David, C. Harvey, P. Resosudarmo, K. Sinclair, D. Kurz, M. McNair, S. Crist, L. Shpritz, L. Fitton, R. Saffouri and R. Blair (1995) "Environmental and Economic Costs of Soil Erosion and Conservation Benefits", *Science* 267: 1117–1122.
Pimentel, David, C. Wilson, C. McCullum, R. Huang, P. Dwen, J. Flack, Q. Tran, T. Saltman and B. Cliff (1997) "Economic and Environmental Benefits of Biodiversity", *BioScience* 47(11): 747–757.
Poincelot, Raymond P. (1986) *Toward a More Sustainable Agriculture*. Westport, CT: AVI Publishing Company, Inc.
Reeves, Wayne D. (1997) "The Role of Soil Organic Matter in Maintaining Soil Quality in Continuous Cropping Systems", *Soil & Tillage Research* 43(1): 131–167.
Reijntjes, Coen, B. Haverkort and A. Waters-Bayer (1992) *Farming for the Future: An Introduction to Low-external-input and Sustainable Agriculture*. London: Macmillan.
Robertson, Philip G., E. A. Paul and R. R. Harwood (2000) "Greenhouse Gases in Intensive Agriculture: Contributions of Individual Gases to the Radiative Forcing of the Atmosphere", *Science* 289: 1922–1925.
Russell, Edward J. (1957) *The World of the Soil*. London and Glasgow: Collins Clear-Type Press.
Smith, Pete, D. Martino, Z. Cai, D. Gwary, H. H. Janzen, P. Kumar, B. McCarl, S. Ogle, F. O'Mara, C. Rice, R. J. Scholes, O. Sirotenko, M. Howden, T. McAllister, G. Pan, V. Romanenkov, U. Schneider, S. Towprayoon, M. Wattenbach and J. U. Smith (2008) "Greenhouse Gas Mitigation in Agriculture", *Philosophical Transactions of the Royal Society B* 363: 789–813.
Sustainable Agriculture Office (2008) *Results of Certification of Ecologically Friendly Farmer*. Available at: <http://www.maff.go.jp/soshiki/nousan/nousan/kanpo/ecofarmer.pdf>.
Tiessen, Holm, E. Cuevas and P. Chacon (1994) "The Role of Soil Organic Matter in Sustaining Soil Fertility", *Nature* 371: 783–785.
Tilman, David., J. Fargione, B. Wolff, C. D'Antonio, A. Dobson, R. Howarth, D. Schindler, W. H. Schlesinger, D. Simberloff and D. Swackhamer (2001)

"Forecasting Agriculturally Driven Global Environmental Change", *Science* 292: 281–284.

Verhoog, Henk, M. Matze, E. Lammerts van Bueren and T. Baars (2003) "The Role of the Concept of the Natural in Organic Farming", *Journal of Agricultural and Environmental Ethics* 16: 29–49.

2-2
Soil quality and sustainable land use

Ademola K. Braimoh

2.2.1 Overview

The global human population was under 1 billion in the middle of the eighteenth century; it was about 2.5 billion by the beginning of the twentieth century. It took only 40 years (1950–1990) for the population to double to 5 billion. It currently stands at 6.5 billion, and is estimated to stabilize at 10 billion in 2100 (Cohen, 2003). The unprecedented growth in human population since the eighteenth century has led to the accelerated consumption of resources, which is manifest in relatively high rates of agricultural output and food production, industrial development, energy production and urbanization. These human enterprises result in changes in local land use and land cover, which, when aggregated, affect different components of the Earth's system. These include climate, hydrology, biogeochemistry, biodiversity and the ability of biological systems to support human needs (Foley et al., 2005; Sala et al., 2000).

Soil is the basic resource of land use. It is the central link in the biosphere, performing multifunctional roles ranging from food production to water purification, biodiversity support and climate regulation. The available soil/land resources have significantly diminished as a result of various anthropogenic impacts. Between 1900 and 1990, the per capita cropland area decreased from 0.75 to 0.35 hectares (ha) (Ramankutty et al., 2008), which is less than the minimum 0.50 ha needed to provide an adequate diet (Lal, 1989). Since the middle of the twentieth

---

Designing our future: Local perspectives on bioproduction, ecosystems and humanity,
Osaki, Braimoh and Nakagami (eds),
United Nations University Press, 2011, ISBN 978-92-808-1183-4

century, human-induced change in land use (an additional challenge to the inherent soil constraints on agricultural productivity) has become so rapid, drastic and pervasive that its impacts are affecting the life support systems of the Earth (Braimoh and Vlek, 2008). Of the total land area of the Earth (134 million km^2), 14 per cent is estimated to have been degraded by anthropogenic activities. The major causes of human-induced soil degradation are overgrazing, which accounts for 35 per cent of degraded land surface around the world, deforestation (29 per cent), agricultural mismanagement (28 per cent), the consumption of wood as fuel (7 per cent) and urbanization and industrial pollution (1 per cent) (United Nations Environment Programme, 2002). Human-induced changes in nutrient cycling in terrestrial ecosystems also affect the sustainability of food production, the state of the natural resource base, and the health of the environment. About 230 Tg (1 Tg = 10^6 metric tons) of plant nutrients are removed yearly from agricultural soils through runoff, erosion, leaching, crop-residue removal and harvested products, whereas fertilizer consumption is only 57 per cent of this, i.e. 130 Tg (Vlek et al., 1997). The twin menaces of the diminution and deterioration of land resources are major challenges to sustainable agricultural development in many parts of the world.

Many properties that contribute to soil quality are mutually influenced by land use. Thus, knowledge of the interaction between soil quality and land use is indispensable for the sustainable use and management of agricultural land. Soil quality refers to the ability of soils to function within natural and managed ecosystems. It influences five functions of the soil including the ability to (i) accept, hold and release nutrients; (ii) accept, hold and release water for plants and for recharge; (iii) sustain root growth; (iv) maintain suitable biotic habitats; and (v) respond to management and resist degradation (Karlen et al., 2001). The "inherent" quality of soils, which is a measure of their natural ability to function, is dependent on soil formation factors (climate, topography, parent material, vegetation and time), whereas the "dynamic" quality of soils refers to their response to use and management.

In this section, two data sets are used to study human impact on soil resources in the Asia Pacific. The data sets included a map of anthropogenic biomes, which reflect direct human interaction with ecosystems, and a map of inherent soil/land quality, which shows the resilience and performance of soils based on pedoclimate information. Spatial patterns of biomes and soil quality were examined, and the association between them was identified. The implication for sustainable use of soil and land was also discussed.

2.2.2 Soil resources and biome data sets

One attempt to map global soil resources involves an overlay of soil climate and soil suborder maps within the Geographical Information System to derive an inherent soil/land quality map (Beinroth et al., 2001). Based on 25 stress factors, the soil and pedoclimate information was used to place the mapping units into nine quality classes integrating soil resilience and soil performance. "Soil resilience" refers to the capacity of the soil to resist change or to recover its functional or structural integrity after a disturbance (Lal, 1997). "Soil performance" refers to the ability to produce the required levels of crop yields under moderate levels of conservation technology, fertilizers and pesticide use.

The inherent soil/land quality map for parts of the Asia Pacific noted in the present study (approximately 24 million km^2) is shown in Figure 2.2.1. Prime lands (Class 1) are not very extensive and occupy only 5 per cent of the entire study area (Figure 2.2.2). They are found mainly in northern China and in the Indo-Gangetic Plain, where the population density is high (>100 persons km^{-2}). Large tracts of Class 2 soils, covering approximately 30 per cent of the study area, are found in India and other tropical areas such as Indonesia. This class of soil is also found in the temperate regions of China and Japan. Class 5, occupying 35 per cent of the study area, is the most extensive. This class occurs mostly in tropical Southeast Asia, where swidden agriculture is rampant (Fox and Vogler, 2005). Classes 1 to 3 lands, covering 36 per cent of the study area, are generally free from constraints for most agricultural uses (Eswaran et al., 1999). Classes 4 to 6, covering about 40 per cent of the study area, have progressively more severe productivity constraints and, therefore, require important inputs for conservation management. Taken together, 76 per cent of the total land area (Classes 1–6) is suitable for crop cultivation, although with varying degrees of constraints. However, this portion of the landscape must also compete with other uses (e.g. urban, recreation, wildlife and forestry). Class 7 (covering about 8 per cent of the land area) should generally not be used for cultivation unless there is intense pressure on land use. Classes 8 and 9, covering about 15 per cent of the study area, are mainly the desert soils of Pakistan, China and Mongolia. They are fragile ecosystems that should be retained in their natural state.

Biomes are the most basic units describing the patterns of ecosystem forms and processes. Hitherto, they have been mapped based on vegetation and climatic differences across locations (Olson et al., 2001). Owing to the ongoing transformation of ecosystems by humans, a new view of the terrestrial biosphere in its contemporary, human-altered form is

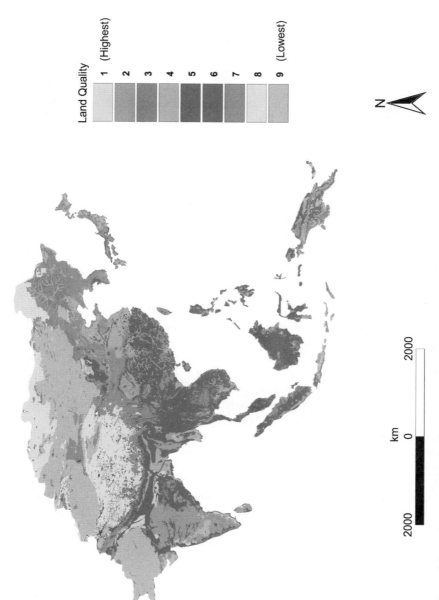

Figure 2.2.1 Inherent soil/land quality map for the Asia Pacific (please see page 473 for a colour version of this figure).
Source: Based on Beinroth et al. (2001).

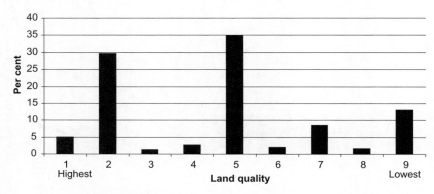

Figure 2.2.2 The distribution of land classes in the Asia Pacific.

required. Ellis and Ramankutty (2008) responded to this need by characterizing terrestrial biomes based on global patterns of sustained direct human interaction with ecosystems. Through the empirical analysis of global land use, land cover and population, they identified 18 anthropogenic biomes and 3 wild biomes.

Figure 2.2.3 shows the distribution of biomes in the Asia Pacific landscape. There are wide differences in these biomes. The six major groupings of biomes and their proportions are displayed in Figure 2.2.4. With the exception of wildlands, all the others are anthropogenic (i.e. human-impacted) biomes. Of the densely populated biomes, villages representing dense agricultural settlements are most extensive, covering 59 per cent of the land area, compared with dense urban areas, which cover 5 per cent. This is not surprising because nearly two-thirds of the >4 billion people residing in Asia live in rural areas and depend on agricultural activities for their livelihood. Croplands (i.e. mosaics of annual crops, including irrigated and rain-fed crops, and other land cover, together with pasture) are the next most extensive, covering about 26 per cent of the land area. Rangelands occurring in arid areas with relatively low population account for about 8 per cent of the study area, whereas forested biomes with minor human populations account for 2 per cent. Wildlands are areas without evidence of human occupation or land use and account for just 0.3 per cent. This implies that virtually all parts of the Asia Pacific are under one form of anthropogenic influence or the other.

2.2.3 The spatial correspondence of biomes and soil quality

The distribution of biomes by different land classes is shown in the left-hand panel of Table 2.2.1. To simplify the analysis, the croplands biome

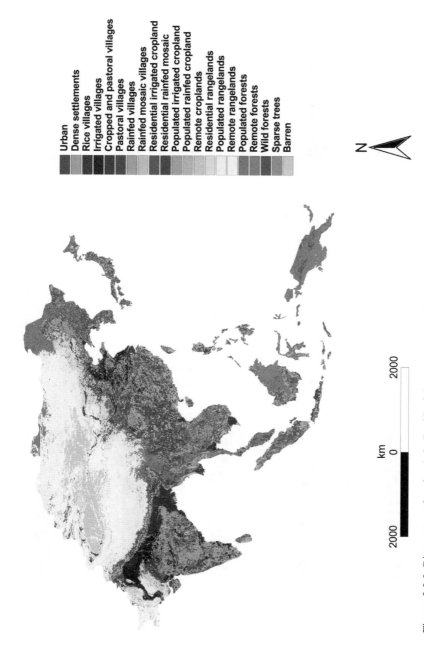

Figure 2.2.3 Biome map for the Asia Pacific (please see page 474 for a colour version of this figure).
Source: Based on Ellis and Ramankutty (2008).

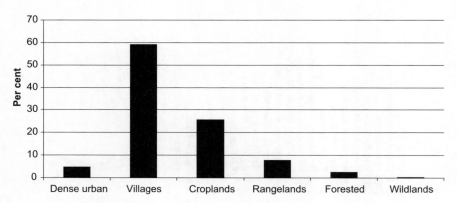

Figure 2.2.4 The distribution of broad biome groupings in the Asia Pacific.

was merged with villages to produce the "agriculture" biome, and the rangelands biome was incorporated into wildlands. With the exception of Class 8 land, the highest proportion of all land classes is occupied by agriculture. Agriculture biomes vary from 66.2 per cent for Class 9 land to 94.1 per cent for Class 2 land. Competition between agriculture and urban uses for the soils most suitable for agriculture (Classes 1 to 6) is apparent. Remarkably, the proportion of primelands (Class 1) supporting urban areas is the highest (11.1 per cent). Cities have traditionally been established in areas with high agricultural potential (Marcotullio et al., 2008). As the world "urbanizes", agricultural land is transformed into urban land, of which a proportion is high-quality farmland. The extent of

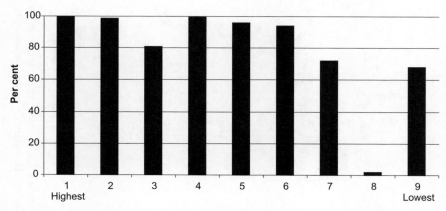

Figure 2.2.5 The extent of urban and agricultural land use on different land classes.

urban and agricultural land use on different lands is shown in Figure 2.2.5. The two uses virtually dominate Classes 1, 2, 4, 5 and 6 soils. This implies that future agricultural expansion will occur on marginal lands. It is also noteworthy that farming has encroached on 66.2 per cent of fragile Class 9 lands (Table 2.2.1). Affected parts of the Asia Pacific include Pakistan, where more than 75 per cent of the >22 million ha of cultivated area is under irrigation (Ashraf et al., 2007). Waterlogging and soil salinity resulting from seepage from canals, poor drainage and poor water management threaten the sustainability of these irrigated lands (Vlek et al., 2008). This leads to increased vulnerability to natural disasters and permanent damage to biodiversity. The need to increase food production will make irrigation indispensable, so efforts to minimize the impacts of irrigation in fragile ecosystems will remain the principal challenge.

Finite land resources are expected to face increasing demands from arable, grazing, forestry, wildlife and urban uses. With the rise in the level of affluence, soil resources are expected to be scarcer, competition among land uses for the available land fiercer, and conflicts over land use more severe. Land-use planning provides the context for the use and development of land, usually through regulatory controls. As a decision-making process, it facilitates land allocation for the use that provides greatest sustainable benefits (United Nations Conference on Environment and Development, 1993). Land parcels exhibit wide variability in quality, which in turn influences their suitability for various uses. Societal values and economic factors also influence land-use decisions. Thus, land-use planning entails the systematic assessment of physical, social and economic factors so as to encourage and assist land users in selecting options that increase their productivity, are environmentally sustainable and satisfy the needs of the society (Food and Agriculture Organization, 1993).

Ideal biome patterns based on soil quality and its potential uses are shown in the middle panel of Table 2.2.1. Under ideal land-use planning, the allocation of land to agriculture decreases progressively from 70 per cent of Class 1 land to 5 per cent for Classes 8 and 9. In practice, deviations from the ideal land-use pattern usually occur as a result of several factors. For instance, economic factors usually influence land-use decision-making through prices, taxes and subsidies on land-use inputs (Braimoh, 2009). In other cases, institutional and other sociocultural factors may have an overriding influence on economic variables in land-use decisions (Nagendra, 2007). The right-hand panel of Table 2.2.1 displays the deviation of the observed biome patterns from the ideal. Urbanization has utilized about 6 per cent above the ideal proportion supported by primelands and Class 4 land. The proportion of urban land use ideally supported by Class 5 lands has also been surpassed, although to a lesser degree (0.3 per cent) compared with Classes 1 and 4. This further confirms

Table 2.2.1 Observed and ideal distribution of biomes by land class (per cent)

Land/soil class	Observed biome pattern				Ideal biome pattern[a]				Difference between ideal and observed pattern			
	Anthropogenic		Natural		Anthropogenic		Natural		Anthropogenic		Natural	
	Urban	Agri-culture	Forested	Wildlands	Urban	Agri-culture	Forested	Wildlands	Urban	Agri-culture	Forested	Wildlands
1 Highest	11.1	88.8	0.0	0.1	5	70	20	5	-6.1	-18.8	20.0	4.9
2	4.8	94.1	0.9	0.2	5	60	30	5	0.2	-34.1	29.1	4.8
3	4.8	76.1	0.2	18.9	5	50	35	10	0.2	-26.1	34.8	-8.9
4	10.7	89.0	0.3	0.1	5	45	40	10	-5.7	-44.0	39.7	9.9
5	5.3	90.8	2.9	0.9	5	40	45	10	-0.3	-50.8	42.1	9.1
6	4.9	89.0	5.4	0.6	5	30	50	15	0.1	-59.0	44.6	14.4
7	1.6	70.5	11.0	16.8	5	10	50	35	3.4	-60.5	39.0	18.2
8	0.0	2.1	1.8	96.0	5	5	60	30	5.0	2.9	58.2	-66.0
9 Lowest	1.7	66.2	0.1	32.0	5	5	30	60	3.3	-61.2	29.9	28.0

[a] Modified from Beinroth et al. (2001).

the preference of human settlement for primelands (Marcotullio et al., 2008). There is still scope for urban expansion on the other land classes, ranging from 0.1 per cent for Class 6 to 5.0 per cent for Class 8. With the exception of Class 8, the ideal proportion of agricultural land use supported by the various land classes has been surpassed by amounts ranging from 18.8 per cent for Class 1 to about 61 per cent for Classes 7 and 9. This mismatch between land quality and land use, coupled with poor agricultural land management, accentuates soil degradation. One estimate (Van Lyden and Oldeman, 1997) indicates that water erosion (the most destructive type of soil degradation) affects 21 per cent of South and Southeast Asia, and is predominant in large parts of China (>180 million ha) and the Indian subcontinent (>90 million ha) and in the sloping parts of Indochina (40 million ha), Indonesia (22.5 million ha) and the Philippines (10 million ha).

The differences between the ideal and the observed forested biomes are positive for all land classes, indicating lower forest cover than is required. This suggests intense pressure on forests in the region. Deforestation rates in Southeast Asia are higher than in other tropical areas of the world. The proximate causes of biodiversity threat in the region include forest conversion to perennial crops, logging, forest fires, hunting for bushmeat and wildlife trade (Sodhi et al., 2004). The underlying factors in deforestation include population pressure, poverty, wealth and entitlement factors, and the property regime.

2.2.4 Future outlook

Civilization will probably depend on soils more crucially than ever in the future owing to the increase in anthropogenic demands on soil resources (Hillel, 2008). Up to the end of the twentieth century, the principal function of soils was the provision of ecosystem services, i.e. the production of food, feed and fibre. However, the demands of modern civilization far exceed these traditional functions, and include regulatory, supporting and cultural ecosystem services, including climate change mitigation, enhancement of biodiversity, desertification control, waste management, meeting energy needs and functioning as an archive of planetary and human history (Lal, 2007). This increasing role of soils has necessitated a renewed commitment to integrated planning and management to conserve soil resources.

The Food and Agriculture Organization framework for evaluating sustainable land management (FESLM) provides a platform to fine-tune strategies for sustainable management of soils vis-à-vis this expanding role. The overall aim of the paradigm is to integrate the sustainability principle into land use and management by harmonizing the complementary goals of providing environmental, economic and social opportunities for the benefit of present and future generations, while maintaining and enhancing soil quality (Smyth and Dumanski, 1993). FESLM is a multi-stakeholder approach based on a systematic procedure (pathway) for the identification and development of indicators and thresholds of sustainability. As a proactive approach for making decisions on land management systems, FESLM is based on five sustainability pillars: the maintenance of productivity; a reduction in the level of production risk; protection against soil degradation; economic viability; and social acceptability. Sustainability within the context of FESLM is a measure of the extent to which the overall objective of sustainable land management can be met by a defined land use on a specific tract of land over a period of time. For instance, if a piece of land is suitable to produce desired levels of crop yields but the envisaged land management system degrades soil quality, then this use is not recommended.

Despite its apparent usefulness with respect to the expanding multi-functional role of soils, FESLM has not been tested in several parts of Asia. Practical tools that could help local stakeholders and multidisciplinary scientists to work together and apply FESLM and associated techniques have been developed recently (Cauwenbergh et al., 2007; Hurni, 2000; Lefroy et al., 2000; Wiek and Binder, 2005). To improve its practical value, the FESLM procedure should be adapted first to include non-agricultural uses and other benefits derived from the soil. Second, stakeholder participation must be integrated in all stages of the trans-disciplinary assessment. Third, interactions between indicators should be considered so that trade-offs can be explicitly analysed. Fourth, the commitment to protect the soil in Asia must address forest conservation challenges. Most of the threats to forests are apparently linked to rural poverty, a dearth of conservation resources and weak enforcement of legislation, so crucial solutions for sustainable land management should include the provision of economic incentives and of appropriate technology for soil conservation.

Acknowledgements

I thank Erle Ellis and Navin Ramankutty for granting access to the biome data, and Paul Reich and Hari Eswaran for the land quality data. I also thank Ayo Ogunkunle for insightful discussions whilst preparing the manuscript.

REFERENCES

Ashraf, M., M. A. Kahlown and A. Ashfaq (2007) "Impact of Small Dams on Agriculture and Groundwater Development: A Case Study from Pakistan", *Agricultural Water Management* 92: 90–98.

Beinroth, F. H., H. Eswaran and P. H. Reich (2001) "Land Quality and Food Security in Asia", in E. M. Bridges, I. D. Hannam, L. R. Oldeman, F. W. T. Pening de Vries, S. Scherr and S. Sompatpanit (eds), *Responses to Land Degradation*. New Delhi: Oxford Press.

Braimoh, A. K. (2009) "Agricultural Land-use Change during Economic Reforms in Ghana", *Land Use Policy* 26: 763–771.

Braimoh, A. K. and P. L. G. Vlek, eds (2008) *Land Use and Soil Resources*. Berlin: Springer.

Cauwenbergh, N. V., K. Biala, C. Bielders, V. Brouckaert, L. Franchois, V. Garcia Cidad, M. Hermy, E. Mathijs, B. Muys, J. Reijnders, X. Sauvenier, J. Valckx, M. Vanclooster, B. Van der Veken, E. Wauters and A. Peeters (2007) "SAFE – A Hierarchical Framework for Assessing the Sustainability of Agricultural Systems", *Agriculture, Ecosystems and Environment* 120: 229–242.

Cohen, Joel E. (2003) "The Human Population: Next Century", *Science* 302: 1172–1175.

Ellis, E. C. and N. Ramankutty (2008) "Putting People in the Map: Anthropogenic Biomes of the World", *Frontiers in Ecology and the Environment* 6(8): 439–447.

Eswaran, H., F. Beinroth and P. Reich (1999) "Global Land Resources and Population Supporting Capacity", *American Journal of Alternative Agriculture* 14: 1129–1136.

Foley, J. A., R. DeFries, G. P. Asner, C. Barford, G. Bonan, S. R. Carpenter, F. S. Chapin, M. T. Coe, G. C. Daily, H. K. Gibbs, J. H. Helkowski, T. Holloway, E. A. Howard, C. J. Kucharik, C. Monfreda, J. A. Patz, I. C. Prentice, N. Ramankutty and P. K. Snyder (2005) "Global Consequences of Land Use", *Science* 309: 570–574.

Food and Agriculture Organization (1993) *Guidelines for Land-Use Planning*. Rome: FAO Development Series 1.

Fox, J. and J. B. Vogler (2005) "Land-use and Land-cover Change in Montane Mainland Southeast Asia", *Environmental Management* 36(3): 394–403.

Hillel, Daniel (2008) *Soil in the Environment*. London: Academic Press.

Hurni, Hans (2000) "Assessing Sustainable Land Management (SLM)", *Agriculture, Ecosystems and Environment* 81: 83–92.

Karlen, D. L., S. S. Andrews and J. W. Doran (2001) "Soil Quality: Current Concepts and Applications", *Advances in Agronomy* 74: 1–40.

Lal, Rattan (1989) "Land Degradation and Its Impact on Food and other Resources", in D. Pimentel (ed.), *Food and Natural Resources*. San Diego: Academic Press, pp. 85–140.

Lal, Rattan (1997) "Degradation and Resilience of Soils", *Philosophical Transactions of the Royal Society of Britain* 352: 997–1010.

Lal, Rattan (2007) "Soil Science and the Carbon Civilization", *Soil Science Society of America* 71: 1425–1437.

Lefroy, R. D. B., H. Bechstedt and M. Rais (2000) "Indicators for Sustainable Land Management Based on Farmer Surveys in Vietnam, Indonesia, and Thailand", *Agriculture, Ecosystems and Management* 81: 137–146.

Marcotullio, P. J., T. Onishi and A. K. Braimoh (2008) "The Impact of Urbanization on Soils", in A. K. Braimoh and P. L. G. Vlek (eds), *Land Use and Soil Resources*. Berlin: Springer, pp. 201–250.

Nagendra, Harini (2007) "Going Beyond Panaceas. Special Feature: Drivers of Reforestation in Human-Dominated Forests", *Proceedings of the National Academy of Sciences* 104(39): 15218–15223.

Olson, D. M., E. Dinerstein, D. Wikramanayake, N. D. Burgess, G. V. N. Powell, J. A. Underwood, I. Itoua, H. E. Strand, J. C. Morrison, O. L. Loucks, T. F. Allnut, T. H. Ricketts, Y. Kura, J. F. Lamoreux, W. W. Wettengel, P. Hedao and K. R. Kassem (2001) "Terrestrial Ecoregions of the World: A New Map of Life on Earth", *Bioscience* 51: 933–938.

Ramankutty, N., J. A. Foley and N. J. Olejniczak (2008) "Land-use Change and Global Food Production", in A. K. Braimoh and P. L. G. Vlek (eds), *Land Use and Soil Resources*. Berlin: Springer, pp. 23–40.

Sala, O. E., F. S. Chapin III, J. J. Armesto, E. Berlow, J. Bloomfield, R. Dirzo, E. Huber-Sanwald, L. F. Huenneke, R. B. Jackson, A. Kinzig, R. Leemans, D. M. Lodge, H. A. Mooney, M. Oesterheld, N. L. Poff, M. T. Sykes, B. H. Walker, M. Walker and D. H. Wall (2000) "Global Biodiversity Scenarios for the Year 2100", *Science* 287(5459): 1770–1774.

Smyth, A. J. and J. Dumanski (1993) *FESLM: An International Framework for Evaluating Sustainable Land Management*. World Soil Resources Report 73. Rome: Food and Agriculture Organization of the United Nations.

Sodhi, N. S., L. P. Koh, B. W. Brook and P. K. L. Ng (2004) "Southeast Asian Biodiversity: An Impending Disaster", *Trends in Ecology and Evolution* 19: 654–660.

United Nations Conference on Environment and Development (1993) *Agenda 21: Earth Summit – The United Nations Programme of Action from Rio*. New York: United Nations Department of Public Information.

United Nations Environment Programme (2002) *Global Environment Outlook 3. Past, Present and Future Perspectives*. London: Earthscan.

Van Lyden, G. W. J. and L. R. Oldeman (1997) *Assessment of the Status of Human-Induced Soil Degradation in South and Southeast Asia*. Amsterdam: International Soil Reference and Information Centre (ISRIC).

Vlek, P. L. G., R. F. Kühne and M. Denich (1997) "Nutrient Resources for Crop Production in the Tropics", *Philosophical Transactions of the Royal Society of Britain* 352: 975–985.

Vlek, P. L. G., D. Hillel and A. K. Braimoh (2008) "Soil Degradation under Irrigation", in A. K. Braimoh and P. L. G. Vlek (eds), *Land Use and Soil Resources*. Berlin: Springer, pp. 101–119.

Wiek, Arnim and Claudia Binder (2005) "Solution Spaces for Decision-making – A Sustainability Assessment Tool for City-Regions", *Environmental Impact Assessment Review* 25: 589–608.

2-3
The role of biological control in sustainable agriculture

Anthony R. Chittenden and Yutaka Saito

2-3-1 Introduction

More than 40 per cent of world food production is lost to insect pests, plant pathogens and weeds every year, despite the annual application of over 3 million metric tons of pesticides. Proven pest management practices developed over several millennia have been supplanted by a vast amount of synthetic chemicals. Although the amount of pesticides applied to crops in the United States has increased 10-fold since 1945, crop losses to insects have doubled over the same period (Pimentel, 2008).

Humans have long utilized biological control methods to manage insect pests. Exotic pests were initially controlled by natural enemy species collected from the same area or country of origin as the pest, an approach known as classical control. In addition, augmentative control techniques have enjoyed success against a wide range of open-field and greenhouse pests, and conservation biological control schemes that use indigenous predators and parasitoids have been developed.

Biological control involves the use of one organism to reduce the population density of another; it can apply to the control of animals, pathogens or weeds. In natural ecosystems, organisms often consume other organisms, thus reducing their populations. This phenomenon is known as *natural control*. Biological control involves selective exploitation of these "natural" relationships by humans to suppress pest species. The objective is the same, namely reduction rather than eradication of pests, such that both the pest and its natural enemies remain in ecosystems at

---

Designing our future: Local perspectives on bioproduction, ecosystems and humanity,
Osaki, Braimoh and Nakagami (eds),
United Nations University Press, 2011, ISBN 978-92-808-1183-4

low densities for long periods. However, repeated releases of natural enemies and/or additional methods are sometimes needed to achieve the desired level of control. Additional methods may include cultural techniques, such as the use of physical barriers or resistant plant strains, or even the judicious application of chemicals, if necessary. This is one of the basic tenets of integrated pest management (Stern et al., 1959).

Biological control has become an integral part of pest management programmes worldwide. The surging development of agriculture witnessed during the twentieth century led to an unprecedented increase in international trade and a concomitant surge in the global transfer of pest species on plants and products. This, in conjunction with human practices, such as the introduction of new crops into new areas, and consumer demands for blemish-free produce, has led to a chronic over-reliance on chemical methods of pest control. The consequences of this dependence have been well described: pest resistance, rising production costs, bioaccumulation in food chains, environmental pollution, loss of biodiversity and human health risks (Bale et al., 2008; Carson, 1962).

Compared with insecticides, biological control techniques enjoy favourable cost–benefit ratios and much lower development costs. Therefore, biological control methods can provide potent solutions to insecticide-resistant pests and allow reductions or even complete cessation of pesticide usage, making them essential components in the establishment of a "systems approach" to integrated pest management. If we are going to safely feed and live on a planet with a population rapidly approaching 7 billion people, the need for clean, sustainable methods of effective pest control is more urgent than ever. Moreover, understanding biodiversity in relation to biological control can play an essential role in local sustainability.

2-3-2 Types of biological control

Biological control using natural enemies can be broadly classified into three main categories: *classical*, *augmentative* and *conservation* control (Van Lenteren, 1993).

Classical control is typically employed against "exotic" pests that have become established in new regions (Type 1 in Figure 2.3.1). Potential natural enemies are sought in the region or country where the pest originated and, once a suitable candidate has been identified, small numbers are collected for rearing. The natural enemy is then released into the new environment where it will ideally build up a long-term level of control. It is most suited to perennial crop and orchard situations, where the stability of the ecosystem allows the pest–natural enemy system to become established over time. One of the earliest successes of classical control occurred in California in 1888, when the predatory coccinellid *Rodolia cardinalis* was imported from Australia for use against the cottony cushion scale *Icerya purchasi* on citrus (Caltagirone, 1981).

Augmentative control refers to a method of biological control in which a large number of commercially produced natural enemies are regularly introduced into crop ecosystems (Type 2 in Figure 2.3.1). It is best suited to short-term annual crops, the cultivation systems of which disrupt predator–prey relationships and prevent natural enemies from establishing themselves. *Inundative* release techniques seek immediate reduction of potentially damaging pest populations, typically through released individuals rather than their offspring. However, the control achieved is often transient and further releases may be necessary. Furthermore, the unpredictable nature of natural enemy activity and the costs of purchasing them limit this approach to certain situations, such as mass releases of *Trichogramma* egg parasites to control lepidopteran pests (Smith, 1996). On the other hand, *seasonal inoculations* involve the release of small numbers of mass-reared natural enemies at prescribed intervals throughout the pest period, beginning when the pest density is very low. The natural enemies are expected to establish themselves to provide long-term control. Continued releases are expected to keep the pest at low numbers and prevent it from approaching an economic injury level. This method is most often utilized when the cropping system prevents the use of classical biological control; for example, using parasitoids and predators to control thrips, aphids and mites in greenhouses (Van Lenteren, 2000).

Conservation control utilizes indigenous predators and parasitoids, primarily against native pests, through the preservation of pest, natural enemy and crop biodiversity (Type 3 in Figure 2.3.1). A raft of measures, direct and indirect, may be employed to enhance the efficacy of natural enemies. These may include manipulating crop microclimates, creating overwintering refuges ("beetle banks"), increasing alternative host and prey availability ("banker plants") and providing essential food resources for adult parasitoids (Gurr et al., 2000).

Although capable of providing long-term regulation of agricultural pests at levels difficult or impossible to attain using traditional chemical means, biological control is not entirely risk free. Introduced natural enemies may negatively affect native ecosystems, directly or indirectly. Recent evidence of non-target host use by bio-control agents has polarized scientists. Once touted as an "environmentally safe" way to control pests without toxic chemicals, biological control is now considered by its opponents to be "a cure worse than the disease". Various nations and organizations have responded by drawing up guidelines to determine the safely of natural enemies, the Code of Conduct for the Import and

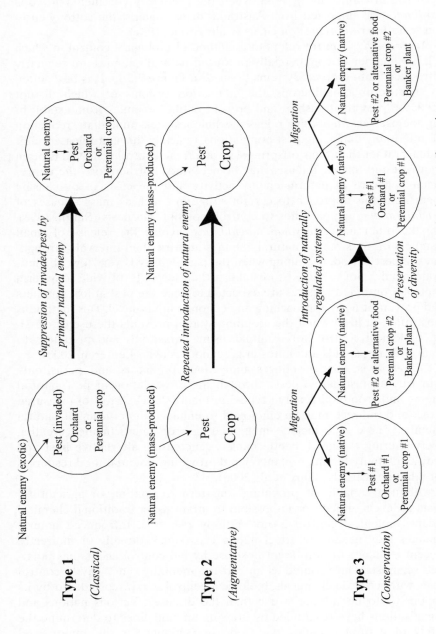

Figure 2.3.1 Three main categories of biological control using natural enemies.

Release of Exotic Biological Control Agents (FAO, 1996) being one such example. One result from this is that polyphagous control agents or those requiring non-targets as part of their life cycle are now rarely considered for release. However, given the complexity of environmental interactions, predicting the outcome of biological control introductions will always be a formidable task.

2-3-3 Two examples of successful biological control

Classical biological control of cassava pests (an example of Type 1)

Cassava (*Manihot esculenta*) is one of the world's most important food crops; the roots and leaves are an essential source of food and income throughout the tropics. About 600 million people worldwide depend on cassava for food and livelihood (IFAD, 2007; and see Table 2.3.1). First introduced into Africa from South America by Portuguese settlers in the sixteenth century, cassava is rich in calories and highly drought tolerant. It thrives in poor soils and is easy to store in the ground. More than 200 million Africans rely on the 101 million metric tons of cassava produced in the "Cassava belt" every year (Radcliffe et al., 2009; IFAD, 2007).

Cassava remained relatively free from arthropod pests until the early 1970s, when *Mononychellus tanajoa* (cassava green mite) and *Phenacoccus manihoti* (cassava mealybug) were both detected in East Africa. These pests spread rapidly and caused up to 80 per cent reduction in cassava root yield in some areas (Hajek, 2004). Two classical biological control programmes were established. The initial step required identifying the pests so that they could be recognized and collected in South America. After identification, the search for potential predators could begin. *Anagyrus lopezi*, an encyrtid wasp collected in Paraguay in 1981, and *Typhlodromalus aripo*, a phytoseiid mite collected in Brazil in 1988, have

Table 2.3.1 World's major cassava producers by area and volume

Country	Area under cultivation (hectares)	Production (metric tons)
Nigeria	3,455,000	34,476,000
Democratic Republic of Congo	1,839,962	14,929,410
Brazil	1,685,275	23,108,076
Indonesia	1,290,000	16,723,257
Thailand	1,030,000	16,870,000

Source: Nasser and Ortiz (2007).

Table 2.3.2 Biological control programmes for two major cassava pests

Pest	Year of first occurrence	Biological control agent	Loss reduction due to biological control[c]
Cassava mealybug P. manihoti	1973[a]	Encyrtid wasp A. lopezi	95%
Cassava green mite M. tanajoa	1971[b]	Phytoseiid mite T. aripo	50%

Sources:
[a] Zeddies et al. (2001).
[b] Neuenschwander (2004).
[c] IITA (2009).

proved very successful, achieving 95 per cent reduction in cassava mealybug damage and 50 per cent reduction in cassava green mite damage, respectively (IITA, 2009; and see Table 2.3.2). These programmes are estimated to have saved African subsistence farmers hundreds of millions of dollars and have secured food supplies for over 200 million people (Radcliffe et al., 2009).

Biological control by restoring biodiversity in moso bamboo plantations (an example of Type 3)

As emphasized previously, increasing biodiversity is fundamental in developing new methods of sustainable pest management. Use of "banker plants" is an effective new method for introducing such biodiversity (Type 3). However, it is still unclear how exactly the conservation of biodiversity is related to effective biological control, especially in large-scale agricultural situations, even though small-scale studies have confirmed it.

The Type 3 method of biological control (Figure 2.3.1) is an example of an "old yet new" concept that attempts to utilize endemic fauna that existed in an area before agricultural practices began. It is difficult, sometimes impossible, to completely restore such fauna, because many of the faunal components have been lost through agricultural activities. Therefore, like the Type 2 method, the search for effective biological control agents sometimes follows a hit-and-miss approach. The possibility of finding such agents increases considerably if plantations where the cultivation style has recently reverted to polyculture are observed. Such a situation is described here and the importance of biodiversity in pest management is indicated.

Moso bamboo (*Phyllostachys pubescens*) is endemic to China. Its culms are utilized in industry, housing and handicrafts, and its shoots are a prized food item. Furthermore, moso bamboo forests play an important role in preventing soil erosion in mountainous areas. Outbreaks of pest mites began in the late 1980s, and many bamboo forests suffered defoliation. Of the approximately 275,000 hectares of moso bamboo in Fujian Province, an estimated 35 per cent suffered moderate to heavy damage from phytophagous mites (Zhang et al., 2000).

Change in cultivation practices is thought to be an important underlying reason for mite outbreaks. Traditionally, Chinese moso bamboo plantations were mixed forests consisting of many bamboo and grass species as well as deciduous and coniferous vegetation types. In the late 1980s, a concerted effort was made to remove all vegetation types, except moso bamboo, to increase yields. Although this shift to monoculture initially resulted in dramatically higher productivity, yields soon plummeted as outbreaks of pest mites took a heavy toll on plant vitality (Zhang et al., 2000). Comparative analysis of both cultivation types revealed that monoculture forests have much lower predatory mite diversity (Zhang et al., 2000). However, precisely how predator diversity was connected to the pest mite outbreaks was not well understood, and a decision was taken to restore a bare minimum of plant diversity (polyculture) back into several moso forests. To realize this, it was essential to identify the key predators and to know how to maintain them in the monoculture moso bamboo plantations. Zhang et al. (2000) reported that the spider mite *Stigmaeopsis nanjingensis* was the most important moso bamboo pest. Furthermore, the intermittent lack of natural enemies, which would have ordinarily suppressed *S. nanjingensis* populations before the change in cultivation practice, was considered responsible for the mite outbreaks. These included *Typhlodromus bambusae*, a phytoseiid mite known to be a specific regulator of *Stigmaeopsis* spp. in Sasa forests in Japan. Figure 2.3.2 shows the population fluctuations of *S. nanjingensis* and *T. bambusae* in heavily damaged (monoculture) moso bamboo forests. During the two-year survey period, *T. bambusae* repeatedly disappeared from bamboo plants. These disappearances were always accompanied by subsequent outbreaks of *S. nanjingensis*.

Discovering the reason for these frequent predator disappearances was the next task. A key point for researchers was that *T. bambusae* occurs only with *Stigmaeopsis* species in Japan. It has adapted to living under *Stigmaeopsis* woven nests to such an extent that it has difficulty reproducing without such nests. Therefore, it was hypothesized that polyculture moso plantations contain several plant species in addition to moso bamboo on which *Stigmaeopsis* species may occur. Following this, Zhang et al. (2004) discovered that *T. bambusae* could prey upon *Stigmaeopsis miscanthi*, a spider mite species occurring on Chinese silvergrass (*Miscanthus sinensis*) as an alternative prey. As expected from its scientific name, this plant species is widely distributed throughout China and other Southeast

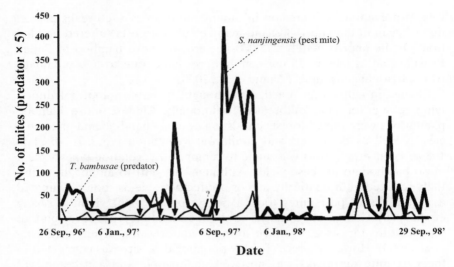

Figure 2.3.2 Population dynamics of *S. nanjingensis* and *T. bambusae* on *P. pubescens* in a monoculture bamboo plantation.
Source: Zhang et al. (2000).

Asian countries. Furthermore, it is a common undergrowth component of polyculture moso bamboo forests in China. *S. miscanthi* never reproduces on moso bamboo leaves, and *T. bambusae* obtained from *M. sinensis* are the same as those obtained from moso bamboo forests, readily able to feed upon *S. nanjingensis* (Zhang, personal communication).

Zhang et al. (2004) discovered that Chinese silvergrass (*M. sinensis*) acts as an important nursery plant where *T. bambusae* can persist. A one-year study of the population dynamics of *S. miscanthi* and *T. bambusae* (all stages were lumped) on *M. sinensis* in a polyculture bamboo forest showed that there is a stable and typical predator–prey interaction, as expected theoretically by Lotka (1925). Saito et al. (2008) thus developed a simulation model to ascertain whether the moso bamboo and Chinese silvergrass systems attain stability through inter-plant migration of predators. Their model showed three possible outcomes: both subsystems become stable; one system becomes stable, but the other does not (both predator and prey gradually increase infinitely); and both systems become unstable. Furthermore, they revealed that migration of the predator species between the two subsystems often stabilizes the system at low levels of prey and predator density (Figure 2.3.3), suggesting that effective control of pest mite outbreaks may be possible by simply recovering a modicum of biodiversity. These findings are thought to partly explain the phenomenon that has occurred in Fujian Province: a change in bamboo

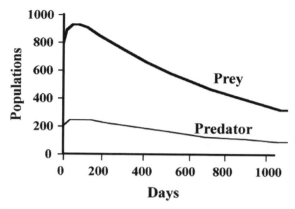

Figure 2.3.3 Stable state of two plants and one common predator in a simulated system.
Source: Saito et al. (2008).

cultivation practice from polyculture to monoculture and the subsequent elimination of Chinese silvergrass vegetation may have triggered the outbreaks of *S. nanjingensis*.

Reintroducing Chinese silvergrass (*M. sinensis*) is necessary to protect moso bamboo plantations in China. Although traditional polyculture moso plantations involve high plant and arthropod diversity, it is noteworthy that even the recovery of a single plant species can greatly improve system stability, i.e. sustainability in agricultural fields. Therefore, conservation of diversified systems must contribute to the stability of bio-systems as well as provide biological control of pest species using endemic natural enemies at the broad-scale level. In other words, biodiversity seems to be an effective regulator of pests.

2-3-4 The sustainability of "*satoyama*"

As shown in the previous subsection, a highly successful method for regulating pest mite outbreaks in Chinese bamboo plantations could be developed by simply reintroducing biodiversity.

Moso bamboo, introduced into Japan approximately 250 years ago from China through the Ryukyu Islands (according to one theory), once played an important role in rural Japan, both as a material for housing and handicrafts and as a source of food. However, moso bamboo is now destroying many traditional Japanese forests (or *satoyama*[1]). Competition from cheaper bamboo products imported from Fujian and other Chinese provinces has resulted in many ageing Japanese bamboo farmers

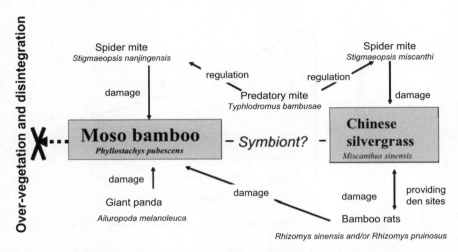

Figure 2.3.4 Hypothetical historical relationships in native moso bamboo forests.
Note: It is assumed that over-vegetation is disadvantageous to the sustainability of moso bamboo forests.
Source: Saito et al. (2008).

abandoning their forests, which often occupy steep hillsides. Poorly maintained bamboo forests rapidly cover over and eliminate native plant species and may subsequently trigger landslides.

Biological pest management, if widely adopted in China, might have further adverse affects on the sustainability of Japanese *satoyama*. This fundamental conflict between Chinese and Japanese bamboo plantations provides a cogent example of how establishing sustainable agriculture in one country may sometimes lead to unexpected problems for sustainability in another.

To resolve this problem, the mechanisms that prevent moso bamboo from crowding out and eliminating other plant species in its native areas, such as Fujian Province in China, need to be identified. During the course of the moso bamboo study, it was learned of two bamboo rat species that feed on moso bamboo shoots. Furthermore, giant pandas were once prolific consumers of bamboo shoots, even in Fujian Province (Figure 2.3.4). How these species regulated moso bamboo in the past is uncertain, but it has been suggested that bamboo rats use Chinese silvergrass for their den sites (Figure 2.3.4; Xu, 1989). However, the idea of introducing exotic pest species (mites or bamboo rats) into Japan remains mere speculation because of the inherent dangers such actions pose to indigenous flora and fauna. Thus, conserving Japanese *satoyama* through the prevention of unchecked moso bamboo growth continues to be a serious and unsolved problem.

2-3-5 The future

Despite the demonstrated ability of biological control programmes to provide successful pest management, the task of persuading farmers to adopt a systems approach to pest management and make greater use of biological control remains a formidable one.

Greater networking among the world's biological control community will facilitate the spread of new techniques, and compilation of easily accessible natural enemy databases will increase the identification rate of new control agents. It is also essential that farmers receive proper training in the adoption of new control techniques and agents as these techniques become available. FAO Farmers Field School (FFS) projects in Asia and Africa have shown that, with appropriate training, farmers quickly learn to select the most appropriate pest management strategy for their crops and many shrug off their reliance on chemicals in favour of cultural and biological methods of pest control (Ooi and Kenmore, 2005). It has even been suggested that the FFS approach to achieving sustainable pest management could be adapted to the developed world as well (Bale et al., 2008). However, until the proponents of biological control generate greater societal awareness about the advantages of sustainable and environmentally friendly pest management, most farmers will continue using conventional chemical control because, despite its many successes in many regions of the world, the incredible potential of biological control remains frustratingly unfulfilled, and its adoption and implementation painfully slow.

It is inevitable that future pest management will be heavily dependent on biological control because it is the most sustainable, cheapest and most environmentally safe system of pest management, with numerous additional benefits for both growers and consumers.

Note

1. See Section 1-1 for a more detailed definition of *satoyama*.

REFERENCES

Bale, J. S., J. C. van Lenteren and F. Bigler (2008) "Biological Control and Sustainable Food Production", *Philosophical Transactions of the Royal Society B* 363: 761–766.

Caltagirone, Leopoldo E. (1981) "Landmark Examples in Classical Biological Control", *Annual Review of Entomology* 26: 213–232.

Carson, Rachel (1962) *Silent Spring*. Harmondsworth, Middlesex: Penguin Books.

FAO [Food and Agriculture Organization of the United Nations] (1996) "Code of Conduct for the Import and Release of Exotic Biological Control Agents". *International Standards for Phytosanitary Measures* no. 3. Rome: FAO.

Gurr, G. M., S. D. Wratten and P. Barbosa (2000) "Success in Conservation Biological Control of Arthropods", in G. Gurr and S. Wratten (eds), *Measures of Success in Biological Control*. Dordrecht: Kluwer Academic Publishers, pp. 105–132.

Hajek, Ann (2004) *Natural Enemies: An Introduction to Biological Control*. Cambridge: Cambridge University Press.

IFAD [International Fund for Agricultural Development] (2007) "Improving Marketing Strategies in Western and Central Africa", IFAD Factsheet, Rome. Available at: <http://www.ifad.org/pub/factsheet/cassava/e.pdf> (accessed 3 March 2010).

IITA [International Institute of Tropical Agriculture] (2009) "Cassava", <http://www.iita.org/cms/details/cassava_project_details.aspx?zoneid=63&articleid=267> (accessed 3 March 2010).

Lotka, Alfred J. (1925) *Elements of Physical Biology*. Baltimore, MD: Williams & Wilkins Co.

Nasser, N. M. A. and R. Ortiz (2007) "Cassava Improvement: Challenges and Impacts", *Journal of Agricultural Science* 145: 163–171.

Neuenschwander, Peter (2004) "Harnessing Nature in Africa: Biological Control Can Benefit the Pocket, Health and the Environment", *Nature* 432: 801–802.

Ooi, P. A. C. and P. E. Kenmore (2005) "Impact of Educating Farmers about Biological Control in Farmer Field Schools", in *Proceedings of International Symposium on Biological Control of Arthropods, 12–16 September 2005, Davos, Switzerland*, pp. 277–289.

Pimentel, David (2008) "Preface Special Issue: Conservation Biological Control", *Biological Control* 45: 171.

Radcliffe, E. B., W. D. Hutchison and R. E. Cancelado (2009) *Integrated Pest Management: Concepts, Tactics, Strategies and Case Studies*. Cambridge: Cambridge University Press.

Saito, Y., N. Tsuji and A. R. Chittenden (2008) "Sustainable Pest Management and Moso Bamboo Forest Conservation – A Preliminary Report", *Proceedings of ICSA* 2008: 309–316.

Smith, Sandy M. (1996) "Biological Control with Trichogramma: Advances, Successes, and Potential of Their Use", *Annual Review of Entomology* 41: 375–406.

Stern, V. M., R. F. Smith, R. van den Bosch and K. S. Hagen (1959) "The Integration of Chemical and Biological Control of the Spotted Alfalfa Aphid – The Integrated Control Concept", *Hilgardia* 29: 81–101.

Van Lenteren, Joop C. (1993) "Biological Control of Pests", in J. C. Zadoks (ed.), *Modern Crop Protection: Developments and Perspectives*. Wageningen, The Netherlands: Wageningen Press, pp. 179–187.

Van Lenteren, Joop C. (2000) "A Greenhouse without Pesticides: Fact or Fantasy?", *Crop Protection* 19: 375–384.

Xu, Longhui (1989) "The Biology of Rhizomys pruinosus", *Acta Theriologica Sinica* 4(2): 100–105.

Zeddies, J., R. P. Schaab, P. Neuenschwander and H. R. Herren (2001) "Economics of Biological Control of Cassava Mealybug in Africa", *Agricultural Economics* 24: 209–219.

Zhang, Y.-X., Z.-Q. Zhang, L.-X. Tong, Q.-Y. Liu and M.-G. Song (2000) "Causes of Pest Mite Outbreaks in Bamboo Forests in Fujian, China: Analyses of Mite Damage in Monoculture versus Polyculture Stands", *Systematic Applied Acarology Special Publication* 4: 93–108.

Zhang, Y.-X., Z.-Q. Zhang, Y. Saito, Q. O. Liu and J. Ji (2004) "On the Causes of Mite Pest Outbreaks in Mono- and Poly-cultured Moso Bamboo Forests", *Chinese Journal of Applied Ecology* 15: 1161–1165 (in Chinese with English abstract).

2-4

The role of biochar in land and ecosystem sustainability

Makoto Ogawa

2-4-1 Biochar supporting sustainable agriculture in East Asia

Charcoal used in agriculture was originally called "charcoal for agricultural use" or "agrichar". In 2006, Lehmann et al. proposed the term "biochar" (i.e. charcoal or biomass-derived black carbon) and they recommend its application not only for the benefit of crop production but also for carbon sequestration (Lehmann et al., 2006). They established an International Biochar Initiative in 2007 and expanded the movement all over the world; the reviews of biochar research were published as a monograph edited by soil scientists (Lehmann and Joseph, 2009). Responding to this activity, the Japan Biochar Association was established in April 2009.

In Asian countries, highly intensive agriculture has been popular since ancient times because of their high population density. Therefore, various traditional cultivation techniques had developed to increase productivity. All types of waste – human and livestock excreta, straw, leaf litter, grass, sewage, ash and rice husk charcoal – have been employed as fertilizers and soil conditioners in agriculture (Ogawa, 1987).

Particularly in the Far East, rice cultivation has enabled sustainable agriculture through a water culture system. Rice husk charcoal can be carbonized by simple methods in the field after harvesting, and has been used as one of commonest and cheapest materials in agriculture. Its application also helps to return silica to paddy-field soil.

It seems that rice husk charcoal has been used since the beginning of rice cultivation in Asia (i.e. for several thousand years), because rice husk with a high silica content is slightly decomposed in soil and compost. Such traditional crop production supported by organic fertilizers and charcoal seems to be a typical sustainable agricultural system, and has a much longer history than that of *terra preta* in the Amazon (Glaser et al., 2002). Even now it is very common to use rice husk charcoal for the improvement of soil properties in arable fields and nursery plots in Southeast Asian countries such as Japan and Korea.

The oldest description of charcoal use in agriculture has been found in a textbook, *Nogyo Zensho (Encyclopedia of Agriculture)* written by Yasusada Miyazaki in 1697 (Miyazaki, 1697). "After carbonization of wastes, condensed human excreta should be mixed with the charcoal and stacked for a while. When you apply this manure to the fields, it improves any crop yield, and is particularly good for legumes. This manure has been called 'charcoal compost'."

2-4-2 Rice husk charcoal specific to East Asia

The practice of using rice husk charcoal mixed with excreta was very common in wheat cultivation until a few decades ago. There are some benefits because porous charcoal has high absorption and retention capacities for nutrients and water. However, this method was too primitive for modern scientists to investigate it. This led to the neglect of the functions of charcoal in agriculture for a long time. After information on wood charcoal use was circulated in the 1980s, the role of rice husk charcoal was recognized again and began to be studied by researchers.

The effects of rice husk charcoal on the formation of arbuscular mycorrhiza of citrus seedlings were reported by Ishii and Kadoya (1994). Other researchers (Ezawa et al., 2002) reported that rice husk itself and the charcoal enhanced arbuscular mycorrhiza formation of some crop plants, improving the soil's physical properties when the material was added to the top soil. It was also suggested that a small amount of rice husk charcoal could boost the growth of *Catharanthus roseus*, but the browning of leaves appeared with excessive application because of the high concentration of potassium and the higher pH of rice husk charcoal than of wood charcoal (Komaki et al., 2002). Takagi and Takanashi (2003) proposed a practical method to reduce the runoff of pesticides and herbicides from paddy fields by utilizing the adsorbing ability of rice husk charcoal. The application of bacteria-enriched charcoal also successfully reduced the runoff of simazine from a golf course (Takagi and Yoshida,

2003). These techniques may be useful for reducing the damage caused by continuous cultivation using chemicals. It was confirmed from these results that rice husk charcoal has a function similar to that of woody charcoal despite their different physical properties and chemical composition.

In Southeast Asia, the use of charcoal in agriculture has been promoted primarily by the Japan International Cooperation Agency. In Indonesia, Dr T. Igarashi experimented with cultivating crop plants with rice husk charcoal in 1989 (Igarashi, 1996). He applied charcoal together with magnesium phosphate and lime in a soybean–maize rotation. The application of charcoal significantly enhanced root nodule formation, plant growth and yields. The effect was also sustained in the second crop of maize without fertilizers, and the residual effect was observed up to the tenth rotation. In particular, the growth and yield of maize treated with charcoal were more than that in the control plot cultivated only with chemical fertilizers. The effects of charcoal under different soil conditions depend on the soil properties and crop types (Igarashi, 2002).

Another research group reported changes in the chemical properties of soil and crop yields with the application of *Acacia mangium* bark charcoal made from pulp waste in Indonesia (Yamato et al., 2006). The yields of maize and peanut increased significantly with charcoal application under fertilized conditions in an acid soil with low fertility. The amount of root and colonization rate of arbuscular mycorrhiza increased especially in maize. Root nodule formation in leguminous plants was also stimulated by charcoal. Application of charcoal improved the chemical properties of soil by neutralizing soil pH and increasing (i) total nitrogen, (ii) available phosphate contents, (iii) cation exchange capacity (CEC), (iv) amount of exchangeable cations, and (v) base saturation. Moreover, it induced a reduction in exchangeable Al ion, which is harmful for root growth. Identical phenomena have been confirmed in other areas (Steiner et al., 2007).

In Thailand, the effects of rice husk charcoal on soybean were investigated in a sandy soil (Oka et al., 1993). Above-ground biomass, root biomass, soybean yield and rate of nitrogen fixation in the soil increased significantly. In fields, the application of 10 tons per hectare (t/ha) was most efficient, and the effect appeared evident in the second (sorghum) and the third (soybean) crops. The soil's physical properties, porosity, water retention capacity, pH and CEC also improved, although the changes varied with soil type.

In the Philippines, the application of charcoal (2.5 t/ha) and lime (1.5 t/ha) increased the number of root nodules and nitrogen fixing rates (Noguchi et al., 1993). The inoculation of root nodule bacteria and application of rice husk charcoal induced the same effect as a single use of lime. Therefore, the use of rice husk charcoal seems to be economic in the Philippines where the use of lime is expensive.

2-4-3 Multiple uses of woody charcoal

Studies on biochar in Japan originally started from woody charcoal made from waste wood (Kishimoto and Sugiura, 1985). In Japan, the domestic supply of energy sources was limited to forest resources until the beginning of the twentieth century. Charcoal production reached a maximum of 2.7 million tons per year in 1947. It is estimated that the wood of domestic broad-leaved trees (approximately 10 million tons) was carbonized only by traditional kilns (Ogawa, 1987). These forests had been maintained in a sustainable manner by coppice regeneration for centuries. These woody charcoals and their wastes were very valuable as a fuel and had never been used in agriculture. Only wood ash containing some cinders was applied in agriculture for soil improvement and mineral supply.

However, owing to the increase in imported fossil fuels in the 1960s, charcoal production decreased abruptly to a low of about 30,000 tons per year in the 1980s. As a result, the forest ecosystem changed completely owing to the abandonment of lowland forest, and induced an outbreak of pine wilting disease throughout the Japanese islands (Ogawa, 2007).

The first reports of biochar experiments were two articles written in 1983 on the effects of bark charcoal in soybean cultivation (Ogawa et al., 1983a) and ectomycorrhiza formation in pine trees (Ogawa et al., 1983b). After the publication of these works, charcoal, including rice husk charcoal, became more actively used. In 1986, the Technical Research Association for Multiuse of Carbonized Materials (TRA) was established with the support of the Japanese government, and studies were undertaken on the multi-use of charcoal. Studies on the effects of charcoal and wood vinegar were performed covering various aspects: the improvement of carbonization technology, soil conditioning, the activation of microorganisms, and water purification. Research results, together with some general comments, were published in 1990 (TRA, 1990). At present, woody charcoal is being used mainly in agriculture, revegetation, tree rehabilitation, house construction, water purification and sewage treatment, and other wastes have been carbonized as biochars (Shinogi et al., 2003). In Japan, total charcoal consumption, not including fuel, has reached approximately 100,000 tons per year, but about half of this charcoal has been imported from Southeast Asia where it is produced from coconut and oil palm shells.

2-4-4 Biochar and micro-organisms

In 1986, Ogawa and Yambe (1986) reported that charcoal powders containing a small amount of chemicals, called "charcoal fertilizer", were effective for the formation of arbuscular mycorrhiza and root nodules of soybean plants. The charcoal fertilizers used in this experiment were bark charcoal with 1 per cent (w/w) of the compound fertilizer N-P-K (8-8-8), urea, super lime phosphate, ammonium sulfate and rapeseed meal, respectively. These charcoal fertilizers were scattered over the soil surface at 500 g/m^2 and 1,500 g/m^2 each before ploughing. The plots to which the compound fertilizer was applied at 100 g/m^2 and 200 g/m^2 (the amounts used in the conventional cultivation of soybeans) each and the one without any treatment were set as controls.

Soybean yields from the plots with charcoal fertilizers of 500 g/m^2 were mostly equal to those from plots with only compound fertilizer. That is, the amount of chemical fertilizers could be reduced remarkably by the application of biochar. Root nodule formation was stimulated by charcoal fertilizers, but it was suppressed by the ones with ammonium sulfate and compound fertilizer. Arbuscular mycorrhiza formation rates and spore numbers also increased in the plots treated with charcoal fertilizers. Charcoal application did not have an effect on plant growth and microbial propagation. Therefore, it is certain that the better growth of the soybean plants resulted from enhanced root growth and the high activity of symbiotic micro-organisms.

Soil microbial flora showed significant changes depending on the charcoal application. Charcoal fertilizers with a pH >8 inhibited the propagation of soil fungi, but enhanced that of bacteria and actinomycetes soon after treatment, and gradually returned to the normal state in two months. It was suggested that carbon dioxide emissions were temporarily increased owing to such high bacterial activity.

The free-living nitrogen-fixing bacteria could be isolated on a nitrogen-free medium. From the inoculation test of the charcoal, which was sterilized and then buried in soil for one week, it was demonstrated that charcoal became a good habitat for the propagation of root nodule bacteria.

These results were subsequently confirmed by TRA. Wood charcoal could improve soil properties, but mixtures with chemical fertilizers, zeolite, wood vinegar and organic fertilizer exhibited better effects than charcoal alone on tea plants, citrus plants and vegetables (Ishigaki et al., 1990), on rice plants and apple trees (Okutu et al., 1990), as well as on some leguminous plants and grass for revegetation (Sano, 1990).

It was found that root nodule bacteria could be immobilized with high frequency in white charcoal produced under a high temperature of 800°C

(Takagi et al., 1990). Fine-pore oak charcoal was more suitable for the immobilization of bacteria and actinomycetes than pine charcoal (Oohira et al., 1992). Meanwhile, the immobilizing ability of pine charcoal was improved by the addition of acetic acid. Arbuscular mycorrhizal fungi showed better growth on black charcoal produced at 400–500°C, and the spores of *Gigaspora margarita* were frequently formed in black soil with a high carbon content (Soda et al., 1990). Application of wood charcoal to Alnus and Myrica associated with Frankia was also effective for tree growth and actinorhiza formation (Aiba, 1990). In Australia, biochar has been applied as a soil conditioner on a large scale (Blackwell and Shea et al., 2007).

Ash or carbonized material was an essential material for accelerating decomposition by stimulating bacterial activity and neutralizing acidity, as well as absorbing smells and liquid. In the compost production process, the greater the amount of charcoal used, the faster decomposition progresses with an exothermic reaction (King Coal Co., 2006). Under aerobic conditions, the Bacillus group, which can produce antibiotics, became dominant, and the growth of the soil-born pathogens Pythium, Rhizoctonia, Phytophtra and Fusarium was greatly inhibited (Kobayashi, 2001). Following this finding, various types of organic compost have been produced from livestock excreta and charcoal and sold commercially (Yoshizawa et al., 2007).

2-4-5 Biochar and reforestation

Ogawa et al. (1983c) promoted the growth of *Pinus thunbergii* and cultivated the mycorrhizal fungus *Rhizopogon rubescens* in a young stand on sand dunes by applying charcoal fertilizer. Charcoal fertilizers with small amounts of phosphate and urea were buried in trenches among the trees. Fresh roots regenerated inside the charcoal layers and mushrooms appeared in abundance along the trenches with phosphate nine months later. After a year, the amounts of pine root and mycorrhiza in the charcoal had increased considerably. The growth of shoots and the colour of the needles were better than before treatment. This may have been caused by the enhanced uptake of nutrients and water absorption through increased mycorrhizal formation. The water content in the charcoal was much higher (40 per cent) than that in sand (5 per cent) in the dry season (Ogawa, 1992). The same method is used by professional gardeners – usually charcoal powder (with a maximum diameter of 1 cm) is buried in a trench or hole together with a small amount of phosphate and the spores of specific mycorrhizal fungi. In nurseries, charcoal fertilizer has been used as a "soil conditioner" for the inoculation of mycorrhizal fungi (Ogawa, 2007).

Pine wilting disease caused by insects and nematodes has become a very serious problem in Japan. Most of the *Pinus densiflora* and *Pinus thunbergii* forests have disappeared during the last 100 years. Coastal pine forests have an important role in preventing natural disasters, so practical rehabilitation methods have been eagerly awaited in rural areas.

Ectomycorrhiza formation is essential for the survival and growth of *Pinus* species. In general, pine trees grow as a pioneer plant in devastated areas with infertile soil, and the mycorrhizal fungi prefer to inhabit the same conditions as the host plant. Therefore, it is well known from experience and research that man-made pine forests should be kept at the primary stage of plant succession by removing all undergrowth and raking out the litter layer. Ogawa (2007) proposes the use of charcoal and mycorrhizal fungi to rehabilitate and reforest the coastal pine forest. In this project, pine seedlings with the mycorrhizal fungi *Rhizopogon*, *Pithoritus* and *Suillus* species were planted together with charcoal powder and phosphate in sand dunes. The survival rates and growth of these seedlings were always much higher than those without mycorrhiza and charcoal.

It was also found that Dipterocarpaceae form ectomycorrhiza with mycorrhizal fungi. Of these, *Scleroderma columnare* was effective in enhancing seedling growth in nurseries. A small amount of charcoal (2 per cent by volume) stimulated the growth of *Shorea* species and mycorrhiza formation (Ogawa, 2006). Practical inoculation methods and the use of rice husk charcoal in nurseries were subsequently developed by others (Kikuti and Ogawa, 1999; Mori and Marjenah, 2000). The nursery technique of inoculating mycorrhizal fungi with charcoal was also used successfully in a pine forestation project in northern China (Takami, 2003). It can be expected that charcoal will be used in dryland farming – for example, date palm plantation and revegetation – to halt desertification (Ogawa, 1998).

2-4-6 The roles of biochar in the future

Because of the population increases in developing countries and the technological development of renewable energy production utilizing biomass, it is postulated that sustainability in agriculture and forestry will become a more serious problem in the near future. The agricultural system will inevitably have to be shifted from the chemical and large-scale systems common in developed countries to an intensive system in which the continuous production of food and feed is maintained, supported by organic farming with compost and biochar (Ogawa, 1994).

Growing green plants and burying biochar in the soil will be significant and practical methods for sequestering carbon, and will simultaneously contribute to the prevention of natural disasters and the conservation of land ecosystems (Okimori et al., 2003). In particular, forestation is essential for fixing carbon dioxide that has already been emitted into the atmosphere. The application of biochar to forest soil that has scarcely been disturbed by human activity could provide a long-term carbon sink, considering the stability of biochar carbon in soil (Kawamoto et al., 2006; Ogawa et al., 2006). These ideas are spreading worldwide as a countermeasure for global warming (Lehmann et al., 2007).

REFERENCES

Aiba, F. (1990) "Effects of the Materials for Greening with Charcoal on the Growth of Herbaceous Plants and Trees (2)", TRA Report, pp. 167–170 (in Japanese).

Blackwell, P. S. and S. Shea, et al. (2007) "Improving Wheat Production with Deep Banded Oil Mallee Charcoal in Western Australia", International Agrichar Initiative Conference, Terrigal, New South Wales, 29 April–2 May.

Ezawa, T., et al. (2002) "Enhancement of the Effectiveness of Indigenous Arbuscular Mycorrhizal Fungi by Inorganic Soil Amendments", *Soil Science and Plant Nutrition* 48(6): 897–900.

Glaser, B., et al. (2002) "Ameliorating Physical and Chemical Properties of Highly Weathered Soils in the Tropics with Charcoal – A Review", *Biology and Fertility of Soils* 35: 219–220.

Igarashi, T. (1996) "Soil Improvement Effect of FMP and CRH in Indonesia", JICA pamphlet.

Igarashi, T. (2002) "Handbook for Soil Amendment of Tropical Soil", AICAFF, pp. 127–134 (in Japanese).

Ishigaki, K., et al. (1990) "The Effect of Soil Amendment Materials with Charcoal and Wood Vinegar on the Growth of Citrus, Tea Plant and Vegetables", TRA Report, pp. 107–120 (in Japanese).

Ishii, T. and K. Kadoya (1994) "Effects of Charcoal as a Soil Conditioner on Citrus Growth and VA Mycorrhizal Development", *Journal of the Japanese Society of Horticultural Science* 63(3): 529–535.

Kawamoto, K., et al. (2006) "Reactivity of Wood Charcoal with Ozone", *Journal of Wood Science* 51: 66–72.

Kikuti, J. and M. Ogawa (1999) "Development of Nursery Techniques Utilizing Microorganisms", RETROF: "Research Report on Rehabilitation of Tropical Forest", pp. 155–182.

King Coal Co. (2006) Hi-pro pamphlet No. 251.

Kishimoto, S. and G. Sugiura (1985) "Charcoal as a Soil Conditioner", *Symposium on Forest Products Research, International Achievement for the Future* 5: 12–23.

Kobayashi, N. (2001) "Charcoal Utilization in Agriculture (1)", *Nogyo Denka* 54(13): 16–19 (in Japanese).

Komaki, Y., et al. (2002) "Utilization of Chaff Charcoal for Medium of Flower Bed Seedlings and Its Effect on the Growth and Quality of Madagascar Periwinkle (*Catharanthus roseus*) Seedlings", *Japanese Society for Soil Science and Plant Nutrition* 73(1): 49–52 (in Japanese).

Lehmann, J. and S. Joseph, eds (2009) *Biochar for Environmental Management, Science and Technology*. London: Earthscan.

Lehmann, J., et al. (2006) "Bio-char Sequestration in Terrestrial Ecosystems – A Review", *Mitigation and Adaptation Strategies for Global Change* 11: 403–427.

Lehmann, J., et al. (2007) "A Handful of Carbon", *Nature* 447: 143–144.

Miyazaki, Y. (1697) *Nogyo Zensho* [Encyclopedia of Agriculture], vol. 1 in *Nihon Nousho Zenshu*, vol. 12 (revised edn). Tokyo: Nousangyoson Bunka Kyokai, pp. 91–104 (in Japanese).

Mori, S. and Marjenah (2000) "A Convenient Method for Inoculating Dipterocarp Seedlings with the Ectomycorrhizal Fungus, *Scleroderma columnare*", in H. Guhardja et al. (eds), *Rainforest Ecosystems of East Kalimanta*, Ecological Studies 140, pp. 251–255.

Noguchi, A., et al. (1993) "Effect of Rice Husk Charcoal Application on the Growth and Nitrogen Fixation of Phaseolus Vulgaris", JICA Internal Report (in Japanese).

Ogawa, M. (1987) *Symbiotic Microorganisms Connecting Soil and Plants, Ecology of Mycorrhiza*. Tokyo: Nosangyoson Bunka Kyokai (in Japanese).

Ogawa, M. (1992) *Cultivation of Wild Mushrooms*, Ringyo Kairyo Fukyu Sosho 110. Tokyo: Zenkoku ringyo Fukyu Kyokai.

Ogawa, M. (1994) "Symbiosis of People and Nature in the Tropics", *Farming Japan* 28(5): 10–34.

Ogawa, M. (1998) "Utilization of Symbiotic Microorganisms and Charcoal for Desert Greening", *Green Age* 14: 5–11.

Ogawa, M. (2006) "Inoculation Method of *Scleroderma Columnare* onto Dipterocarps", in K. Suzuki et al. (eds), *Plantation Technology in Tropical Forest Science*. Tokyo: Springer Verlag, pp. 185–197.

Ogawa, M. (2007) *Reviving Pine Tree with Charcoal and Mycorrhiza*. Tokyo: Tsukiji Shokan (in Japanese).

Ogawa, M. and Y. Yambe (1986) "Effects of Charcoal on VA Mycorrhiza and Root Nodule Formations of Soybean. Studies on Nodule Formation and Nitrogen Fixation in Legume Crops", *Bulletin of Green Energy Program Group II*, No. 8, Japanese Ministry of Agriculture, Fisheries and Forestry, pp. 108–134 (in Japanese with English summary).

Ogawa, M., Y. Yambe and G. Sugiura (1983a) *Effects of Charcoal on the Root Nodule Formation and VA Mycorrhiza Formation of Soybean*, Third International Mycological Congress (IMC3) Abstract.

Ogawa, M., Y. Yambe and G. Sugiura (1983b) *Cultivation of the Hypogenous Mushroom, Rhizopogon rubesscen*, IMC3 Abstract.

Ogawa, M., et al. (1983c) "Charcoal and the Mushroom Rhizopogon rubescens", *Forestry and Forest Products Research Institute News* 223(2): 1–3.

Ogawa, M., et al. (2006) "Carbon Sequestration by Carbonization of Biomass and Forestation: Three Case Studies", *Mitigation and Adaptation Strategies for Global Change* 11: 429–444.

Oka, H., et al. (1993) "Improvement of Sandy Soil in the Northeast by using Carbonized Rice Husks", JICA Technical Report 13, pp. 42–40 (in Japanese).

Okimori, Y., et al. (2003) "Potential of CO_2 Emission Reduction by Carbonizing Biomass Wastes from Industrial Tree Plantation in South Sumatra, Indonesia", *Mitigation and Adaptation Strategies for Global Change* 8: 261–280.

Okutu, M., et al. (1990) "The Effect of the Soil Amendment Materials with Charcoal and Wood Vinegar on the Growth of Rice Plant, Apple Tree and Vegetables", TRA Report, pp. 121–131 (in Japanese).

Oohira, T., et al. (1992) "Function of Charcoal as Microbial Carrier in Soil", *Journal of Antibacterial and Antifungal Agents* 20(10): 511–517 (in Japanese with English summary).

Sano, H. (1990) "Effects of the Materials for Greening with Charcoal on the Growth of Herbaceous Plants and Trees (1)", TRA Report, pp. 155–165 (in Japanese).

Shinogi, Y., et al. (2003) "Basic Characteristics of Low-Temperature Carbon Products from Waste Sludge", *Advances in Environmental Research* 7: 661–665.

Soda, R., et al. (1990) "Spore Propagation of VA Mycorrhizal Fungi", TRA Report, pp. 199–212.

Steiner, C., et al. (2007) "Long Term Effect of Manure, Charcoal and Mineral Fertilization on Crop Production and Fertility on a Highly Weathered Central Amazonian Upland Soil", *Plant and Soil* 294: 275–290.

Takagi, K. and S. Takanashi (2003) "Development of a Technique for Reducing Herbicide Runoff from Paddy Fields using the PCPP-1 Model and Rice Husk Charcoal Powder", *Proceedings of the Third International Conference on Contaminants in the Soil Environment in the Australasia–Pacific Region, Beijing, China*.

Takagi, K. and Y. Yoshida (2003) "In Situ Bioremediation of Herbicides Simazine-polluted Soils in a Golf Course Using Degrading Bacteria-Enriched Charcoal", *Proceedings International Workshop on Material Circulation through Agro Ecosystems in East Asia and Assessment of Its Environmental Impact, Tsukuba*, pp. 58–60.

Takagi, S., et al. (1990) "Immobilization Method of Root Nodule Bacteria within Charcoal and Effective Inoculation Method for Legumes", TRA Report, pp. 229–248 (in Japanese).

Takami, K. (2003) *Apricot Bore Fruit in Our Village*. Tokyo: Nihon Keizai Shinbun (in Japanese).

TRA [Technical Research Association for Multiuse of Carbonized Materials], ed. (1990) *Research Report on New Uses of Wood Charcoal and Wood Vinegar*. Tokyo (in Japanese).

Yamato, M., et al. (2006) "Effects of the Application of Charred Bark of *Acacia mangium* on the Yield of Maize, Cowpea and Peanut, and Soil Chemical Properties in South Sumatra, Indonesia", *Soil Science and Plant Nutrition* 52: 489–495.

Yoshizawa, H., et al. (2007) "Proliferation Effect of Aerobic Microorganism during Composting of Rice Bran by Addition of Biomass Charcoal", *Proceedings of the International Agrichar Conference, Terrigal, NSW Australia*, May.

3
How to make food, biological and water resources sustainable

3-1
Biomass as an energy resource

Fumitaka Shiotsu, Taiichiro Hattori and Shigenori Morita

3-1-1 The creation of a sustainable society and renewable energy

The creation of a sustainable society and reducing dependence on fossil fuels

In the twentieth century, large amounts of energy were used to maintain a lifestyle of mass production, mass consumption and mass disposal. During this time, total worldwide energy consumption doubled. In recent years, this type of lifestyle has been rethought, building a strong desire for the creation of a sustainable society.

One factor necessary for the creation of a sustainable society is decreasing (or at least not increasing) the total amount of energy used. Another factor is to reduce dependence on petroleum as a source of energy. More than half of the world depends on fossil energy; and petroleum accounts for almost half of the entire energy source. Burning fossil energy (including petroleum) is responsible for increasing the concentration of carbon dioxide in the atmosphere and can cause global warming. Thus, reducing petroleum dependency is necessary to combating global warming.

These are not the only reasons the dependency on petroleum must be reduced. Although large quantities of petroleum were consumed in the twentieth century, it was possible to delay the depletion of petroleum by simultaneously discovering new oil fields. The discovery of new oil fields

---

Designing our future: Local perspectives on bioproduction, ecosystems and humanity,
Osaki, Braimoh and Nakagami (eds),
United Nations University Press, 2011, ISBN 978-92-808-1183-4

has dropped sharply since the late 1960s and is expected to "zero-out" in the first half of the twenty-first century. This does not mean that petroleum will completely disappear at that point, but it does mean that there will be an oil peak sooner or later and ultimately the unavoidable depletion of petroleum will have to be faced in the near future. This is the major reason the dependence on petroleum must be reduced.

The development of renewable energy

The development and use of renewable energy as an alternative to fossil fuel energy, especially petroleum, are necessary. The major renewable energy sources are solar power, wind-generated power, tidal power, geothermal power, small- and medium-scale hydroelectric power and biomass energy. The development and use of most renewable energies are strongly influenced by the related natural conditions in a particular location. For example, strong and hot winds blow all year in the desert area of Xinjiang-Uygur Autonomous District of China, and this wind generates electric power. This is a typical and effective development and utilization of renewable energy based on the natural conditions there.

Because the development and use of renewable energy are usually restricted to particular regions, it is not easy to construct a universally applicable model. The development of solar power and biomass energy have been encouraged, because they are available worldwide, in contrast to other renewable energy sources. Biomass energy, in particular, has an additional advantage. Biomass can produce several types of energy, including liquid fuel, whereas most other renewable energy sources such as solar power are usually converted into electrical energy. That is, it is possible to mix bioethanol (one form of biomass energy) with petroleum or use it in place of petroleum. This section will discuss the possible utilization of biomass energy in Japan and other Asian countries by focusing on bioethanol.

3-1-2 Types and use of biomass energy

Biomass energy is energy generated from biomass, which refers to renewable organic resources originating from biological materials excluding fossil energy. Specifically, biomass includes crops and their residues, livestock excreta, foodstuff waste, forest thinning debris and used paper (Komiyama et al., 2003).

There are several types of biomass energy, depending on the raw materials and the process of generation. This section will focus on the manner of use, i.e. whether it can be burned directly or converted into gaseous or liquid fuel. The direct combustion of biomass, such as firewood and charcoal, is the most traditional use and has a long history. Recently, unused wood biomass has been converted into pellets and briquettes to be burned.

Gaseous fuel is produced from biomass by gasification or methane fermentation. In the gasification method, mixing with vapour in a hypoxic state and steaming generates a compound gas containing methane, hydrogen and carbon monoxide. In the methane fermentation method, fermenting material with a high moisture content, such as sewage sludge or livestock manure, produces biogas.

Bioethanol and biodiesel are major forms of liquid fuel. Bioethanol is obtained by preprocessing and fermenting the raw materials of sugar, starch or cellulose. It is then ethanol-fermented with yeast, distilled and dehydrated. Adding methanol to vegetable oil as the raw material yields biodiesel (fatty acid methyl ester) and glycerin.

3-1-3 Biomass energy and bioethanol

"Carbon neutral" in biomass energy

As noted above, there are several types of biomass energy source. No single type is completely inexhaustible, but they are all renewable. Unlike petroleum, most of them can be produced repeatedly as long as the sun and Earth exist. In addition, because biomass energy is produced from plant biomass, the carbon dioxide that is emitted during combustion was originally generated from the atmospheric carbon dioxide that was assimilated into the dry matter by photosynthesis. Therefore, as long as a cycle is maintained in which the raw material plants continue to grow, the entire process is carbon neutral even though carbon dioxide is released when the biomass energy is burned. However, energy is also required in the process of creating gaseous or liquid fuel from biomass. This cannot be called carbon neutral in the strictest sense if fossil energy is used in this process. Besides, although it is relatively easy to create a universal model for generating energy using biomass, the distribution of biomass is not concentrated but instead scattered; it is also time-consuming, laborious and costly to gather biomass, which is another disadvantage to its use.

Nonetheless, the advantages of biomass energy outweigh its disadvantages. The most important advantage is that it can be used as a liquid fuel, i.e. it can be mixed with gasoline or used instead of gasoline. Bioethanol is a typical biomass energy that can be used in this way. Thus, in addition to having the potential to work as a countermeasure to global warming, it has a great advantage from the perspective of energy security.

In essence, it also assists in guaranteeing food security. However, global concerns regarding competition between food and biofuels have arisen, possibly because of the rapid increase in global bioethanol production, which has doubled since 2000 (Licht, 2007).

Major issues in bioethanol production

The increase in bioethanol production may be a factor in the increasing international market price of cereal crops that are also raw materials for bioethanol. As a result, competition between food and fuel in relation to bioethanol production has been a problem, especially in developing countries. This has led to the use of raw material crops rich in cellulose (switchgrass, napier grass, *Erianthus* spp., *Miscanthus* spp.) instead of raw material crops rich in sugar (sugarcane, sugar beets) or rich in starch (maize, wheat, cassava) to produce bioethanol. Innovations in technology will be required to convert cellulose raw materials into bioethanol. Moreover, existing farmland cannot be used for growing such raw material crops rich in cellulose because of possible competition with food production. The basic lack of understanding of this perspective is a major problem.

Another criticism is that expanding bioethanol production will lead to an increasing load on the environment. For example, forests have been cut down to create new farmland for growing raw material crops. Such deforestation increases carbon dioxide emissions into the atmosphere and simultaneously decreases the ability to fix carbon dioxide into forests. Furthermore, intensive crop cultivation uses a large amount of energy for machinery, fertilization and agricultural chemicals. If raw material crops for bioethanol are grown in the same high-input manner, as much energy will be consumed as in the process of producing and using bioethanol. Thus, it is necessary to take into consideration the greenhouse gas balance as well as the energy balance. The same problems exist for growing sunflower and rapeseed for biodiesel and for producing pellets and charcoal from wood biomass.

Low-input sustainable cultivation of raw material crops

Although the energy balance and the greenhouse gas balance are extremely important concerns in the production and use of bioethanol, a problem is generally perceived only with respect to the process of converting biomass into ethanol. The process of cultivating the raw material crops has not received much attention. A great deal of energy goes into most crop cultivation processes through the use of machinery, fertilizers and agricultural chemicals (Pimentel et al., 1973). As a result, although crop yields have definitely been growing year on year, the ratio of the energy yield from harvested crops compared with the input energy is steadily decreasing. This is not always a problem from the perspective of food production because the high input of energy has been increasing the efficiency of solar energy assimilation into crops. However, the production and utilization of bioethanol will no longer be a countermeasure against global warming if the output/input energy ratio and greenhouse gas balance are poor in the cultivation of raw material crops for bioethanol.

Thus, regarding growing raw material crops for bioethanol, very careful consideration must be given to which raw material crops should be grown where and how (Morita et al., 2009). Taking into account the literature on crop science, perennial gramineous plants of the C4 type with high biomass productivity and stress tolerance will be strong candidates for raw materials. Such plants must be grown in a low-input, sustainable manner to save energy, and, to prevent food–fuel competition, they must be grown in areas that are not suitable for agriculture, instead of using existing farmland or new farmland created by cutting down forests. Land use (including soil management) is particularly important from the perspective of the carbon cycle with reference to global warming.

3-1-4 Utilization of biomass energy in Asia

The leading countries in bioethanol production are the United States and Brazil, followed by China and India. Particularly in China and India, recent rapid economic growth and population expansion are causing a huge increase in the demand for fossil energy (mainly petroleum). Most of the petroleum in China and India must be imported from other countries, so a stable supply of energy is an important issue as regards energy security. This has resulted in both countries having a growing interest in renewable energy, including the development and use of bioethanol. The current state of traditional biomass use and modern bioethanol production and use in China and India will be introduced, and then future prospects discussed.

China

Bioethanol production in China began with maize and wheat as raw materials in Heilongjiang and Jilin provinces in the late 1990s. These materials were surplus from previous overproduction of crops that could not be eaten after being stored for a long period. From March 2002, the production, storage, transportation, supply, sale and use of E10 gasoline (gasoline mixed with 10 per cent ethanol) for automobile fuel were carried out on a trial basis targeting the cities of Zhengzhou, Luoyang and Nanyang in Henan Province, and of Harbin and Zhaodong in Heilongjiang Province. In 2004, this expanded to the five provinces of Heilongjiang, Jilin, Anhui, Henan and Liaoning, and to 27 cities in the provinces of Hubei, Shandong, Hebei and Jiangsu. Currently approximately 1.25 million kL of bioethanol is produced annually in China (NEDO, 2007).

However, since 2005, along with a rapid increase in bioethanol production, the rise in the international market price of maize (a major raw material for bioethanol) and a reduction in the year-end inventory level have made supplying raw materials difficult. To constrain the dramatic increase in grain prices, in December 2006 the National Development and Reform Commission (NDRC) issued an emergency notice regulating new bioethanol production projects using maize as a raw material. Because new factories for bioethanol production continued opening in each region even after the notice, the NDRC made another announcement in October 2007, entitled the "Guidance opinion to promote the healthy development of the corn-processing industry", which prohibited the introduction and expansion of corn-processing projects during the period covered by the Eleventh Five-Year Plan (2006–2010). From the perspective of food security, it is advocated that bioethanol materials should change from food crops to non-food crops and to alternative raw materials rich in cellulose, such as rice and wheat straw, in the near future.

Additionally, desertification and environmental degradation in China are reducing the area of cultivated land, which is becoming a serious issue for maintaining arable land for food production. Thus, using land unsuitable for farming (such as regions with alkaline soils) is being considered as an option for growing raw material plants for bioethanol production.

The Chinese government has issued two targets for biofuel production: the NDRC's Long-Term Renewable Energy Development Plan, and the Agriculture and Forestry Biomass Project of the National Long-Term Science and Technology Development Plan. The plans list the final target of annual production as approximately 12.5 million kL and 15.0 million kL, respectively, by 2020. To achieve these targets, future research to develop new technologies to produce bioethanol from raw materials rich in cellulose is vital. Promising candidates for energy crops in China include cassava and sorghum with a high tolerance for biotic and abiotic stresses (China has widespread arid and semi-arid regions where there are stresses of drought or salt accumulation).

India

In India, the use of abundant biomass resources is being promoted as a national policy through the ongoing development of various measures and projects. In particular, bioethanol production from molasses obtained during the production of sugar from sugarcane is being promoted from the perspective of the effective utilization of waste material. Sweet sorghum and sugar beet are also raw materials for bioethanol production.

The Biofuel Ethanol Plan (Ethanol Programme) was launched in 2001. It was initiated by the Ministry of Petroleum and Natural Gas (MoPNG), which announced the phased popularization of E5 gasoline, with subsequent developmental expansion towards E10 gasoline and a blend of 5 per cent ethanol in diesel fuel.

The government is developing bioethanol production and use throughout the country via a three-phase plan. Phase 1, which aimed for E5 gasoline use in 9 Indian states and 4 cities, began in January 2003. However, the popularization schedule was delayed and implementation did not start until July 2003 owing to a delay in increasing the dehydration facilities and the preparation of ethanol and gasoline mixing facilities. Phase 2, which started in October 2006, called for a nationwide expansion of the limited supply region of Phase 1. As of February 2007, bioethanol use had been implemented in 20 states and 4 cities. Phase 3 started in June 2007, whereby the ethanol content was projected to increase from 5 per cent to 10 per cent throughout India. However, there may be an insufficient supply of molasses (used as a raw material for ethanol) to introduce E10 (NEDO, 2007).

Biomass energy other than bioethanol also has incredible potential. It has a wide diversity of applications, scales, purposes (compatibility with the micro-demand involved in the development of small rural communities, in-house power generation, power sales, fuel supply) and energy sources (agricultural residue, energy crops, livestock manure, waste materials). Various technological set-ups are available, from those already established domestically (such as biogas and biomass power generation and cogeneration) to formats at the test-development phase and progressing towards practical application. Biomass energy use is limited in scale and application, but government expectations for possible expansion are extremely high (MNES, 2006). Policy and preparation for an investment environment to promote the popularization of biomass energy with diverse applications and scales are needed to progress.

China and India are focusing on bioethanol production plants that are decentralized and small scale in rural communities. Large-scale bioethanol production of the type seen in the United States or Brazil is not realistic for Asia, where farming has traditionally been based on small-scale and intensive rice cultivation. In Asia, it is preferable to consider comparatively small-scale bioethanol production, and for the community to choose the biomass raw materials. The next subsection will consider this point in relation to the "biomass town" concept in Japan.

3-1-5 The biomass town as an Asian model

The Biomass Nippon Strategy decided by the Japanese cabinet in 2002 and revised in 2006 is the foundation of a policy for promoting the use of biomass within Japan. The pillars of this strategy are promotion of the use of biofuel for transportation and promotion of the use of regional biomass resources through creating biomass towns throughout the country.

Regarding biofuel for transportation, the "Significant Expansion of Domestic Biofuel Production" announced in 2007 contained a numerical target calling for the supply of 6 million kL of bioethanol annually by approximately 2030. It is specified that the raw materials should be at least 70 per cent cellulosic (e.g. rice straw, energy crops and woody biomass). This means that production technology for bioethanol derived from cellulose must be established to promote the use of biofuel for transportation.

A biomass town is a community-based system where people in the local community join together to process biomass efficiently, from generation to use, and to facilitate stable and appropriate application. Here, "community" specifically refers to local government at the city, town and village level. The Japanese Ministry of Agriculture, Forestry and Fisheries intends to actively support cities, towns and villages that put forward their own biomass town plan. As of the end of May 2009, 212 communities nationwide had publicly announced biomass town plans. Biomass town plans were expected to be announced in 300 cities, towns and villages nationwide by the end of fiscal year 2010.

The promotion of the use of biofuel for transportation and the plans for biomass towns should not be thought of as independent goals. Instead, they are goals that should be achieved in harmony. That is, promotion of the use of biofuel for transportation not only relies on the development of manufacturing technology for bioethanol derived from cellulose, but also requires the construction of community systems that promote the efficient harvesting of biomass resources and their active use as biofuel. The biomass town system is a system that aims for the comprehensive use of biomass in the community, and the biomass use formats involved should not be limited to biofuel alone. It is preferable to implement "cascading use" for the "5 Fs" of food, fibre, feed, fertilizer and fuel from high to low added value. Biofuel, which is at the bottom of the cascading use list, has an important position in promoting the use and application of community biomass resources.

Test projects related to biofuels

The Japanese Ministry of Agriculture, Forestry and Fisheries is currently supporting several test projects nationwide. These entail the bringing together of individuals in the community involved in processes from the procurement of biomass materials to the manufacture and sale of biofuel. Regarding bioethanol, from fiscal year 2007 verification projects using high-yielding rice varieties or irregular agricultural products that cannot be accepted in the market as raw materials to produce biofuel for community utilization started in two areas in Hokkaido Prefecture and one area in Niigata Prefecture. Regarding the systems of producing bioethanol from cellulosic material, research and development programmes have been launched as projects to establish soft cellulose use and application technology in Hokkaido, Akita and Hyogo prefectures from fiscal year 2008. Although the biofuel test project as a community model aims for annual production of 31,000 kL of bioethanol, the other projects for establishing soft cellulose use are aiming for annual production of only 24 kL in three areas. However, because several test projects using soft cellulose have already started, technological innovations are expected to develop in the course of the production and use of bioethanol from cellulosic raw materials. In addition, The University of Tokyo has started a research project on biomass utilization, including bioethanol production in Shinano Town of Nagano Prefecture, which aims to create a community self-sufficient fuel system using biomass resources from the community such as rice, rice straw and husk, and the prunings from apple trees as raw materials (Igarashi and Saiki, 2008).

Regarding biodiesel, test projects were launched in three areas in Ibaraki, Tokyo and Fukuoka prefectures in fiscal year 2007 and in nine areas including Akita and Chiba prefectures in fiscal year 2008. Currently, they have expanded to 12 areas nationwide. Focusing on edible oil waste, sunflower and rapeseed oils are being used as raw materials for biodiesel, and the annual target for the total production quantity for all areas is approximately 10,000 kL.

In tandem with the promotion of test projects, as seen with the May 2008 approval of the Act on the Promotion of Producing Biofuels from Biomass of Agricultural, Forestry and Fisheries (enforcement began in

October of the same year), government support has increased even further for activities implemented jointly between entities in the agriculture, forestry and fishery industries and biofuel manufacturing entities, and for research and development programmes geared towards streamlining biofuel manufacture. As the application and use of biomass at the community level are being promoted, the importance of biofuel has been increasing. In the future, there is the possibility of a shift from fossil energy to alternatives energies including fuel cells and hydrogen. Because it will take time to complete the shift, it is necessary to develop and utilize various renewable energies for a soft landing. Biofuel is a possible candidate and its production and utilization in Japan have another important role in supporting weakened regional agriculture and forestry.

Unlike the United States or Brazil, where large-scale bioethanol production is possible, Japan has few large areas and flat farmland for producing biomass. The "starting point" (raw material production) in the system for the production and use of biofuel is completely different from that in the United States and Brazil. The model of the use and application of biomass focusing on farming development at the community level (i.e. the biomass town model) is Japan's latest contribution. Moreover, this biomass town model has great potential for application in many Asian regions that depend on small-scale and intensive agriculture based on rice farming.

The problem of the sharp division between urban and rural areas (which is particularly serious in China) has become notable in recent years, and ways are being sought to achieve sustainable development based on linking these two types of areas (Morita, 2007). Although there is no "miracle cure" to resolve this issue immediately, the ongoing creation of a "recycling society", including the production and use of biofuels, is an effective solution. At the 8th ASEAN +3 Meeting of the Ministers of Agriculture and Forestry in October 2008 (Japan, China and South Korea participated in addition to the 10 ASEAN countries), Japan proposed a new project that would utilize the knowledge and knowhow that Japan had accumulated through its efforts to develop biomass town plans. The goal is to promote the creation of biomass towns in East Asian countries. Specifically, support would be offered for human capacity-building and basic research geared to drawing up biomass town concepts in East Asia. This proposal was accepted by each country at the meeting and was granted permission to be put forward in the future under the ASEAN +3 framework. Future movements leading to an Asian model for the sustainable use and application of biofuels that is different from those in the United States and Brazil should be studied closely.

REFERENCES

Igarashi, Y. and T. Saiki (2008) *Production of Bioethanol from Rice Straw*. Tokyo: Japan Association of Rural Resource Recycling Solutions.

Komiyama, H., A. Sakoda and Y. Matsumura (2003) *Biomass Nippon – Aiming at the Regeneration of Japan*. Tokyo: Nikkan Kogyo Shinbun.

Licht, F. O. (2007) "World Fuel Ethanol Production", *World Ethanol and Biofuels Report* 5(17).

MNES [Ministry of New and Renewable Energy] (2006) *XIth Plan Proposals for New and Renewable Energy*. India: Government of India.

Morita, S., et al. (2007) "Establishment of Sustainable and Symbiotic Recycling-Based Society in Tianjin City of China as an Example of Mega City with Suburb Rural Area", AGS Research Report 2007 (3), <http://en.ags.dir.u-tokyo.ac.jp/repo_2007_3#2007_3_1>.

Morita, S., T. Hattori and F. Shiotsu (2009) "Viewpoint Regarding Growing Raw Material Crops for Bioethanol", *Agriculture and Horticulture* 84(7): 687–691.

NEDO (2007) *Report of Summary of Information Related to Biomass and Examination of Business Related to Biomass in Asian Countries*. Tokyo: New Energy and Industrial Technology Development Organization.

Pimentel, D., L. E. Hurd, A. C. Bellotti, M. J. Forster, I. N. Oka, O. D. Sholes and R. J. Whitman (1973) "Food Production and the Energy Crisis", *Science* 182: 443–449.

3-2
Worldwide cross-country pattern of ecosystem and agricultural productivities

Takashi S. Kohyama and Akihiko Ito

In this section it is pointed out that country-based agricultural productivity, population density and economic productivity do not reflect the expected primary productivity of potential vegetation. This is not explained simply by overseas trade or by uneven economic development. Cereals, or seed grains of annual plants, require enduring fertile soil and are naturally adapted to extratropical ecosystems. Crops from perennial plants such as tree fruits, roots and tubers are adapted to tropical ecosystems. The geographical patterns of human population and society also reflect the properties of natural ecosystems. It is suggested that sustainable land use needs to be evaluated from the properties of natural ecosystems.

3-2-1 Human dependence on net primary productivity

The human population (6.5 billion, increasing at 80 million per year) predominates in, and substantially modifies, the global landscape (Foley et al., 2005; Vitousek et al., 1997). Human activities and their consequences are constrained by the Earth's system. As heterotrophs, humans exclusively depend on the primary production of the Earth's system for their food supply. Net primary production (NPP) is a measure of the production rate of organic matter by autotrophs (i.e. plants). It is the gross rate of carbon fixation as organic matter by photosynthesis minus the rate of consumption of organic matter for the life of autotrophs through their respiration. NPP therefore represents the portion available for ecosystem

Designing our future: Local perspectives on bioproduction, ecosystems and humanity,
Osaki, Braimoh and Nakagami (eds),
United Nations University Press, 2011, ISBN 978-92-808-1183-4

development in terms of increasing the organic carbon pool and for heterotrophs' consumption. The production and life of heterotrophs (including the human population) are ultimately dependent on NPP. The human dependence on NPP through the consumption of organic carbon is composed of food and non-food organic matter that supports food production (such as the non-edible parts of crop plants and pasture to feed livestock, as well as timber, fibre and fuelwood). Because of this NPP dependence, the human population has modified about 40 per cent of land (Vitousek et al., 1997) and now consumes about 20 per cent of annual NPP (Imhoff et al., 2004).

There are substantial differences between the geographical distribution of NPP and the organic matter consumed by the human population, or the "human appropriation of NPP (HANPP)" (Haberl et al., 2007; Imhoff et al., 2004). This difference is caused by two components: (1) geographical variation in the proportion of NPP for human use via agriculture, and (2) the anthropogenic trade in organic matter from source to sink regions. The second component, referred as "food/fibre mileage", causes additional consumption of organic carbon as fuel for transportation, and also emphasizes geographical transportation of inorganic nutrients (such as fertilizer) and water (contained in crops) from source to sink regions, which can bring about the decline of potential primary productivity in source regions. The first component, which causes spatial heterogeneity of HANPP/NPP, or the *in situ* ratio of anthropogenic net primary production (ANPP) to total NPP, can be caused by various natural and social factors. For the maintenance of ANPP, it is essential to analyse and understand the structure of ANPP/NPP. This section presents an approximate description of ANPP/NPP based on world countries as macroscopic sociopolitical units. A political unit is used rather than a geographical or spatial unit with the same physical size because societal and agricultural statistics are usually compiled at the scale of political units. Societal quantities that characterize the local human population often reflect the present policy and the change in policy at the scale of each country. An identical procedure can be applied to a smaller scale of municipal units within countries to characterize the social structure in terms of human dependence on NPP.

3-2-2 Country-level data on ecosystem productivity

The land area mesh-based estimates of NPP, plant carbon and soil carbon from Sim-CYCLE[1] (Ito and Oikawa, 2002) are aggregated into the land area of each country. Sim-CYCLE estimates are based on the simulation of biogeochemical carbon cycles in terrestrial ecosystems, using ecophysiological relationships with a variety of vegetation types, and are good estimators of potential ecosystem productivity and biomass accumulation under present climatic conditions. These country-based ecosystem measures are merged with records of agricultural production in 2005, obtained from the statistical database of the Food and Agriculture Organization of the United Nations (FAOSTAT),[2] and demographic and economic statistics for 2005, obtained from GeoHive.[3] To avoid error in aggregation, countries with a land area of >104 km^2 were chosen, excluding zero-NPP Sim-CYCLE estimates and countries with unreliable primary industry statistics in FAOSTAT; 124 countries are listed in the full data set of land ecosystem measures and human societal measures. The data set is composed of NPP per area, the residence time of plant carbon (plant organic carbon divided by NPP), the residence time of soil carbon (soil organic carbon divided by NPP), annual production per area of cereals and of roots and rhizomes, human population density, the per capita rate of increase in the human population and per capita gross domestic product (GDP). The original data set is available upon request.

3-2-3 Latitudinal pattern of natural and social productivities

To illustrate the geographical pattern of NPP and biomass turnover, these measures are plotted along the latitudinal location of countries (Figure 3.2.1), which is represented by the latitude of the capital city. Latitudinal location is important because the thermal environment and seasonality are primarily explained by latitude. A dominant pattern is that NPP per area has its peak in always hot tropical regions (20°N–20°S) (Figure 3.2.1, top panel). Though the detail is not described here, a wide variation in NPP in the same latitudinal zone reflects geographical variation in precipitation. A humid environment is required for high NPP, and arid tropics that accelerate evapotranspiration are characterized by low NPP. Plant mass that contributes to primary production also has its peak in tropical regions. A peak of plant mass, typically in tropical rain forests, is not simply explained by high primary productivity; it is also due to the high longevity of plants in terms of the residence time of plant organic carbon (plant-C/NPP) (Figure 3.2.1, middle). This pattern corresponds to the fact that the most productive vegetation is found in tropical rain forests with long-lived trunk woods. In contrast, soil organic carbon (soil-C) is long-lived in high-latitude regions, where the decomposition rate of organic carbon, through soil respiration, is slow owing to low temperatures and short summers (Figure 3.2.1, bottom). These climatic conditions in high-latitude regions enhance the accumulation of humus and organic soil.

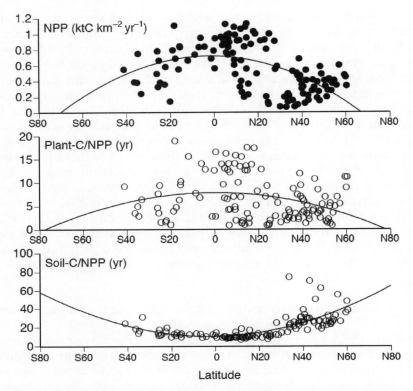

Figure 3.2.1 Latitudinal pattern of country-based net primary productivity (NPP) and residence time of plant carbon and of soil carbon, estimated by Sim-CYCLE.
Source: Ito and Oikawa (2002).
Notes: Symbols correspond to the geographical location of the country, represented by the latitude of the capital city. Quadratic curves fitted to data.

Country-based agricultural productivity shows a common trend in that the maximum per-area productivity mostly occurs in the temperate region of the Northern Hemisphere (30°N–40°N) accompanied by a large between-country variation (Figure 3.2.2). This pattern is most pronounced for the production of cereals and meat; a tropical region with high NPP shows markedly low agricultural productivity. An exceptionally high record of cereal production (Figure 3.2.2, top panel) corresponds to Bangladesh, with maximum rice production supported by extremely fertile river deposits from the inner Indian subcontinent. The productivity of roots and tubers (Figure 3.2.2, middle) shows a less marked trend of low tropical productivity. The three most productive countries in the tropics are, in decreasing order, Nigeria, Ghana and Benin, in West Africa, with

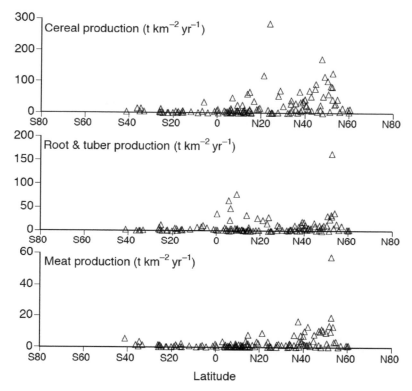

Figure 3.2.2 Latitudinal pattern of country-based annual production in 2005 of cereals, roots and tubers, and meat.
Source: Based on FAOSTAT.
Notes: Symbols correspond to the geographical location of the country, represented by the latitude of the capital city.

the main crop there being cassava. The highest productivity in the Northern Hemisphere is in the Netherlands, which is characterized by the high productivity of potatoes and sugar beet. The Netherlands also has the highest meat productivity with intensive dairying.

By comparing Figure 3.2.2 with Figure 3.2.1, it is obvious that a high NPP of vegetation does not correspond to high agricultural productivity. The latter is instead related to the long residence time of soil organic carbon. It also depends on the type of crop. Typical tropical crops such as cassava, taro and yam are vegetative storage organs of perennial plants. As shown in Figure 3.2.1, tropical vegetation is characterized by an emphasis on aboveground plant biomass with poor soil organic mass. Consequently, nutrients are accumulated in plants rather than in soil, and

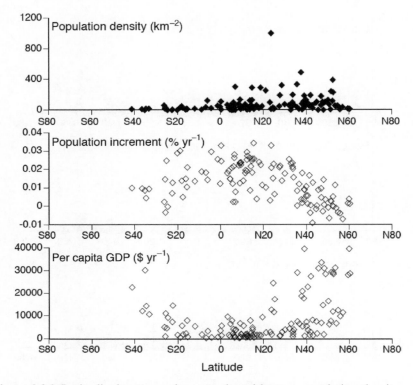

Figure 3.2.3 Latitudinal pattern of country-based human population density, annual population increase and per capita GDP.
Notes: Symbols correspond to the geographical location of the country, represented by the latitude of the capital city.

vegetation removal often brings about the outflow of nutrients. In such an environment, the partial harvest of perennial crop plants is adopted to prevent nutrient loss (compared with the harvest of cereals, which removes the whole plant biomass).

Human population density shows a latitudinal pattern similar to that of agricultural productivity (Figure 3.2.3, top). The highest density is again recorded for Bangladesh, where the highest rice production is not sufficient to feed this dense population. There is a good correlation between human population density and cereal productivity (Figure 3.2.4). These two parameters are interdependent: human population density provides labour intensity for cereal production, and cereal grain is a major type of human food. The exceptional case of the United Arab Emirates is an example of a country dependent on food importation, which is paid for

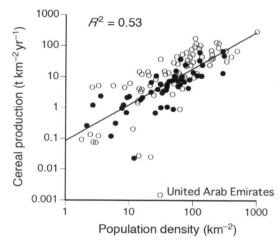

Figure 3.2.4 Correlation between human population density and annual production of cereals.
Notes: Closed circles show tropical countries (20°S–20°N by latitude of the capital city), open circles show extratropical countries. An exceptional country with low cereal productivity relative to population density is United Arab Emirates.

by fossil fuel, its major economic resource. The annual rate of population increase is generally high in tropical countries (Figure 3.2.3, middle) and is opposite to the geographical pattern of per capita GDP (Figure 3.2.3, bottom). The high population growth rate in tropical countries suggests increasing crop productivity in areas with high potential ecosystem productivity through increasing fertilization or a "green revolution" (Tilman et al., 2002). High per capita GDP in extratropical countries is partly a result of historically stable agricultural production with sustained cropland fertility.

3-2-4 The need for a unique strategy for tropical development

The brief sketch presented above suggests that the low agricultural and economic productivity in tropical countries is not easily explained by social historical factors. The historical rise and fall of local societies and civilizations are dependent on regional climate, land fertility, landscape heterogeneity and natural and/or social factors enhancing the maintenance of wild vegetation (Diamond, 2005; Rolett and Diamond, 2004). To support human society, wild vegetation with high and persistent NPP such as forests not only provides land for crop production but also provides supporting and regulative functions for the maintenance of civilization. From this perspective, one might think that the relative poverty of tropical regions, which are still largely covered by wild vegetation characterized by a huge stock of biomass, high NPP and biodiversity, simply represents a delay in the time-course of economic development, and one might conclude that support by developed countries in temperate regions based on their successful experiences of land use would be sufficient to enhance the development of tropical countries.

This "delayed tropics" view may be questioned. Additional factors are important for land-use planning in tropical regions. Humid tropical environments provide no more than high primary productivity of vegetation. High temperatures enhance soil respiration, resulting in a short residence time of soil organic matter that would retain nutrients returning from plants. The excess of precipitation causes leaching of nutrients, and regularly submerged tropical flatlands result in reductive inorganic soil (tropical heath) or wood-peat deposition (tropical peat). Tropical vegetation is adapted to such environments: nutrients are stored in live-plant biomass with a long residence time. By converting natural perennial vegetation to artificial annual vegetation for cereal cropland, nutrients are easily removed from soil, and it is difficult to replicate crop production and maintain cropland sustainability. Shifting cultivation (intermittent burning and crop production in woodland), perennial crops and agroforestry (agriculture in persistent woodland and forest) are adaptive crop production in tropical regions. The required proportion of forested area and the favourable spatial pattern of reserved forest patches to sustain human society will be different between tropical and extratropical environments. Tropical models that support sustainable human society and natural ecosystems in an ideal balance need to be proposed and applied.

Notes

1. A simulation model of the carbon cycle in land ecosystems.
2. See <http://faostat.fao.org>.
3. See <http://www.geohive.com>.

REFERENCES

Diamond, Jared (2005) *Collapse: How Societies Choose to Fail or Succeed*. New York: Viking Press.

Foley, Jonathan A., R. DeFries, G. Asher, et al. (2005) "Global Consequences of Land Use", *Science* 309: 570–574.

Haberl, Helmut, K. H. Erb, F. Krausmann, et al. (2007) "Quantifying and Mapping the Human Appropriation of Net Primary Production on Earth's Terrestrial Ecosystems", *Proceedings of the National Academy of Sciences* 104: 12942–12947.

Imhoff, Marc L., L. Bounoua, T. Ricketts, et al. (2004) "Global Patterns in Human Consumption of Net Primary Production", *Nature* 429: 870–873.

Ito, Akihiko and T. Oikawa (2002) "A Simulation Model of the Carbon Cycle in Land Ecosystems (Sim-CYCLE): A Description Based on Dry-matter Production Theory and Plot-scale Validation", *Ecological Modelling* 151: 147–179.

Rolett, Barry and J. Diamond (2004) "Environmental Predictors of Pre-European Deforestation on Pacific Islands", *Nature* 431: 443–446.

Tilman, David, K. G. Cassman, P. A. Matson, R. Naylor and S. Polasky (2002) "Agricultural Sustainability and Intensive Production Practices", *Nature* 418: 671–677.

Vitousek, Peter M., H. A. Mooney, J. Lubchenco and J. M. Mellilo (1997) "Human Domination of Earth's Ecosystems", *Science* 277: 494–499.

3-3
Sustainable forest management and evaluation of carbon sinks

Noriyuki Kobayashi

3-3-1 Introduction

This section considers sustainable forest management initiatives in relation to the current economic methods for exploiting carbon sinks as a measure against global warming on a worldwide scale. References are also made to the Japanese context.

Forests play an important role in the prevention of global warming by acting as sinks for the removal and storage of carbon dioxide. It has been well documented that the world's forests are diminishing and that the emission of greenhouse gases from forests is equivalent to 20 per cent of total global emissions. Stopping deforestation and making the most of the value of forests as carbon sinks are two of the most important solutions to the problem of global warming. As a way of correcting the problem, global warming countermeasures need to be implemented that achieve sustainable forest management and improve the lives of local people at the same time. One way to achieve this dual benefit is by making the CO_2 absorbed by the forests financially and socially valuable, thereby establishing financial incentives and bringing income to local people and forest stewards. A number of measures have already been put in place based on the Kyoto Protocol; for example, providing incentives for emission reduction projects such as forestation under the clean development mechanism (CDM); emissions trading and the United Nations Collaborative Programme on Reducing Emissions from Deforestation and Degradation in Developing Countries (UN-REDD), which is being developed as part of

---

Designing our future: Local perspectives on bioproduction, ecosystems and humanity,
Osaki, Braimoh and Nakagami (eds),
United Nations University Press, 2011, ISBN 978-92-808-1183-4

the next-generation framework; and the carbon-offsetting Japan Verified Emission Reduction (J-VER) system.

3-3-2 The value of the forests through history

Historical perspective

The function of forests can be broadly separated into economic and environmental purposes as well as purposes for the public good. How these functions are evaluated varies depending on the era and region. When natural forest resources were in abundance or forests were being developed locally, the emphasis was on economic functions, such as the use of forests as a source of timber or the conversion of these lands for farming or other purposes. However, as a result of the global environmental impact of the reduction of forests, the environmental function has been brought to the fore. The basic concept of forest management is closely linked to the evaluation of these functions.

International attention was drawn to the destruction of global forests in *The Planet in 2000* – an analysis of global forest resources released in 1980 by the Carter administration in the United States (Council on Environmental Quality and US Department of State, 1980). It warned that the most recent data at that time showed a dramatic reduction in tropical forests.

In the second half of the 1980s, the decrease in tropical forests was striking, but attention was drawn to this issue by movements in developed countries protesting against the destruction of rainforests in the Amazon basin of Brazil and also the Sarawak region of Malaysia. What is more, in developed countries there were boycotts of products from the rainforest as awareness of this environmental problem increased. Against this background, the Tropical Forestry Action Plan (TFAP) was launched by the Food and Agriculture Organization of the United Nations (FAO), the United Nations Development Programme (UNDP), the World Bank and the World Resources Institute. In addition, the International Tropical Timber Organization (ITTO), a United Nations body, was established in 1986 in Yokohama to promote appropriate trade in tropical forest timber. During the 1980s, forest initiatives were mainly implemented by international bodies targeting tropical forests.

A turning point for actions in support of the world's forests came in 1992 at the United Nations Conference on Environment and Development (UNCED), which was held in Rio de Janeiro, Brazil. At this summit, the issue of forests and, in particular, tropical rainforests was discussed as an important global environmental issue, and a critical topic of these discussions was the symbolic North–South divide concerning environmental protection and economic development. Summit participants agreed to include not only tropical forests but all global forests in the global environment problem, the Non-Legally Binding Authoritative Statement of Principles for a Global Consensus on the Management, Conservation and Sustainable Development of All Types of Forests was adopted, and in *Agenda 21* section II, part 11, a plan of action was laid down relating to the reduction of forests. Under the Statement of Forest Principles, basic rules for sustainable forest management were proposed for the administration of global forests. Standards and guidelines to achieve sustainable forest management were drawn up between the United Nations and various governments, and within regions.

In addition to the initiatives between the United Nations and the participating governments noted above, the global summit served as an opportunity for public-level initiatives to pick up speed. Most representative are the global initiatives for forest certification and environmental labelling. The first global forest certification system was the Forest Stewardship Council (FSC), established in 1993. In 1998, the International Organization for Standardization (ISO) issued ISO/TRI 146061, which applied to forest management under the ISO 14001 Environmental Management System, and which represented the ISO's forest management certification. Following these steps, in 2003 the Programme for the Endorsement of Forest Certification (PEFC) was established as a mutual certification system. Currently, the FSC and the PEFC are the two most common global forest certification systems. In 2003, Japan initiated its own Sustainable Green Ecosystem Council (SGEC); the SEGC Forest Certification System was set up along with the FSC as well-established forest certification systems in Japan.

Towards the end of 1990, the role of forests in the prevention of global warming was becoming recognized as important in the evaluation of forests. Under the Kyoto Protocol adopted in 1997 (United Nations, 1998), it was acknowledged that the amount of carbon dioxide absorbed by carbon sinks was limited, but it was included in the calculation of reduction targets. Furthermore, forest carbon sinks were a subject of the Kyoto mechanism general regulations. The fine details such as the range of forests that were counted as carbon sink forests, the estimated capture volume and the monitoring process were determined under modalities and procedures of the Kyoto Protocol. It can be said that the Kyoto Protocol ushered in an internationally recognized quantitative evaluation method of CO_2 capture by forests.

The perception of the value of forests has evolved on a decade-by-decade basis in the 1980s, 1990s and 2000s; however, the major landmarks were the 1992 UNCED summit and the adoption of the Kyoto Protocol in 1997. The following subsections discuss sustainable forest management and forest certification systems in order to further clarify the value of forests.

Definition, criteria and indicators of sustainable forest management

Sustainable forest management is defined as follows in the global forest management basic regulations proposed in the Statement of Forest Principles adopted by UNCED in 1992:

> Forest resources and forest lands should be sustainably managed to meet the social, economic, ecological, cultural and spiritual needs of present and future generations. These needs are for forest products and services, such as wood and wood products, water, food, fodder, medicine, fuel, shelter, employment, recreation, habitats for wildlife, landscape diversity, carbon sinks and reservoirs, and for other forest products. Appropriate measures should be taken to protect forests against harmful effects of pollution, inducing air-borne pollution, fires, pests and diseases, in order to maintain their full multiple value. (United Nations, 1992a: Article 2(b))

This definition advocates appropriate measures to be taken to protect the forests in order to preserve their multiple values, including as carbon sinks and reservoirs.

Global understanding of the conditions for attaining sustainable forest management is essential. Therefore, in the United Nations Forum on Forests, each country reports annually on the status of its achievements. Forest policy in Japan is founded on sustainable forest management. The basic concept of exploiting the multi-functionality of the forests and maintaining sustainable yet revitalized forestry industries was the underlying principle of the 2001 Forest and Forestry Basic Act. In addition, the Basic Plan for Forest and Forestry established in September 2006 considers forests to be "resources for a green society", promotes long-term forest creation developments and sets targets for forest creation plans to demonstrate the multi-functionality of the forests (Kobayashi, 2008a).

It is necessary to consider what criteria should be set for the development of sustainable forest management and to establish indicators to show whether these aims have been achieved. Japan was a participant in the 1995 Montreal Process meeting in Santiago, Chile, which endorsed a statement of political commitment, known as the "Santiago Declaration", along with a comprehensive set of 7 criteria and 67 indicators.[1] Criterion 5 is "Maintenance of forest contribution to global carbon cycles". The criteria and indicators are determined by bodies such as the United

Nations' Intergovernmental Panel on Forests or the ITTO (Kobayashi, 2001).

Current situation and issues of forest certification

The global total area of forest covered by the FSC is around 107 million hectares (ha), and by the PEFC is around 191 million ha; however, the proportion of the global forest covered is no more than approximately 3 per cent by the FSC and 5 per cent in the case of PEFC. If one considers forested areas by region, certified areas are clustered mainly in Europe, the United States and Canada, where the FSC covers 81 per cent and the PEFC 94 per cent. The PEFC provides no data on certified areas by type of forest but, according to FSC reports, subarctic forests comprise 47 per cent and temperate forests 39 per cent, whereas tropical and subtropical forests comprise no more than 14 per cent (Tachibana, 2009).

In Japan, the area of certified forest was around 270,000 ha covered by the FSC as of December 2008 and 780,000 ha by the SGEC as of March 2009. These figures represent a total area of 4 per cent of the 2.5 million ha of forest in Japan, which is extremely low compared with the rates in Europe.

Global forest certification systems such as the FSC started originally as a response to the destruction of tropical forests. The FSC and the PEFC have both spread mainly around Europe, and the coverage of tropical and subtropical forests is extremely limited. The challenge for global forest certification schemes is to spread to tropical and subtropical forests in developing countries.

It is believed that the lack of proliferation to these regions is because the economic benefit of gaining certification is small compared with the financial outlay and workforce required for the custodians of the forests. In short, it is not cost-effective. Japan could be said to be facing the same challenge. The main key to proliferation is concern about forest certification among consumers of timber and paper products.

3-3-3 Climate change and the forest carbon sink market mechanisms

Evaluation of forest carbon sinks by the IPCC

In the Special Report on *Carbon Dioxide Capture and Storage* (Metz et al., 2005) by the Intergovernmental Panel on Climate Change (IPCC), experts analysed the average annual budget of CO_2 in the 1990s. The annual average volume of emissions resulting from the reduction of forest and land-use change was 1.6 Gt of carbon ± 0.8 Gt of carbon, which is thought to be mainly due to deforestation of tropical forests. Furthermore, net capture by land ecosystems was estimated at 0.7 Gt of carbon ± 0.8 Gt of carbon, but this was considered mainly the result of an increase in forest stock in the Northern Hemisphere (Kobayashi, 2005: 30).

In the IPCC's Fourth Assessment Report (AR4) of 2007, it is assumed that the main causes of the emission and capture of CO_2 by forests during the last decade of the twentieth century were deforestation in the tropics and regrowth in the temperate zones. Emissions due to deforestation during the 1990s were estimated at 5.8 Gt of CO_2 annually (Barker et al., 2007: 67). Multiplying that 5.8 Gt of CO_2 by 12/44 from the IPCC's Special Report on *Carbon Dioxide Capture and Storage* gives 1.6 Gt of carbon – the same value.[2]

The IPCC also assesses the function of forests and timber in global warming prevention, although in the IPCC's Third Assessment Report (AR3) the balancing function of increased CO_2 from Land Use, Land-Use Change and Forestry (LULUCF) is called "biological mitigation", and the report outlines a strategy for promoting "biological mitigation" measures (Albritton et al., 2001). In AR4, officials continue to consider the AR3 mitigation measures, and set out options in four categories for the reduction of emissions and the increase of carbon capture in forests (Barker et al., 2007: 68):

- maintaining or increasing the forest area;
- maintaining or increasing the site-level carbon density;
- maintaining or increasing the landscape-level carbon density and
- increasing off-site carbon stocks in wood products and enhancing product and fuel substitution.

It could be said that the most effective measure for the prevention of global warming is appropriate sustainable forest management: "In the long term, a sustainable forest-management strategy aimed at maintaining or increasing forest carbon stocks, while producing an annual yield of timber, fibre or energy from the forest, will generate the largest sustained mitigation benefit" (Barker et al., 2007: 69). Mitigation measures in the area of forests are evaluated as follows: "Forestry can make a very significant contribution to a low cost global mitigation portfolio that provides synergies with adaptation and sustainable development" (Barker et al., 2007: 70).

The position of forest carbon sinks in the United Nations Framework Convention on Climate Change and the Kyoto Protocol

The IPCC highly values the role of forests in the prevention of global warming based on scientific analysis, but it is in the United Nations Framework Convention on Climate Change (UNFCCC) and the Kyoto Protocol that this assessment is manifested in treaties that reflect system designs for international initiatives for the prevention of global warming. In this subsection, the position of forest carbon sinks in the agreements is analysed (Kobayashi, 2009).

Under the UNFCCC, in the Preamble the authors mention "awareness of the role and importance of capture and storage of greenhouse gases (GHG) by ecosystems on land or in the oceans", and they define the role of forests and so forth as ecosystems on land for the capture and storage of GHG. A sink is defined in Article 1, paragraph 8, as "any process, activity or mechanism which removes a greenhouse gas, an aerosol or a precursor of a greenhouse gas from the atmosphere" (United Nations, 1992b). Included in this definition are the action of removing CO_2 by plants and trees or activities and systems to increase the capture of CO_2.

In Article 4, paragraph 1(a) of the text regulating the pledges of the signatory countries to the UNFCCC, each country is required to submit to the Convention office a national inventory of anthropogenic emissions by sources and removals by sinks. In addition, it is stipulated in paragraph 1(b) that each country must submit a plan for mitigation of global warming and removal by sinks, including carbon sinks. Furthermore, in 1(c) forestry is included in the area of technological development and promotion of technological cooperation, and in 1(d) it is stipulated that each country pledge cooperation in promoting sustainable management of forests and preserving and strengthening sinks and storage.

From these provisions it is clear that, in the UNFCCC, capture by carbon sinks is ranked on the same level as emissions from pollution sources. The importance of carbon sinks can be seen in the provisions, which demonstrate that it is essential that each country act to promote sustainable forest management and put in place policies against global warming.

The Kyoto Protocol, adopted in December 1997 at the Third Conference of the Parties to the Framework Convention on Climate Change held in Kyoto, came into effect in February 2005. This agreement determined the duties of each country to achieve the aims of the Framework Convention. Regarding forest sinks, basic initiatives are set out under the UNFCCC, and only the theory is regulated; however, matters regarding the detailed duties of each country are determined by the Protocol.

The Kyoto Protocol (United Nations, 1998) recognizes that the capture of CO_2 by forests has a role to play in the prevention of global warming, and that the volume of CO_2 absorbed by forests (forest carbon sinks) may be included in the reduction target in order that developed countries bear responsibility for the reduction of greenhouse gases and reach their reduction targets. Article 3, paragraph 3, of the Kyoto Protocol allows that limited human activity (afforestation, reforestation and deforestation after 1990) targeting the capture of greenhouse gases be used to meet the reduction targets of Annex I parties (developed countries). Furthermore, Article 3, paragraph 4, also takes into consideration additional human activity leading to changes in emissions and the capture of GHG in the areas of agricultural land, land-use changes and forestry. Additionally, in Article 2, paragraph 1 (a)(ii), which relates to policies and measures, it stipulates that the practice of sustainable forest management and the promotion of afforestation and reforestation be carried out in accordance with the circumstances in each country (Kobayashi, 2008a).

The Marrakesh Accords and forest carbon sinks

Under the modalities and procedures of the Kyoto Protocol, there were many issues surrounding the treatment of forest carbon sinks based on scientific evidence, but in the IPCC's Special Report on *Carbon Dioxide Capture and Storage* (Metz et al., 2005), international negotiations moved forward because more scientific evidence had come to light. The legal text of the modalities and procedures of the Marrakesh Accords was adopted at the Seventh Conference of the Parties (COP 7) in 2001 (UNFCCC Conference of the Parties, 2002: Decision 11/CP.7). Under the Marrakesh Accords, in addition to the rules of operation such as definitions and measurements of volume of capture, activities subject to the Kyoto Protocol, Article 3, paragraph 4, were determined, including forest management, cropland management, grazing land management and revegetation.

Forest management is defined as "a system of practices for stewardship and use of forest land aimed at fulfilling relevant ecological (including biological diversity), economic and social functions of the forest in a sustainable manner". The definition does not go further than the commonly held view, and, although it does stipulate further details clearly, individual countries evaluate and apply forest management "human-induced activities" noted in Article 3, paragraph 4, to the forest management situation in those countries. The definitions of afforestation and reforestation refer to the three points of land-use change, when there was forest previously and when it was planted (Kobayashi, 2008b). In addition, upper limits were determined for the application (calculation) of capture volume for the commitment targets of individual countries. For Japan, these limits were set at 13 million tonnes of carbon (t-C). Methods of calculating forest capture rates used the IPCC's "Good Practice Guidance for Land Use, Land-Use Change and Forestry" (IPCC NGGIP, 2003), adopted at the Tenth Session of the Conference of Parties (COP 10) in 2004.

3-3-4 Japan's framework for the achievement of the Kyoto Protocol targets and forest carbon sink measures

Status of forest carbon sinks under the framework for the achievement of the Kyoto Protocol targets

The act that forms the foundation of Japan's global warming countermeasures is the 1998 Law Concerning the Promotion of the Measures to Cope with Global Warming (hereinafter referred to as the "Global Warming Countermeasures law"). Article 8 of this law stipulates the establishment of a framework for achieving Kyoto Protocol targets. Article 28 of the same act specifies plans for maintaining and strengthening greenhouse gas capture activities to achieve the commitment targets determined in the framework for the achievement of Kyoto Protocol targets.

The Forest Agency, established in 2002, is implementing a "10-year Action Plan on the Mitigation of Global Warming by Forest Carbon Sinks" in accordance with the Basic Plan for Forest and Forestry (stipulated in the Forest and Forestry Basic Act, Article 11, para. 1), which provides the framework for the achievement of the targets of the Global Warming Countermeasures Act, Article 28. Japan is committed to reducing greenhouse gases by 6 per cent over 1990 levels (the baseline year), but a large proportion – 3.8 per cent (or 13 million t-C) – comes from forest carbon sink countermeasures.

From among the human-induced activities noted in Article 3, paragraph 4, of the Kyoto Protocol, Japan has decided to implement a plan to guarantee its 3.8 per cent reduction through the use of "forest management" as stipulated in the Marrakesh Accords. Japan has intensified forest management by implementing the thinning of 550,000 ha of artificial forest annually, putting in place forest protection directives on natural forests and enhancing preservation and protection measures.

The challenge of forest carbon sink measures

Tree thinning is promoted as a measure for forest sinks but, as a result of the reduction in the profitability of the forestry industry, tree thinning in the privately owned forests that make up 70 per cent of Japanese forest area did not go according to plan, although implementation of tree thinning was previously an important policy in the forestry industry. An aid scheme was established to encourage tree thinning so that, in general, the costs of forest maintenance projects such as thinning were divided up, with national government aid covering around 50 per cent, prefectural governments providing approximately 20 per cent and forest owners approximately 30 per cent. How to reduce costs that fall outside government assistance and how to proceed with tree thinning are major issues, because many local governments are under financial pressure and tree thinning has no economic merits for the individual forest owner. Furthermore, as there is no benefit to be gained from tree thinning and removal of wood, "cut and leave thinning", whereby the cut material is abandoned in the forest, is increasing. A reduction in "cut and leave thinning" and an increase in the removal of thinned wood are important challenges in terms of the cycle of resources (Kobayashi, 2008b). In addition, owing to a rise in domestic demand in recent years, areas of clear-cut harvest without reforestation are growing but, judging by the forestry industry's worsening profitability, it is possible that there will be an increase in non-reforestation, not only affecting the local environment, society and economy but also having an impact on sustainable forest management. Moreover, there are fears that the problem will affect global warming.

Local governments' new initiatives for forest sinks

Forward-thinking local governments and forest owners in Japan are enthusiastically stepping up to the challenges to increase even slightly the value of forests and forestry. As a first step, forest certification systems and schemes such as partnership forests and enterprise forests are being introduced as new initiatives to increase the value of forests. In recent years, there have been a number of initiatives to reduce carbon emissions through forest CO_2 capture certification schemes and wood-based biomass use. These strategies have been launched by various local governments, but here two representative examples are examined: Shimokawa-cho, Hokkaido, and Kochi Prefecture.

Shimokawa-cho, Hokkaido

The basic policy of forest management in Shimokawa is "sustainable cyclical forest management" (Shimokawa Town Office, 2009; Shimokawa Regional Studies Association, 2003). In 1960, Shimokawa town council was already planting and harvesting between 40 and 60 hectares annually. Now it has established a sustainable forest management method whereby 40–50 ha are repeatedly harvested and reforested on an approximately 60-year cycle. The aim of this approach is to maintain the forest while recycling resources and at the same time ensuring continuity of employment and wood production. The most important aim of the municipal forest management is the knock-on effect on the local economy; it is no exaggeration to say that the municipal forest plays a vital role in slowing down depopulation and supporting the town's economy.

According to the Shimokawa town council, the municipal forest is estimated to have a stabilized stock volume of approximately 1 million tonnes of CO_2 in 2007, and to capture approximately 4,200 tonnes of carbon per annum (Shimokawa Town Office, 2009).

Shimokawa leads the way for local governments across the nation in the capture and stabilization of CO_2 by the forest, and, in August 2002, the town looked into a way of adding economic and social value to the volume of CO_2 capture. In other words, it considered the possibility of trading the volume of captured CO_2 on the international emissions trading market. Although initially this idea was not realized, it was reported in the national edition of the *Nihon Keizai Shimbum* (Japan Economic News) and was discussed across the nation (Kobayashi, 2008b). Following that effort, Shimokawa continued to consider ways of adding social and economic value to the volume of CO_2 captured by the forest, and the four towns of Ashoro, Shimokawa, Takinoue and Bihoro established the Association for the Promotion of Carbon Capture by Forest Biomass. The purpose of this Association was to stimulate the local community through the exploitation of CO_2 capture, to enhance the potential of the locally available resource of forest biomass and to construct systems for effective CO_2 reduction by switching from fossil fuels (Shimokawa Town Office, 2009). To set up the systems, a committee of seven renowned forest scientists, economists and lawyers was established.

Kochi Prefecture

In many prefectures, new taxes are being put forward as a way of maintaining and preserving forests and as a measure against global warming. In April 2003, Kochi Prefecture introduced the pioneering "forest environment tax", which forms an important fund for activities such as promoting the thinning of plantations. As of 2008, 29 prefectures are introducing a taxation system for the maintenance of forests.

With a view to a future emissions trading system, Kochi Prefecture set up the "Partnership forest" in September 2005. This project aims to encourage regeneration of forests and involvement of the local community through sustainable forest environment management and partnerships with businesses that have an interest in environmental issues. As of March 2009, 39 businesses had joined the agreements, and total contributions were in excess of ¥240 million (Uchimura, 2009). There are similar systems across the country, but this one is thought to be the earliest and the most widespread.

Kochi Prefecture developed the 2006 "Partnership forest" project further and introduced the Partnership forest CO_2 capture certification system, which certifies the amount of CO_2 captured by the forests. This initiative was the first in the country to be managed by a local government. The number of CO_2 certification certificates issued has risen year on year from 3 in 2007, to 20 in 2008, to 28 in 2009.

In 2007, Kochi Prefecture launched the Kochi CO_2 reduction certification system, coordinated by the CO_2 Reduction Special Committee. The first initiative under this system was the establishment of a wood-based biomass mixed incinerator facility at the Sumitomo Osaka Cement Company, Kochi Plant (in Suzaki). This project was the first registered emissions reduction project under the J-VER system – Japan's carbon-offsetting system discussed below. The first third-party certified J-VER in Japan came into effect on 10 March 2009 (Uchimura, 2009).

Creating new value by utilizing forest resources in Japan

The emissions trading "trial run" by the Ministry of Economy, Trade and Industry (METI) and the Japan Verified Emission Reduction (J-VER) carbon-offsetting credits system established by the Ministry of the Environment are new systems implemented by the government that offer financial incentives. Under the J-VER system, credits are offered for the volume of carbon sink capture and for emissions reduction through biomass energy. The "trial run", however, does not include carbon sink credits. Therefore, at present in Japan, the only system under which forest carbon sinks can be used for credits is the J-VER system.

3-3-5 Carbon offsetting: J-VER system and forest projects

About the carbon-offsetting system

Carbon offsetting is a method for preventing global warming that incorporates a wide range of initiatives related to the everyday activities of ordinary citizens. It is becoming more widespread in the United States and Europe. Carbon offsetting is defined as:

> a system whereby all members of society: citizens, businesses, NPOs/ NGOs, local government, national government, etc., are aware of their own emissions of greenhouse gases and make an effort to reduce them autonomously by purchasing the reduction or capture of emissions occurring in another location to cover the portion which is difficult to reduce (known as credits) or, partially or

wholly compensating for emissions by establishing activities or projects to reduce or absorb emissions in another location. (Ministry of the Environment, 2008)

Carbon offsetting is a voluntary system that was first used commercially in 1997 by the United Kingdom company Future Forests. However, emissions trading is intended to enable the fulfilment of emissions reduction obligations under the Kyoto Protocol or other local or national systems. The credits used under carbon offsetting are called verified emission reduction (VER). The scale of the market is increasing, and it is expected to be 400 million tonnes of CO_2 worldwide in 2010 (Takeda, 2009).

Japan carbon credit system initiative

To promote the spread of carbon offsetting, the Ministry of the Environment set up the Commission for the Status of Carbon Offsetting in September 2007, and in February 2008 released the results as "Guidelines for Carbon Offsetting in Japan". Furthermore, in March 2008, the Commission for Certification Criteria for VER Using Carbon Offsetting was established. This body considered system design issues of VER and third-party verification, and in November 2008 released the "Implementation Rules for Offsetting and Carbon Credit (J-VER)". The implementation rules stipulated highly transparent monitoring, calculation and verification rules and a framework for the issuance and management of carbon-offsetting credits.

Overview of J-VER

Figure 3.3.1 shows a chart of J-VER operations, in which the process flow is shown by the lines. The party in charge of the proposed project (A) creates a project application (the project design document, PDD):
(1) the application is submitted to the J-VER executive board (C) of the Certification Center on Climate Change, Japan (4CJ);[3]
(2) this application is investigated by the J-VER executive board and registered at 4CJ;
(3) the applicant carries out the project;
(4) the emissions and captured volume are monitored, and the results are verified by a third-party organization (D);
(5) the results of the verification are submitted to the J-VER executive board and the reduction removal by sink or capture of emissions is certified.

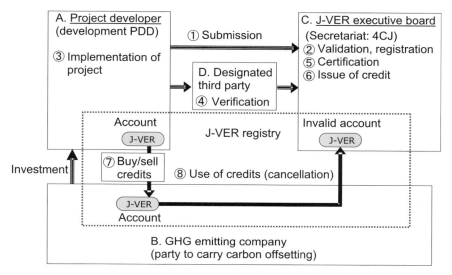

Figure 3.3.1 Process and organization of J-VER.
Source: Adapted from Ministry of the Environment materials.

In terms of the Kyoto Protocol clean development mechanism (CDM), (2) is equivalent to validation, (4) to verification and (5) to certification.

Registration of the credits (J-VER) and the management process are as follows:

(6) The J-VER executive board registers and issues the J-VER credits to the party carrying out the project in accordance with the verified reduction or removal;
(7) the party undertaking the project sells the credits to companies that are emitting greenhouse gases (B); and
(8) the purchasing business can use the credits for carbon offsetting; at the time of carbon offsetting, the credits that balance out the emissions and removal are cancelled on the J-VER register.

The J-VER executive board (C), which at present comprises eight members employed by the Ministry of the Environment, plays a very important role in certification, issuance, management and the positive list (explained later) of J-VER, as well as in discussion of policy.

The third-party certification organization (D) is vital for maintaining the credibility of J-VER, and eligibility is strictly limited to organizations that have been authorized under ISO 14065. As of July 2009, 11 organizations were provisionally registered as authorized.

J-VER forest sinks credit certification criteria

For forest projects subject to the J-VER system, a forest sink working group (WG) was established under the Commission for Certification and Criteria for VER, using carbon offsetting, and from October 2008 to March 2009 it deliberated on the certification criteria for greenhouse gas removal expansion projects under the J-VER system and current forest management projects. The certification criteria it stipulated were intended to determine the standard evaluation of the volume of CO_2 captured by individual forest management projects in Japan. The primary matters for consideration by the forest WG were the types of projects, the method of calculating the volume of removal, and monitoring methods. The types of forests or projects are detailed in the positive list conditions of eligibility. In addition, the methodology for calculating capture and the monitoring methods that have been determined are available in the positive list, as are summaries of methodology guidelines and guidelines for monitoring.

The positive list

The positive list refers to a list of the types of greenhouse gas reduction or capture projects to which the system applies; under the system, the "criteria of eligibility" are determined for individual project types. Furthermore, the "criteria of eligibility" refer to the fulfilment of necessary requirements at the time of project application and, if the criteria are met, additionality is verified. The applicants complete their application in accordance with the positive list. When an applicant proposes a project that does not appear on the positive list, the applicant must supply certification of additionality and obtain authorization from the J-VER executive board.

As of July 2009, three types of forest management projects were on the positive list: forest management projects promoting tree thinning (VER 0002-1), projects contributing to sustainable forest management (VER 0002-2) and tree-planting projects (VER 0003). The eligibility conditions for individual projects are discussed below.

Maintaining the credibility of credits

To ensure the credibility of the credits, it is necessary to guarantee the permanency of capture volume and countermeasures against risk in forest management projects. The following points are used under the J-VER system.

The permanency of forest management is assured in Articles 1 and 3 of VER 0002-1 and 0002-2 and in Article 3 of VER 0003 as conditions of eligibility under the positive list. In addition, responsibility for the management of a project is borne by the party in charge of it from its registration until 31 March 2035. There are measures in place if the parties administering these projects do not strictly adhere to these conditions. Risk countermeasures are procedures such as the establishment of a "buffer management account" – a pool comprising 3 per cent of the removal volume of each project, which acts as a buffer system to cope with forest fires or other natural disasters.

Monitoring and calculation formula

The government employs the gross net formula for calculating total capture by domestic forest management activities during the first commitment period, in accordance with Article 3, paragraph 4, of the Kyoto Protocol. This formula, used by the J-VER system, is a method of calculating the increases in carbon capture for applicable projects based on fluctuations in annual total stored carbon. The calculation method is shown on the positive list for each project type under the heading "methodology". Therefore, the party in charge of a project simply needs to use the capture volume calculation formula shown there.

Monitoring is defined as gathering essential information and data in order to regulate the volume of removal or emission of greenhouse gases, as well as measuring, calculating and recording them. "The J-VER Monitoring Guidelines" detail how to obtain the necessary data and information and also how to execute the work (Ministry of the Environment, 2009). The party carrying out the project monitors it in accordance with these guidelines, collects the data and calculates the results. The project supervisor then creates a monitoring report that undergoes third-party verification.

3-3-6 Conclusion

Evaluation of the potential of forests is changing with the times. When the 1992 Statement of Forest Principles was adopted, the emphasis was on evaluating the multi-purpose functionality of forests, and this document proposed sustainable forest management. More recently, evaluation of forests has focused on the capture and storage of CO_2 and on biodiversity.

It is thought that the role of forests in the prevention of global warming is vital but, in order to achieve that role, basic initiatives for sustainable forest management in the world's forests are essential. Three proposed functions of forests are considered here: the outlook for forest carbon sinks; the economic and social added value of forest carbon sinks; and domestic carbon sink initiatives in Japan under a next-generation Kyoto Protocol (post-Kyoto).

Outlook for post-Kyoto forest carbon sinks

A challenge for forests, post-Kyoto, is reducing emissions from deforestation and forest degradation (REDD) and the treatment of domestic forest carbon sinks. The developing countries are encouraged to form cooperating bodies with developed countries under REDD, or afforestation and reforestation CDM. It is thought that, under the extension of the Kyoto Protocol, Article 3, paragraph 4, the forests of developed countries can be incorporated as domestic forest sinks in individual countries.

Deforestation and degradation of forests in developing countries account for a large part of the global volume of greenhouse gas emissions and are recognized as an important challenge in combating global warming. In December 2005, at COP 11, Papua New Guinea and Costa Rica raised REDD on the formal agenda as a proposal. Consideration of measures and positive incentives regarding REDD were included in the "Bali Road Map" (UNFCCC Conference of the Parties, 2007: para. 1(b)(ii)), which won agreement in December 2007 at COP 13.

Bringing together global warming countermeasures and developing countries, particularly the control of the deforestation and degradation of tropical forests, is not without objections, but there are many issues that could be resolved at the time of implementation of the specific measures. The principal issues for consideration that relate to REDD can be broadly summarized under the following two points:

1. Methodological issues concerning the development of methods of measurement and the observation of reductions in emissions caused by deforestation and the degradation of forests, as noted in paragraph 7 of the resolution on reducing emissions from deforestation in developing countries passed at COP 13 (UNFCCC Conference of the Parties, 2008: 9).
2. Initiatives to provide incentives (investment and support) to developing countries to control GHG emissions from deforestation and forest degradation.

The technical problems include how to draw up a reference level that will form the criteria for predicted deforestation and forest degradation, and so forth, and how to develop measuring and monitoring methods for greenhouse gas emissions caused by deforestation and degradation of forests. Strategic problems include the anticipated divergence between the developing countries, which expect the financial burden to be passed to developed countries under the positive incentive system, and the developed countries, which are trying to avoid the costs as much as possible. It will be difficult for international negotiators to decide how best to balance out these advantages and disadvantages to each side, and there were misgivings about the agreement format for the REDD framework in COP 15 in Copenhagen in December 2009 (Kobayashi, 2009).

Under the next-term framework in developed countries, discussion of the treatment of domestic sinks is one of the agenda items at the Ad Hoc Working Group on Further Commitments for Annex I Parties under the Kyoto Protocol (AWG-KP), entitled "Analysis of stages to attain emissions reduction targets further after 2013 by Annex I Parties", and discussions are ongoing. Also, in the resolution document regarding potential reductions by developed countries and stages for attaining reduction (UNFCCC AWG-KP, 2008: para. 10) – bearing in mind that the use of sinks should continue in the next-generation framework through LULUCF – it was reconfirmed that the treatment of developed countries' forest carbon sinks would be included under the next-generation framework (Kobayashi, 2009).

It will be a huge step forward in the incorporation of targets for the prevention of global warming and for global sustainable forest management if the initiative regarding REDD and domestic carbon sinks can be agreed to under the post-Kyoto next-generation framework. When implementing post-Kyoto forest carbon sink initiatives, it is essential to have strategies that link with the United Nations Forum on Forests for the achievement of sustainable forest management and comply with the Convention on Biological Diversity.

Social and economic added value of forest sinks

Forest certification systems are methods of objectively evaluating the value of a forest. Despite the challenge of spreading this approach to developing countries, developed nations have had good results. Initiatives involving the added social and economic value of forest sinks provide a new method of evaluating the environmental value of forests socially and economically. CDM credits and emissions trading under the Kyoto Protocol provide a first step, as do the forest carbon sink credits through the VER carbon-offsetting system.

REDD has the potential to create a new method of evaluation in relation to tropical forests and global warming. The "positive incentive" system will accord social and economic value to tropical forest carbon sinks by offering an economic incentive to developing countries. The World Bank's Forest Carbon Partnership Facility (FCPF) was established in

2007 as a method of providing economic incentives to leading businesses through REDD. FCPF – a combination of basic policy and the utilization of emissions rights – is a method whereby investment is made in projects that avoid deforestation of the tropical forests. The investment return comes in the form of credits for the reduced GHG emissions through this reduction control. It can be said that this method incorporates the role of the social and economic added value of tropical forest sinks (Kobayashi, 2008b).

It is essential to involve local people in sustainable forest management systems such as REDD and FCPF, and to offer these people economic benefits. Such methods can potentially achieve the dual benefits of global warming countermeasures through the attainment of sustainable forest management and improvement of the lives of local people. The examples of CO_2 sink certification systems and carbon-offsetting systems that have been implemented by local governments in Japan have been a source of inspiration for initiatives at a local level in developing countries that are aimed at the dual benefits.

Suggestion for initiatives for Japanese domestic sinks

Reduction in demand over the years and downward pressure on prices have had a negative impact on the profitability of forest management for Japan's forestry industry. This has led to problems such as a slowdown of forest maintenance by tree thinning or cut and leave thinning, or non-reforestation. These issues have led the industry to have misgivings about sustainable forest management.

Further anti-global warming initiatives are to be expected, in addition to the introduction of new initiatives such as the CO_2 sink certification systems promoted by local governments, the J-VER system of carbon offsetting by the Ministry of the Environment or forest maintenance by individual investment, along with assistance in the promotion of forest maintenance, such as tree thinning or usage of domestic timber. Furthermore, these methods also emphasize ordinary citizens' deepening appreciation of the role played by forests in the prevention of global warming.

It is a precondition of sound forest management based on the principles of sustainability that the role of forests in tackling global warming is evaluated and that economic and social value is given to forest carbon sinks through CO_2 sink certification or the J-VER system. The current status of forestry management in Japan limits the extent to which it should be entrusted to individual forest owners. Instead, a "green social system" needs to be constructed. One idea is that relationships of confidence and collaboration should be built between the various stakeholders and local forest management should be implemented as a technique of forest governance (Kakizawa, 2007). The various initiatives to utilize forest resources in Kochi and Shimokawa, which forged links between various stakeholders, and the idea of forest governance can be seen as the embodiment of this social system (Kobayashi, 2008b).

Post-Kyoto domestic carbon sink initiatives after 2013 will be considered in accordance with the rules under the next-generation framework for the treatment of forest sinks, but the next step needs to be considered in relation to the domestic system. First, it is essential to establish a policy that makes the most of original ideas from local governments or forest owners that will lead to the revitalization of forests and to the added social and economic value of forest sinks. Second, under the next-generation framework, carbon sink credits need to be utilized to achieve Japan's reduction commitment; any future emissions trading system in Japan must include carbon sink credits. The proliferation of carbon sink credits under the J-VER system is vital as a test case.

It is to be hoped that policymakers, researchers, local and national governments and business people who are concerned about the prevention of global warming and about sustainable forest management will take note of the remarks in this section.

Notes

1. The Montreal Process is the Working Group on Criteria and Indicators for the Conservation and Sustainable Management of Temperate and Boreal Forests, formed in 1994.
2. $CO_2 * 12/44 = C$.
3. See <http://www.4cj.org/english/4CJ-Eng.pdf>.

REFERENCES

Albritton, D. L., T. Barker, I. A. Bashmakov, O. Canziani, R. Christ, U. Cubasch, O. Davidson, H. Gitay, D. Griggs, K. Halsnaes, J. Houghton, J. House, Z. Kundzewicz, M. Lal, N. Leary, C. Magadza, J. J. McCarthy, J. F. B. Mitchell, J. R. Moreira, M. Munasinghe, I. Noble, R. Pachauri, B. Pittock, M. Prather, R. G. Richels, J. B. Robinson, J. Sathaye, S. Schneider, R. Scholes, T. Stocker, N. Sundararaman, R. Swart, T. Taniguchi and D. Zhou (2001) *Climate Change 2001: Synthesis Report. Contribution of Working Groups I, II and III to the Third Assessment Report of the Intergovernmental Panel on Climate Change*, ed. R. T. Watson. Cambridge and New York: Cambridge University Press. Available at: <http://www.grida.no/publications/other/ipcc_tar/> (accessed 5 April 2010).

Barker, T., I. Bashmakov, L. Bernstein, J. E. Bogner, P. R. Bosch, R. Dave, O. R. Davidson, B. S. Fisher, S. Gupta, K. Halsnæs, G. J. Heij, S. Kahn Ribeiro, S. Kobayashi, M. D. Levine, D. L. Martino, O. Masera, B. Metz, L. A. Meyer, G.-J. Nabuurs, A. Najam, N. Nakicenovic, H.-H. Rogner, J. Roy, J. Sathaye, R. Schock, P. Shukla, R. E. H. Sims, P. Smith, D. A. Tirpak, D. Urge-Vorsatz and D. Zhou

(2007) "Technical Summary", in *Climate Change 2007: Mitigation. Contribution of Working Group III to the Fourth Assessment Report of the Intergovernmental Panel on Climate Change*, ed. B. Metz, O. R. Davidson, P. R. Bosch, R. Dave and L. A. Meyer. Cambridge and New York: Cambridge University Press.

Council on Environmental Quality and US Department of State (1980) *The Global 2000 Report to the President. Vol. I: Entering the Twenty-First Century. Vol. II: The Technical Report*. Gerald O. Barney, Study Director. Washington DC: US Government Printing Office.

IPCC NGGIP [Intergovernmental Panel on Climate Change, National Greenhouse Gas Inventories Programme] (2003) "Good Practice Guidance for Land Use, Land-Use Change and Forestry", ed. Jim Penman et al. Kanagawa, Japan: Institute for Global Environmental Strategies.

Kakizawa Hiroaki (2007) "For the Establishment of Forest Governance", *Sanrin* (Journal of the Japan Forestry Association), No. 1478: 2–9.

Kobayashi, Noriyuki (2001) "The Latest Trend of ISO in the Field of Forestry Management and Related Industries", IGES Policy Trend Report 2001: 81–94.

Kobayashi, Noriyuki (2005) *Climate Change and Forestry Business*. Japan Forestry Investigation Committee, pp. 30–34.

Kobayashi, Noriyuki (2008a) "A Legal and Social System for Sustainable Forest Management", *Nihon University Law Review* March.

Kobayashi, Noriyuki (2008b) *Global Warming and Forest-Protection of Global Benefit*. Japan Forestry Investigation Committee, pp. 40–43, 54–57, 234–239.

Kobayashi, Noriyuki (2009) "Climate Change and Role of Forest", *Environmental Law Journal* No. 33: 110–136.

Metz, B., O. Davidson, H. de Coninck, M. Loos and L. Meyer, eds (2005) *Carbon Dioxide Capture and Storage*, Special Report Prepared by Working Group III of the Intergovernmental Panel on Climate Change. Cambridge: Cambridge University Press. Available at <http://www.ipcc.ch/pdf/special-reports/srccs/srccs_wholereport.pdf> (accessed 5 April 2010).

Ministry of the Environment (2008) "Study Group of Carbon-offset 1.22".

Ministry of the Environment (2009) "J-VER Monitoring Guidelines (Forest Management) (Ver. 2.0)".

Shimokawa Regional Studies Association (2003) "Forest of Shimokawa".

Shimokawa Town Office (2009) "Toward Low Carbon Community by Forest".

Tachibana, Satoshi (2009) "The Trends of Forest Certification System of Recent Years", *Sanrin* (Journal of the Japan Forestry Association), No. 1499: 60–68.

Takeda Masahiro (2009) *OECC Journal* (Overseas Environmental Cooperation Centre, Japan), No. 56: 9.

Uchimura, Naoya (2009) "J-VER and Wood Biomass of Kochi Prefecture", *Gendai Ringyo* July: 18–21.

UNFCCC AWG-KP [Ad Hoc Working Group on Further Commitments for Annex I Parties under the Kyoto Protocol] (2008) "Means, Methodological Issues, Mitigation Potential and Ranges of Emission Reduction Objectives, and Consideration of Further Commitments". Available at: <http://unfccc.int/resource/docs/2008/awg6/eng/l18.pdf> (accessed 9 March 2010).

UNFCCC Conference of the Parties (2002) "The Marrakesh Accords", in *Report of the Conference of the Parties on Its Seventh Session, Held at Marrakesh from*

29 October to 10 November 2001. Addendum – Part Two: Action Taken by the Conference of the Parties, Volume I, FCCC/CP/2001/13/Add.1. United Nations, 21 January. Available at: <http://unfccc.int/resource/docs/cop7/13a01.pdf> (accessed 8 March 2010).

UNFCCC Conference of the Parties (2007) "The Bali Road Map", *Revised Draft Decision -/CP.13: Ad Hoc Working Group on Long-term Cooperative Action under the Convention*. Conference of the Parties, Thirteenth session, Bali, 3–14 December 2007, FCCC/CP/2007/L.7/Rev.1, 14 December 2007. Available at: <http://unfccc.int/files/meetings/cop_13/application/pdf/cp_bali_act_p.pdf> (accessed 9 March 2010).

UNFCCC Conference of the Parties (2008) *Report of the Conference of the Parties on Its Thirteenth Session, Held in Bali from 3 to 15 December 2007. Addendum: Part Two: Action Taken by the Conference of the Parties at Its Thirteenth Session*, FCCC/CP/2007/6/Add.1*, 14 March. Available at: <http://unfccc.int/resource/docs/2007/cop13/eng/06a01.pdf> (accessed 5 April 2010).

United Nations (1992a) "Non-Legally Binding Authoritative Statement of Principles for a Global Consensus on the Management, Conservation and Sustainable Development of All Types of Forests", *Report of the United Nations Conference on Environment and Development (Rio de Janeiro, 3–14 June 1992)*, Annex III, A/CONF.151/26 (Vol. III), 14 August.

United Nations (1992b) *United Nations Framework Convention on Climate Change*, FCCC/INFORMAL/84, <http://unfccc.int/resource/docs/convkp/conveng.pdf> (accessed 8 March 2010).

United Nations (1998) *Kyoto Protocol to the United Nations Framework Convention on Climate Change*, <http://unfccc.int/resource/docs/convkp/kpeng.pdf> (accessed 8 March 2010).

3-4
Ocean ecosystem conservation and seafood security

Masahide Kaeriyama, Michio J. Kishi, Sei-Ichi Saitoh and Yasunori Sakurai

3-4-1 Introduction

Marine food should be a renewable resource for humans. However, world fish catches have peaked since the 1990s, despite increases in aquaculture production. Fish provide more than 2.9 billion people with at least 15 per cent of their animal protein intake (FAO Fisheries and Aquaculture Department, 2009). Tuna (*Thunnus* spp.) stocks have decreased severely from overfishing since the 1980s (Myers and Worm, 2003). Bluefin tuna (*T. thynnus*) is already listed as a "critical species" by the IUCN (International Union for Conservation of Nature). Although aquaculture production is increasing worldwide, many aquaculture programmes such as shrimp farming have caused the destruction of mangrove forests over the past 20 years in East Asia (Primavera, 2005), and marine pollution has also been reported for farmed Atlantic salmon (Hites et al., 2004).

Traditional fisheries science considers only fisheries, some of whose consequences include fishing down marine food webs (Pauly et al., 1998), the overfishing of tuna, the tragedy of the commons, food miles, ecosystem crashes and food pollution. A paradigm shift is needed from traditional fisheries science to a new ecological fisheries science and oceanography for the protection of ocean ecosystems and human seafood resources for the well-being of future generations. This section discusses sustainable conservation based on the ecosystem approach as a new paradigm for risk management in fisheries. The NEMURO ocean ecosystem model and the application of remote sensing and marine Geographical

Designing our future: Local perspectives on bioproduction, ecosystems and humanity, Osaki, Braimoh and Nakagami (eds),
United Nations University Press, 2011, ISBN 978-92-808-1183-4

Information Systems (GIS) are discussed, as well as a case study of fisheries management at the Shiretoko World Natural Heritage Site in northern Japan.

3-4-2 Sustainability issues for ocean ecosystem conservation and seafood security?

Carrying capacity and long-term climate change

The population dynamics of Pacific salmon (*Oncorhynchus* spp.) are directly affected by a number of stresses (climatic and human impacts) that need to be considered within an ecosystem context. A significant positive correlation was observed between the Aleutian Low Pressure Index (ALPI) and carrying capacity at the species level. Residual carrying capacity was significantly positively correlated with body size and negatively related to age at maturity in chum salmon (*O. keta*) as an example of a density-dependent effect (Kaeriyama, 2008). On the other hand, the biomass of wild chum salmon populations in the 1990s decreased to 50 per cent of 1930s levels, despite a significant increase in hatchery populations (Figure 3.4.1). This phenomenon suggests that the increased number of hatchery salmon affected the reproductive value (e.g. fecundity) of wild populations.

Global warming is affecting marine organisms. How will Pacific salmon be affected by global warming in the near future? Using the A1B

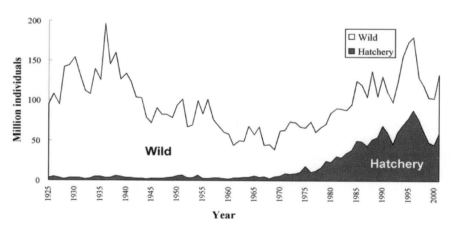

Figure 3.4.1 Temporal change in biomass of wild and hatchery populations of chum salmon in the North Pacific, 1925–2001.
Source: Kaeriyama (2008).

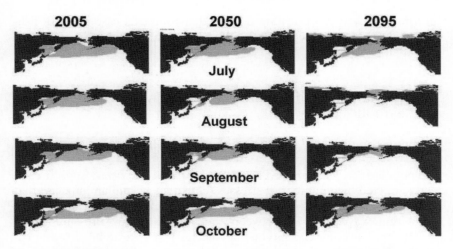

Figure 3.4.2 Prediction of the global warming effect on chum salmon in the North Pacific Ocean.
Source: Kaeriyama (2008), based on the SRES-A1B scenario of the IPCC.

scenario of the Intergovernmental Panel on Climate Change (IPCC) Special Report on Emissions Scenarios (SRES), it is possible to infer the global warming effect on chum salmon based on their optimal temperature for growth (8–12°C; Kaeriyama, 1986; Ueno and Ishida, 1998). This suggests that chum salmon would be brought into direct competition with other salmon populations, leading to a decrease in the survival rate and population density-dependent effects because of a reduction in the distribution area, displacement to the north (e.g. the Arctic Ocean) and loss of migration routes (e.g. the Okhotsk Sea; see Figure 3.4.2; Kaeriyama, 2008).

Sustainable conservation management based on the ecosystem approach

The structure of ecosystems includes interaction between the abiotic environment and the organism, and the function of ecosystems includes interaction between the abiotic environment and biodiversity. The aquatic ecosystem is subject to disturbance by natural factors and human impact. Recently, human impact has strongly affected the aquatic ecosystem; for example, global warming, overfishing, habitat loss, creation of artificial river channels, dam construction and the negative effects of aquaculture and hatchery programmes (Kaeriyama and Edpalina, 2004). It is necessary to identify the limitations of fisheries management, focused at the population level, and to establish sustainable conservation and management based on the integration of population-level approaches within a wider ecosystem-based approach.

Definitions of an ecosystem-based approach to fisheries management have been proposed by several authors (FAO Fisheries Department, 2003; McLeod et al., 2005; Marasco et al., 2007; Murawski and Matlock, 2006; NRC, 1999; Witherell et al., 2000). The definition of McLeod et al. (2005: 1) is used in this section:

> Ecosystem-based management is an integrated approach to management that considers the entire ecosystem, including humans. The goal of ecosystem-based management is to maintain an ecosystem in a healthy, productive and resilient condition so that it can provide the services humans want and need. Ecosystem-based management differs from current approaches that usually focus on a single species, sector, activity or concern; it considers the cumulative impacts of different sectors.

Sustainable conservation management based on the ecosystem approach (SCMEA) for Pacific salmon should be a part of sustainability science of fisheries and oceans. Three aspects of the structure and function of the ocean ecosystem should be monitored: (1) spatial and temporal changes, including carrying capacity, food web and trophic level; (2) climatic oceanic conditions, including global warming and regime shifts; and (3) biological interactions between wild and hatchery populations, density-dependent effects and inter- and intra-specific competition. For the SCMEA of Pacific salmon, adaptive management and precautionary principles are important. In particular, adaptive management should be conducted based on the feedback of monitoring, modelling and adaptive learning, which includes learning by conducting risk analyses and consensus-building (Figure 3.4.3; Kaeriyama, 2008).

3-4-3 The NEMURO model: An example of an end-to-end food web model and its application to marine ecosystem investigations

The PICES CCCC (North Pacific Marine Science Organization, Climate Change and Carrying Capacity Program) MODEL Task Team achieved a consensus on the structure of a prototype lower trophic level ecosystem model for the North Pacific Ocean and named it "NEMURO" (Kishi et al., 2007). Its extension "NEMURO.FISH" (Megrey et al., 2007; Figure 3.4.4) includes a fish growth model and represents a step towards end-to-end modelling (Travers et al., 2007). Since 2000, the year of the first NEMURO workshop, an extensive dialogue between modellers,

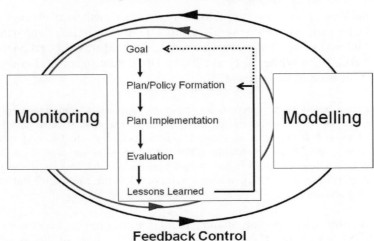

Figure 3.4.3 Scheme for adaptive management of sustainable fisheries of Pacific salmon based on the ecosystem approach.
Source: Kaeriyama (2008).

plankton biologists and oceanographers has occurred, producing over 30 NEMURO-related papers in the peer-reviewed literature. Many of these were conducted for distinct geographical regions and specific marine organisms.

Many papers on NEMURO and NEMURO.FISH are published in the *Ecological Modelling* special issue (vol. 202, 2007), and these are highlighted here because they relate to end-to-end modelling applications. Previous publications that used NEMURO.FISH included Ito et al. (2004), who developed a fish bioenergetics model coupled to NEMURO to analyse the influence of climate change on the growth of Pacific saury. In their model, the biomass of three functional groups of zooplankton (ZS, ZL and ZP) was supplied to the bioenergetics model as prey for saury.

NEMURO.FISH provides estimates of fish growth and weight-at-age of either Pacific herring or saury in the cases considered here and can be run in two modes. In the uncoupled mode, the growth and weight of an individual fish were computed, but the total number of fish was not followed, and thus there was no effect of fish predation on zooplankton concentrations. In the coupled mode, the fish component included calculation of bioenergetics and the total number of fish was followed, enabling fish consumption to impose a mortality term on zooplankton, in

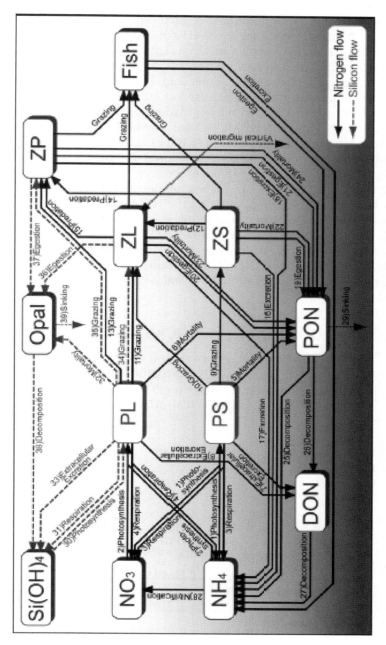

Figure 3.4.4 Schematic view of NEMURO.FISH flow chart.

effect allowing density-dependent growth of the fish. However, following the number of fish is an additional complication because the fish's life cycle must be closed and a new year-class needs to be added each year. NEMURO.FISH in the coupled mode permits long-term simulations by enabling the generation of new fish recruits within the simulation model. A spawner–recruit relationship, dependent on the prevailing environmental and climatic conditions, was used to estimate the number of new 1-year-old individuals added to the population each year from the spawning biomass a year earlier. NEMURO has provided beneficial results on the northern Pacific and coastal area (Kishi et al., 2009, 2010).

NEMURO is one of the pre-eminent lower trophic level models for the northern Pacific Ocean, although the iron effect is not included. NEMURO started from a single box model and extended to three-dimensional models. Figure 3.4.5 shows the NEMURO family model distribution on a space–biological resolution plane. Several other ecosystem models were plotted as references. 0D-NPZD denotes a box model of NPZD, and 3D-NPZD denotes a three-dimensional NPZD model. NEMURO.SAN indicates "NEMURO sardine and anchovy", which includes population dynamics of sardine and anchovy (Rose et al., 2007);

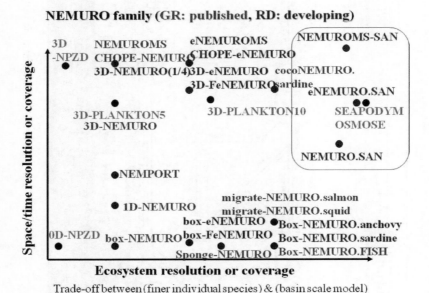

Figure 3.4.5 Summary of the features of NEMURO.

e-NEMURO is "extended NEMURO", which includes subtropical species of plankton in addition to NEMURO. Furthermore, Fe-NEMURO, which includes iron limitation in NEMURO, has been developed. The next targets are the following: (1) a multi-species model (such as NEMURO.SAN), (2) inclusion of predators, (3) coupling with population dynamics, (4) inclusion of a dynamic link between prey plankton and fishes, (5) inclusion of fisheries effects and (6) interaction with human activities (management, economy, food supply, culture, etc.).

3-4-4 Operational use of remote sensing and marine GIS for sustainable fisheries and aquaculture

Satellite remote sensing for fisheries has been developing since the 1980s, and its effectiveness on a synoptic scale and synchronized observation has been demonstrated in the scientific literature (Kiyofuji and Saitoh, 2004; Saitoh et al., 1986). Over the past 10 years, GIS has also been recognized as an important tool for visualizing and analysing the spatio-temporal distribution of fisheries resources. Recently, GIS has been widely adopted in fisheries and oceanographic research (Nishida et al., 2001, 2004). A newly developed ubiquitous fisheries information system and service for Japanese waters is described in this subsection, with the aim of responding to the need for sustainable development of fisheries resources. Target species for this research and development are the Japanese common squid (*Todarodes pacificus*), Pacific saury (*Cololabis saira*), skipjack tuna (*Katsuwonus pelamis*) and albacore tuna (*Thunnus alalunga*), which are a popular and important food source in Japan.

TOREDAS Fisheries Information Service

The TOREDAS (Traceable and Operational Resource and Environment Data Acquisition) system was developed to provide and transfer information to the public through the Internet and to support fishing operations or resource management using satellite connections. The important goals of the system are to (1) develop a system for near real-time data transfer through the Internet and satellite connections during fisheries operations, (2) estimate or predict optimal fishing areas based on scientific findings and (3) provide high value-added fisheries oceanographic information (Kiyofuji et al., 2007; Figure 3.4.6).

In the near future, research and development factoring in spatio-temporal analysis techniques should be included for better assessment of stocks within a multi-species approach (Mugo et al., 2008). This could

Figure 3.4.6 Onboard PC system and display of an SST image from the onboard GIS.
Source: Saitoh et al. (2009).
Note: Estimated potential fishing grounds shown by grey polygon.

also give a new scientific perspective to the field of fisheries science. Stock management methods based on an ecosystem approach are gaining momentum at the global scale. The TOREDAS system and service could contribute to the development of such methodology in Japan. As a result, the system will play a part in reduced fuel consumption and time spent searching for suitable fishing areas (Figure 3.4.7), which will foster effective fisheries by reducing input costs and improving energy efficiency (Saitoh et al., 2009).

Food resources for Japanese scallop culture

The culture of scallops through suspension and bottom seeding is an important fishery activity in northern Japan. Drawing together biophysical, logistic and other limiting data, Radiarta et al. (2008) undertook constraint mapping of scallop culture areas in Funka Bay to quantify site selection. The biophysical data consisted of sea surface temperature (SST), chlorophyll (OC4v4), turbidity (water-leaving radiance at 555 nm) and

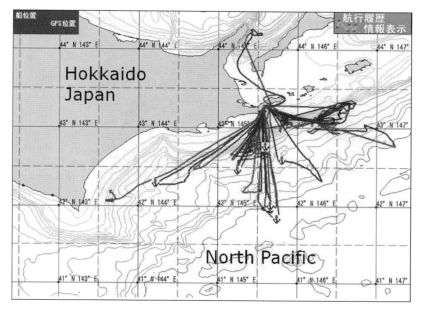

Figure 3.4.7 Trajectories of Pacific saury fishing fleets showing a direct approach to fishing grounds from the mother port off eastern Hokkaido, Japan, for the period 1 September to 30 October 2006 (please see page 475 for a colour version of this figure).
Source: Saitoh et al. (2009).

bathymetry, using MODIS data for SST and SeaWiFS (Sea-viewing Wide Field-of-view Sensor) for the two water quality variables. There are few examples of constraint mapping applied to aquaculture, and their application was made more quantitative by a weighting scheme for the relative importance of environmental variables. Results indicate that 88 per cent of the bay area is suitable for scallop culture and that 56 per cent of this area had a high suitability rating (Figure 3.4.8). This type of thematic mapping, employing data from a variety of remote sensors coupled with decision support through GIS spatial analysis, provides more rigour and insight into aquaculture planning.

Radiarta and Saitoh (2008) examined temporal variation in chlorophyll, turbidity and temperature in Funka Bay, Japan. These data were used to explain seasonal trends in the spring bloom and relate them to scallop production. However, it is obvious that a time series of chlorophyll, turbidity and temperature can be generated by this approach, which would provide the boundary conditions for simulation models of the ecosystem. This is particularly useful in cases where there is not much field data for model parameterization.

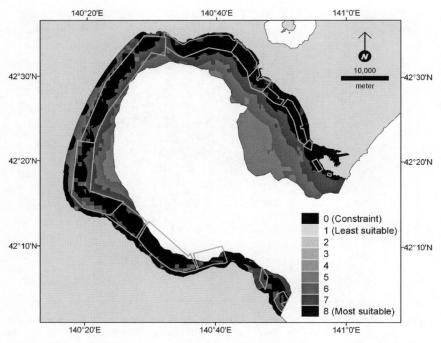

Figure 3.4.8 Overall site selection map, masked to depths in excess of 60 metres, for Japanese scallop aquaculture potential in Funka Bay, south-western Hokkaido (please see page 476 for a colour version of this figure).
Note: Grey polygons represent existing Japanese scallop aquaculture area.

3-4-5 A case study of an ecosystem approach to fisheries and the adaptive management of the Shiretoko World Natural Heritage Site

Human dimensions of global change in marine ecosystems

Marine ecosystems are affected by global-scale climate variability and change as well as by human activities such as intensive fishing that may occur on more immediate time scales than those of climate change. Recent international ocean research programmes such as Integrated Marine Biogeochemistry and Ecosystem Research (IMBER) focus on the human dimensions of global change in marine ecosystems using the concept of coupled marine social–ecological systems. Efforts are now being made to measure and alleviate the ecosystem effects of fishing (Hall, 1999), and the focus is mainly on how an ecosystem approach to fisheries (EAF) may be implemented (Garcia and Cochrane, 2005). Furthermore, a full social–ecological system approach to the management of marine

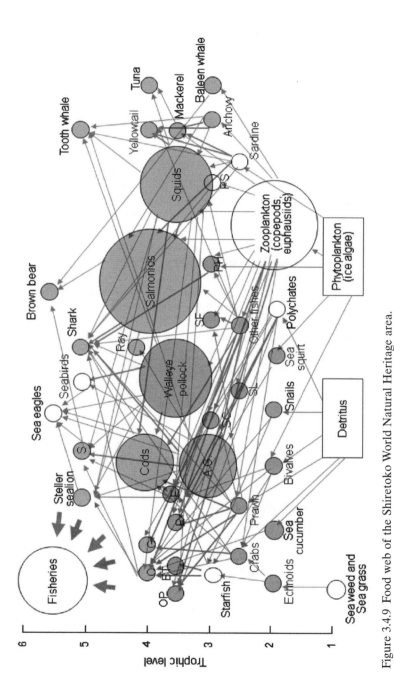

Figure 3.4.9 Food web of the Shiretoko World Natural Heritage area.
Source: Makino et al. (2008).
Notes: AG: arabesque greenling; BT: bighand thornyhead; F: flatfishes; G: greenlings; O: octopus; OP: ocean perch; PH: Pacific herring; PS: Pacific saury; R: rockfish; S: seals; SC: saffron cod; SF: sandfish; SL: sand-lance.

resources must involve multiple-scale objective setting (from government to local fishing sectors) based on societal choices, including ecological, economic and social considerations. Operational objectives need to be established, requiring identification of indicators and reference points for sector impacts.

EAF and adaptive fisheries management of the Shiretoko World Natural Heritage Site, Hokkaido, Japan

For long-term sustainable utilization of marine resources in marginal seas, local fishers must be made aware of EAF and adaptive fisheries management. Adaptive management predicts and monitors changes in the ecosystem and subsequently reviews and adjusts the management and use of natural resources (Matsuda et al., 2009). Such predictions and monitoring are best accompanied by feedback controls, such as verification of hypotheses based on the results of monitoring to review and modify management activities.

Fisheries management in Japan is characterized by seeking balance between sustainable use and ecosystem conservation and involves the co-management of fishers' organizations (Makino et al., 2008). Fisheries in Japan face several challenges, including: (1) exclusive use by fishers with fishery rights/licences (there are a few exceptions for free-fisheries and recreational angling), (2) lack of full transparency in management procedures, (3) lack of objective benchmarks or numerical goals in management plans and (4) strong dependency on political pressure from abroad. This subsection elaborates on these characteristics and issues of marine ecosystem management in Japan (Matsuda et al., 2009).

The Marine Management Plan for the Shiretoko World Natural Heritage Site is an important case study of adaptive marine ecosystem management and co-management of coastal fisheries (Figure 3.4.9). The Oyashio shelf region and seasonally ice-covered areas north of Hokkaido, including Shiretoko, are highly productive, supporting a wide range of species, such as marine mammals, seabirds and commercially important species in the western subarctic Pacific (Sakurai, 2007). Shiretoko is the third World Natural Heritage Site registered in Japan. It is characterized by the formation of seasonal sea ice, high biodiversity and the occurrence of many globally threatened species. The natural resource management plan of the Shiretoko site is characterized by transparency and consensus-building, because UNESCO and IUCN require that the plan be sustainable and the Government of Japan has guaranteed local fisheries that there will be no additional regulations included in the plan. The Marine Management Plan describes which species and factors are monitored, how these data are evaluated and how the benchmarks specified by ecosystem management are determined. The Plan will provide a valuable example for the establishment of "environment-friendly fisheries" in Japan and other countries because it includes voluntary activities by resource users, which are suitable for use in a local context, flexible to ecological/social fluctuations and efficiently implemented through increased legitimacy and compliance. This approach is appropriate for developing coastal countries where a large number of small-scale fishers catch a variety of species using various types of gear (Matsuda et al., 2009).

REFERENCES

FAO [Food and Agriculture Organization] Fisheries Department (2003) *Fisheries Management – 2. The Ecosystem Approach to Fisheries*. FAO Technical Guidelines for Responsible Fisheries No. 4, Suppl. 2. Rome: Food and Agriculture Organization of the United Nations.

FAO Fisheries and Aquaculture Department (2009) *The State of World Fisheries and Aquaculture: 2008*. Rome: Food and Agriculture Organization of the United Nations.

Garcia, Serge M. and K. L. Cochrane (2005) "Ecosystem Approach to Fisheries: A Review of Implementation Guidelines", *ICES Journal of Marine Science* 62: 311–318.

Hall, Jason S. (1999) *The Effects of Fishing on Marine Ecosystems and Communities*. Oxford: Blackwell Science.

Hites, Ronald A., J. Foran, D. O. Carpenter, M. C. Hamilton, B. A. Knuth and S. J. Schwager (2004) "Global Assessment of Organic Contaminants in Farmed Salmon", *Science* 303: 226–229.

Ito, Shin-ichi, M. J. Kishi, K. Kurita, Y. Oozeki, Y. Yamanaka, B. A. Megrey and F. E. Werner (2004) "A Fish Bioenergerics Model Application to Pacific Saury Coupled with a Lower Trophic Ecosystem Model", *Fisheries Oceanography* 13 (suppl. 1): 111–124.

Kaeriyama, Masahide (1986) "Ecological Study on Early Life of the Chum Salmon, *Oncorhynchus keta* (Walbaum)", *Scientific Reports of the Hokkaido Salmon Hatchery* 40: 31–92 (in Japanese, with English abstract).

Kaeriyama, Masahide (2008) "Ecosystem-based Sustainable Conservation and Management of Pacific Salmon", in K. Tsukamoto, T. Kawamura, T. Takeuchi, T. D. Beard, Jr and M. J. Kaiser (eds), *Fisheries for Global Welfare and Environment*. Tokyo: TERRAPUB, pp. 371–380.

Kaeriyama, M. and R. R. Edpalina (2004) "Evaluation of the Biological Interaction between Wild and Hatchery Population for Sustainable Fisheries Management of Pacific Salmon", in K. M. Leber, S. Kitada, H. L. Blankenship and T. Svasand (eds), *Stock Enhancement and Sea Ranching, Second Edition: Developments, Pitfalls and Opportunities*. Oxford: Blackwell Publishing, pp. 247–259.

Kishi, M. J., M. Kashiwai, D. M. Ware, B. A. Megrey, D. L. Eslinger, F. E. Werner, M. N. Aita, T. Azumaya, M. Fujii, S. Hashimoto, D. Huang, H. Iizumi, Y. Ishida, S. Kang, G. A. Kantakov, H. C. Kim, K. Komatsu, V. V. Navrotsky, L. S. Smith,

K. Tadokoro, A. Tsuda, O. Yamamura, Y. Yamanaka, K. Yokouchi, N. Yoshie, J. Zhang, Y. I. Zuenko and V. I. Zvalinsky (2007) "NEMURO: Introduction to a Lower Trophic Level Model for the North Pacific Marine Ecosystem", *Ecological Modelling* 202: 12–25.

Kishi, M. J., K. Nakajima, M. Fujii and T. Hashioka (2009) "Environmental Factors Which Affect Growth of Japanese Common Squid, *Todarodes pacificus*, Analyzed by a Bioenergetics Model Coupled with a Lower Trophic Ecosystem Model", *Journal of Marine Systems* 78: 278–287.

Kishi, M. J., M. Kaeriyama, H. Ueno and Y. Kamezawa (2010) "The Effect of Climate Change on the Growth of Japanese Chum Salmon (*Oncorhynchus keta*) Using a Bioenergetics Model Coupled with a Three-dimensional Lower Trophic Ecosystem Model (NEMURO)", *Deep Sea Research Part II: Topical Studies in Oceanography* 57 (13–14): 1257–1265.

Kiyofuji, H. and S. Saitoh (2004) "Use of Nighttime Visible Images to Detect Japanese Common Squid, *Todarodes pacificus*, Fishing Areas and Potential Migration Routes in the Sea of Japan", *Marine Ecology Progress Series* 276: 173–186.

Kiyofuji, H., F. Takahashi, D. Tachikawa, M. Abe, K. Tateyama, M. Hiraki and S. Saitoh (2007) "A Ubiquitous Information System for the Offshore Fisheries Activities around Japan", in T. Nishida, P. J. Kailola and A. E. Caton (eds), *GIS/Spatial Analyses in Fishery and Aquatic Sciences*, vol. 3. Saitama, Japan: Fishery-Aquatic GIS Research Group, pp. 313–324.

McLeod, K. L., J. Lubchenco, S. R. Palumbi and A. A. Rosenberg (2005) *Scientific Consensus Statement on Marine Ecosystem-based Management. Prepared by Scientists and Policy Experts to Provide Information about Coasts and Oceans to U.S. Policymakers*. Communication Partnership for Science and the Sea (COMPASS), <http://compassonline.org/pdf_files/EBM_Consensus_Statement_v12.pdf> (accessed 9 March 2010).

Makino, M., H. Matsuda and Y. Sakurai (2008) "Expanding Fisheries Co-management to Ecosystem-based Management: A Case in the Shiretoko World Natural Heritage Area, Japan", *Marine Policy* 33: 207–214.

Marasco, R. J., D. Goodman, C. B. Grimes, P. W. Lawson, A. E. Punt and T. Quinn Jr II (2007) "Ecosystem-based Fisheries Management: Some Practical Suggestions", *Canadian Journal of Fisheries and Aquatic Sciences* 64: 928–939.

Matsuda, H., M. Makino and Y. Sakurai (2009) "Development of Adaptive Marine Ecosystem Management and Co-management Plan in Shiretoko World Natural Heritage Site", *Biological Conservation* 142: 1937–1942.

Megrey, B. A., S. Ito, D. E. Hay, R. A. Klumb, K. A. Rose and F. E. Werner (2007) "Basin-scale Differences in Lower and Higher Trophic Level Marine Ecosystem Response to Climate Impacts Using a Coupled Biogeochemical-fisheries Bioenergetics Model", *Ecological Modelling* 202: 196–210.

Mugo, R., S. I. Saitoh, A. Nihira and T. Kuroyama (2008) "Exploiting the Edge: Evaluating Predator–Prey Interactions between Skipjack Tuna, Pacific Saury and Squid Using Satellite Remote Sensing and GIS". Proceedings of 5th World Fisheries Congress, CD-ROM, 1a_0005_234.

Murawski, S. A. and G. C. Matlock (2006) *Ecosystem Science Capabilities Required to Support NOAA's Mission in the Year 2020*. US Department of Commerce NOAA Technical Memorandum NMFS-F/SPO-74.

Myers, R. A. and B. Worm (2003) "Rapid Worldwide Depletion of Predatory Fish Communities", *Nature* 423: 280–283.
Nishida, T., C. E. Hollingworth and P. J. Kailola (2001) *GIS/Spatial Analyses in Fisheries and Aquatic Sciences (Volume 1)*. Proceedings of the First International Symposium on GIS/Spatial Analyses in Fishery and Aquatic Sciences. Kawagome, Saitama, Japan: Fishery-Aquatic GIS Research Group.
Nishida T., P. J. Kailola and C. E. Hollingworth (2004) *GIS/Spatial Analyses in Fisheries and Aquatic Sciences (Volume 2)*. Proceedings of the Second International Symposium on GIS/Spatial Analyses in Fishery and Aquatic Sciences, Fishery-Aquatic GIS Research Group, Kawagome, Saitama, Japan.
NRC [National Research Council] (1999) *Sustaining Marine Fisheries*. Washington DC: National Academy Press.
Pauly, D., V. Christensen, J. Dalsgaard, R. Froese and F. Torres Jr (1998) "Fishing Down Marine Food Webs", *Science* 279: 860–863.
Primavera, Jurgenne H. (2005) "Mangroves, Fishponds, and the Quest for Sustainability", *Science* 310: 57–59.
Radiarta, I. N. and S. Saitoh (2008) "Satellite-derived Measurements of Spatial and Temporal Chlorophyll-a Variability in Funka Bay, Southwestern Hokkaido, Japan", *Estuarine, Coastal and Shelf Science* 79: 400–408.
Radiarta, I. N., S. Saitoh and A. Miyazono (2008) "GIS-based Multi-criteria Evaluation Models for Identifying Suitable Sites for Japanese Scallop (*Mizuhopecten yessoensis*) Aquaculture in Funka Bay, Southwestern Hokkaido, Japan", *Aquaculture* 284: 127–135.
Rose, K. A., F. E. Werner, B. A. Megrey, M. Noguchi-Aita, Y. Yamanaka, E. E. Hay, J. F. Schweigert and M. B. Foster (2007) "Simulated Herring Growth Responses in the Northeastern Pacific to Historic Temperature and Zooplankton Conditions Generated by the 3-Dimensional NEMURO Nutrient–Phytoplankton–Zooplankton Model", *Ecological Modelling* 202: 184–195.
Saitoh, S. I., S. Kosaka and J. Iisaka (1986) "Satellite Infrared Observations of Kuroshio Warm-core Rings and Their Application to Study of Pacific Saury Migration", *Deep Sea Research Part A* 33: 1601–1615.
Saitoh, S., E. Chassot, R. M. Dwivedi, A. Fonteneau, H. Kiyofuji, B. Kumari, M. Kuno, S. Matsumura, T. Platt, M. Raman, M. Sathyendranath, H. Solanki and F. Takahashi (2009) "Remote Sensing Applications to Fish Harvesting", in M.-H. Forget, V. Stuart and T. Platt (eds), *Remote Sensing in Fisheries and Aquaculture*, Reports of the International Ocean-Colour Coordinating Group, No. 8, IOCCG, Dartmouth, Canada.
Sakurai, Yasunori (2007) "An Overview of the Oyashio Ecosystem", *Deep Sea Research Part II* 54: 2525–2542.
Travers, M., Y. J. Shin, S. Jennings and P. Cury (2007) "Towards End-to-End Models for Investigating the Effects of Climate and Fishing in Marine Ecosystem", *Progress in Oceanography* 75: 751–770.
Ueno, Y. and Y. Ishida (1998) "Summer Distribution and Migration Routes of Juvenile Chum Salmon (*Oncorhynchus keta*) Originating from Rivers in Japan", *Bulletin of National Research Institute of Far Seas Fisheries* 33: 139–147.
Witherell, D., C. P. Pautzke and D. Fluharty (2000) "An Ecosystem-based Approach for Alaska Groundfish Fisheries", *ICES Journal of Marine Science* 57: 771–777.

4
Regional initiatives for self-sustaining models

4-1

Biomass town development and opportunities for integrated biomass utilization

Ken'ichi Nakagami, Kazuyuki Doi, Yoshito Yuyama and Hidetsugu Morimoto

This section focuses on the development of a Diagnostic Evaluation Model for Biomass Circulation (DEMBC), which allows the researcher to diagnose regional resources and evaluate opportunities and alternative policies for utilizing biomass. There are three parts to the research. First, the DEMBC was developed with a sub-model, which analysed energy, economic efficiency and greenhouse gases (GHG). Second, the model was applied to 71 municipalities with a regional variety of resources. In addition, environmental effects and eco-efficiency were diagnosed and evaluated in the regions affected by the political development of biomass systems. Last, 24 municipalities designated as "Japanese biomass towns" were interviewed. Based on the results of this evaluation with DEMBC, an understanding was achieved of those biomass towns from the perspective of eco-efficiency, and the towns were categorized into groups using principal component analysis. Finally, some common tendencies and particular parameters of the social conditions for planning, of incentives to introduce the business, and of the decision-making process were extracted.

4-1-1 Diagnostic Evaluation Model for Biomass Circulation

Basic structure of the model

The Diagnosis Model for Biomass Resources Circulation (DMBRC) was the basis of the DEMBC (Morimoto et al., 2009). The DEMBC was developed by the systemization sub-team of the Bio-recycle Project in the National Institute for Rural Engineering (Systemization Sub-team, Bio-recycle Project, 2006; Yuyama, 2004). The DEMBC can analyse the flows of biomass, cash, energy and GHG. This model has a structure that includes the site at which biomass is generated (such as livestock, food industry, farmland, forests, forestry, lumber industry and living environment), some biomass conversion centres and locations outside of the area. These domains are called "compartments". The amount of materials (nitrogen, phosphorus, potassium, carbon), cash, energy and GHG that moves between these compartments is expressed as flows (Figure 4.1.1).

Subjects of biomass

DEMBC builds some parameters involving conversion and product utilization through biomass generation into the database. The typical sources of biomass that the database targets are listed in Table 4.1.1. Moreover, this system can easily add other sources of biomass to the list.

Evaluation factors

Biomass utilization is expected to generate various benefits. The calculation of these can also consider environmental impact mitigation and the economic impact effect. On the other hand, economic costs, such as institutional development and management, need to be included. In the model, a number of criteria were established to consider trade-offs based on scientific evaluation criteria. Environmental impact mitigation includes reduction of fossil energy consumption and GHG emissions, the amount of cyclic usage of nitrogen and organic material, and the ratio of carbon utilization. Economic balance is used as an economic evaluation criterion, as shown in Table 4.1.2. In this research, the evaluation index is calculated by dividing these estimates by the population and the farmland area. Although research and development of a technique for unifying multiple evaluation criteria have been carried out elsewhere, it has not been adopted in this model. The criteria for biomass utilization are listed in Table 4.1.3.

Boundary conditions

For both waste and unused biomass, the analysis boundary of environmental impact assessment and economic evaluation was limited to transportation and conversion up to the production stage. For energy crops, the cultivation process is also added. We assume that the heat from fossil

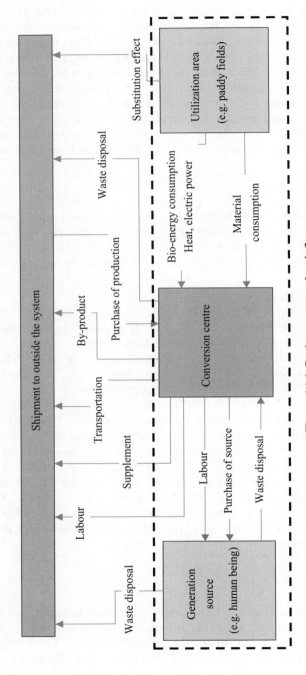

Figure 4.1.1 Basic structure of cash flow.

Table 4.1.1 The main sources of biomass in the database

Class	Species
Livestock waste	Beef, cow, pig, chicken
Leftover food	Food industry
Domestic waste	Kitchen waste, waste paper, waste cooking oil
Woody waste	Sawmill waste, construction debris, prunings (trees, garden shrubs)
Sludge	Raw, digestion tank, human waste, community sewage
Agricultural by-products	Straw, husk, crop residue
Forestry by-products	Thinned wood, forestry residues
Energy crop	Crop produced as energy resource

Table 4.1.2 Main condition setting for economic efficiency evaluation

Economic balance	*Income − Expenditure = Product Sales − (Amortisation + Running + Supply)*
Amortization cost	Duration: 15–20 years^a
	Interest rates: 4%, non-subsidy
Running costs^a	Electricity: 15 JPY/kWh
	Fuel: heavy oil 40 JPY/L, light oil 70 JPY/L
	Labour: 4.5 million JPY/person
Supply cost (JPY/t)	Forestry residues: 17,000^b
▲: Inverse onerous contract	Sawmill residues, straw, chaff: 1,000
	Livestock manure: ▲500^a
	Kitchen waste: ▲15,000^a
	Sludge: ▲15,000^a
	Prunings: ▲20,000
	Construction debris: ▲5,000
	Energy crops^c
Product sales (JPY/L, JPY/MJ, JPY/kWh)	Charcoal: 1,000; Compost: 5,000; Liquid fertilizer: 1,000; Biodiesel fuel: 50; Heat: 2; Electricity: 15; Pellets: 33,000

Sources:
^{a} Yuyama et al. (2006).
^{b} Forestry Machine Association (2003).
^{c} Ministry of Agriculture, Forestry and Fisheries (2006).

energy origins is reduced when heat from biomass origins is utilized. The environmental impacts from machine production or structure construction used for collection, transportation and conversion are not considered. The dashed line in Figure 4.1.1 represents the border of the evaluation range.

Table 4.1.3 Evaluation criteria and index

Evaluation items	Unit of analysis	Evaluation criteria
Economic balance	Million JPY	1,000 JPY/person
Waste biomass (carbon ton)	t-C	kg-C/person
Consumption of fossil energy (heavy oil equivalent)	kL	L/person
GHG emissions (CO_2 equivalent)	t-CO_2eq	kg-CO_2eq/person
Carbon efficiency	%	%
Carbon usage	t-C	kg-C/person
Cyclic usage of nitrogen for fertilization (upland field area)	t-N	kg-N/ha

Data collection

Interviews and public hearings were conducted in 24 areas classified as designed biomass towns. The purpose was to determine management objectives and to collect data on biomass generation and utilization, including economic efficiency for the institutions involved. It also aimed to describe the decision-making process in the configuration of biomass utilization systems through the social analysis of advanced areas. Factors such as the incentive of the design decision, a decision-making system, personnel training, organizational set-up, governance and finance were investigated. When performing scenario evaluation from a macro perspective, it is helpful to use many examples in the evaluation. Therefore many existing biomass town designs were analysed using the DEMBC. The index was proposed for evaluating other scenarios. In fact, it is a technique for examining the rank of different scenarios for planning purposes that takes into consideration the needs and realities of existing biomass town designs.

4-1-2 Analysis of Japanese biomass towns

The DEMBC was used to determine the changes arising from the introduction of the biomass utilization system in 71 biomass towns. It is necessary to know the amount of biomass utilization included in the design. When designs are not defined clearly, an estimate that uses a general numerical value is taken. In addition, several conditions or assumptions for analysis need to be set up. The assumptions are as follows:
- livestock manure, kitchen waste and so forth are used to make fully ripened compost, dried using electric power or other fuel;
- the waste heat from carbonization is used effectively;
- biomass products are fully used;

- in wood pelletizing, the pellet boiler is simultaneously introduced as a product use institution;
- charcoal is used as a soil improver;
- methane fermentation allows power generation and heat use;
- the methane fermented liquid is used as a liquid fertilizer; and
- small-scale gasification allows power generation and uses heat.

4-1-3 Development of Diagnostic Evaluation Model for Biomass Circulation

Materials, energy flow units and cost functions

Standard physical units and the costs of energy transmission or materials used in collection, transportation, conversion, production and conveyance to build the model were investigated. The cost function was determined based on the assumption that the relation between materials and costs is an involution function.

Conversion technology

Conversion processes incorporate the technologies available in the current practical use stage as indicated in the biomass town designs. These are shown in Table 4.1.4. The model enables the addition of conversion technologies by clients. The model is shown in Figure 4.1.2.

Table 4.1.4 Transformation technologies

Conversion	Classification	Products
Compost	Fully ripened compost, dried using energy	Compost
Methane fermentation (bio-gasification)	Wet type / liquid fertilizer	Electricity, heat, liquid fertilizer
	Wet type / solid–liquid separation	Electricity, heat, compost
	Dry type	Electricity, heat, compost
Biodiesel fuel ("esterization")	Alkali-catalysed	Biodiesel fuel
Carbonization	Woody biomass only	Charcoal, heat
Direct combustion power generation	Woodchip boiler, electric steam generator	Electricity, heat
Small-scale gasification	Distributed	Electricity, heat
Pelletization & pellet boiler	Distributed (pellet boiler)	Heat
Ethanolization	Cellulose fermentation	Bioethanol

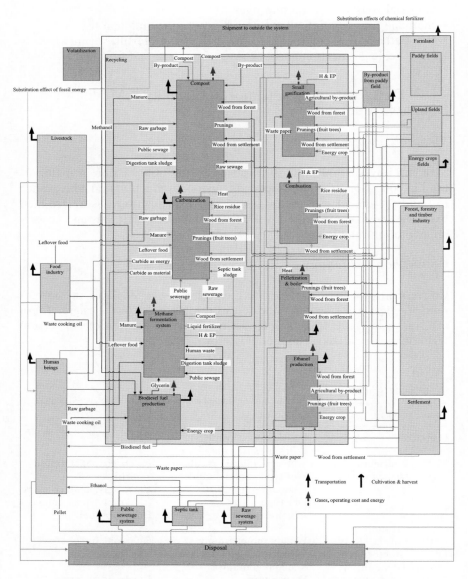

Figure 4.1.2 Diagnostic Evaluation Model for Biomass Circulation.

4-1-4 Results

Incentives and biomass utilization data

As of March 2008, 138 areas had put forward plans. According to interviews in 24 areas, the majority of which have already planned to introduce biomass, the conversion institution drew up the designs in order to receive a subsidy for the introduction. In some cases, private sector companies request the government to draw up the designs, because they too can receive a subsidy as a start-up enterprise. On the other hand, few places created the design from the very beginning. However, although data on the approximate utilization amounts are indicated, most areas had not described in detail which biomass, how much, and which conversion institution would be used.

Incentives and management expenses of business introduction

If the costs of introducing methane fermentation of garbage and sludge are equivalent to or less than conventional garbage-processing costs, then the GHG emissions reduction and reclamation are considered acceptable. In the methane fermentation system for livestock manure, it seems that the collection cost is not included in the overall operational expenses. Thus, communities are offering support in the form of economic subsidies. The conventional garbage-processing cost is about 20,000 JPY/t. The expenses related to collecting manure from livestock farmers are 500–1000 JPY/t.

Quantity of biomass utilized

The index of biomass utilization is expressed as annual carbon usage divided by the total population in the area. Figure 4.1.3 shows the areas ranked according to the amount of carbon used. Of the 71 areas analysed, 49 (69 per cent) are in the range of 0–200 kg-C/person/year. Rural areas that use a lot of livestock manure and where the population is small occupy higher ranks.

Consumption of fossil energy

The inputs of energy for biomass utilization and the outputs of the energy produced were calculated. The energy balance is converted into the amount of heavy oil consumed per person per year. Figure 4.1.4 shows the areas ranked according to the varying amounts of fossil energy consumption. Of the 71 areas, 26 (37 per cent) show high positive fossil

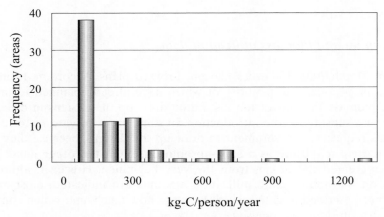

Figure 4.1.3 Biomass utilization.

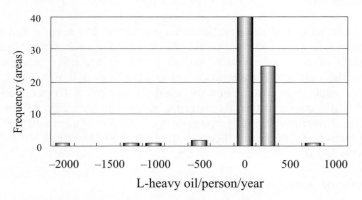

Figure 4.1.4 Consumption of fossil energy.

energy consumption; in 68 areas consumption was below an annual rate of 100 L-heavy oil/person, and was in the range of –100 to 100 L-heavy oil/person in 75 per cent of all areas.

Greenhouse gas emissions

The index of GHG emissions is expressed as the amount of discharge (t-CO_2eq) divided by the population in the area. Figure 4.1.5 shows the areas ranked according to the amount of GHG emissions from biomass utilization. Of the 21 areas where GHG emissions are reduced by biomass utilization, 85 per cent (18 areas) show discharges of not more than 100 kg-CO_2eq/person.

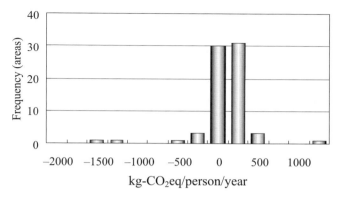

Figure 4.1.5 Greenhouse gas emissions.

Economic efficiency

Figure 4.1.6 shows the areas classified by their economic balance: 10 areas score positively. In contrast, 49 areas have a negative balance of 0–25,000 JPY/person; and 15 per cent had an excess of expenditure over income of more than 30,000 JPY/person. Both construction waste and sawmill residues provide biomass on an inverse onerous contract, under which the facilities that dispose of waste are charged (see Table 4.1.2). Then, because these materials contain little moisture, they have the advantage of being able to produce a lot of heat or electricity. Therefore, a biomass town that utilizes inverse onerous contract woody biomass can reduce its fossil energy consumption and turn its economic balance positive.

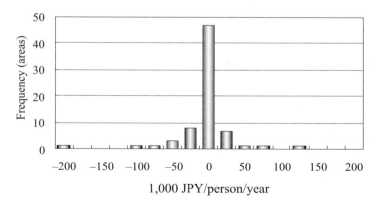

Figure 4.1.6 Economic balance.

Table 4.1.5 Evaluation standard for scenario assessment

	Rank			
	A	B	C	D
Economic balance (1,000 JPY/person/year)	>−2.0	−6.4 to −2.0	−20.4 to −6.4	<−20.4
Waste (kg-C/person/year)	<219	219–397	397–613	>613
Fossil energy consumption (L-heavy oil/person/year)	<−52.6	−52.6 to −9.9	−9.9 to −1.9	>−1.9
GHG emissions (kg-CO_2eq/person/year)	<−47.4	−47.4 to −15.1	−15.1 to 19.5	>19.5
Cyclic usage of nitrogen as fertilizer (t-N/ha/year)	<0.1	0.1–0.2	0.2–0.5	>0.5
Carbon usage (kg-C/person/year)	>234	92–234	54–92	<54

Systematization of the local diagnosis by creating an evaluation-based scenario

In the results, the values were calculated in quartiles (rank A, B, C, D) according to the evaluation criteria (Tables 4.1.5 and 4.1.6). Each rank indicates a different biomass utilization scenario and its effects based on the existing biomass town scenario. It is important to utilize these values for decision-making and policy-making purposes. The authors analysed 103 existing and 71 future biomass towns. Besides the advantage of utilizing

Table 4.1.6 Evaluation standard for present condition

	Rank			
	A	B	C	D
Economic balance (1,000 JPY/person/year)	>0.36	0 to 0.36	−0.05 to 0	<−0.05
Waste (kg-C/person/year)	<396	396–751	751–1,632	>1,632
Fossil energy consumption (L-heavy oil/person/year)	<−16.7	−16.7 to −6.5	−6.5 to −1.4	>−1.4
GHG emissions (kg-CO_2eq/person/year)	<−36.1	−36.1 to −8.8	−8.8 to −0.7	>−0.7
Cyclic usage of nitrogen as fertilizer (t-N/ha/year)	<0.1	0.1–0.3	0.3–0.7	>0.7
Carbon usage (kg-C/person/year)	>198	81–198	29–81	<29

very cheap biomass, even if biomass utilization involves the use of inverse onerous contract resources, there are a few cases in which the economic balance is positive. If the use of biomass is economically advantageous, one can proceed. On the other hand, if the use is disadvantageous, then action must not be taken. Thus, it is possible that the use of several types of biomass in combination is not viable.

Classification of the biomass towns by resource distribution

Potential biomass abundance has three sources: domestic waste, waste from agriculture and livestock, and forestry residues. Figure 4.1.7 shows the types or categories of division of potential biomass abundance in the 71 municipalities. There is only one area that has mainly life system biomass. There is no area for a hybrid category of forestry and life. However, areas with domestic waste are mainly urban areas, and will be able to expect an increase in biomass towns, including big cities, in the future. As shown in Figure 4.1.7, the three types that are the most prevalent are daily life, agriculture and forestry, and forestry.

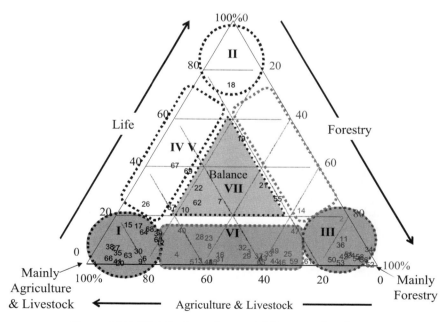

Figure 4.1.7 Distribution of potential biomass abundance.

Selection of biomass utilization system

First, the detailed methods, circumstances, measurement processes, existing institutions and so forth of designed biomass utilization were reviewed. Second, biomass utilization systems were classified into four stages: no plan; exploratory stage; in operation; and mostly successful, as shown in Figure 4.1.8. A total of approximately 340 systems in the 71 biomass towns are attempting to use one or more of the biomass utilization systems. Although composting is at present practised widely, it is thought that in the future the ratio will fall owing to the rapid diffusion of other conversion techniques. On the other hand, use of both biodiesel fuel (BDF) and pelletization will expand from the present number of 10 systems. At present, no areas are producing ethanol, but several mention the possibility in their plans. Although the proportion of methane fermentation will expand in the future, carbonization and direct combustion will tend to fall. There tend to be greater possibilities for introducing methane fermentation in highly populated areas and in places where product shipment is important. However, its introduction is not related to finances and area. BDF has greater usage both in areas with high agricultural

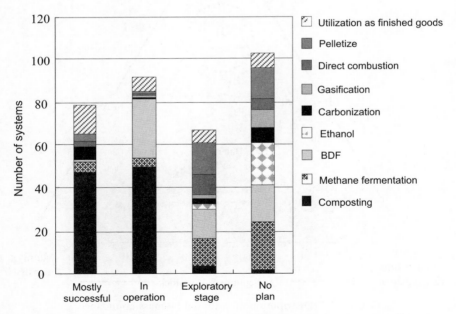

Figure 4.1.8 Biomass utilization systems classified by introduction stage.

production (Figure 4.1.7, I) and in balanced areas (Figure 4.1.7, VII), each having average biomass.

Pellets are extensively used in forested areas and in areas that are poor or where the product shipment value is low. There are few cases of carbonization and direct combustion. These results imply that it is necessary to combine several conversion technologies in creating a utilization scenario. Especially in rural areas, the combined use of BDF, pellets, composting and methane fermentation to generate ethanol is effective from the utilization system point of view.

4-1-5 Conclusion

In this research, a model for evaluating the potential utilization of biomass was developed. Several biomass utilization systems were built into the model as benchmarks. The model shows the flow of material, its ingredients, circulation, economic efficiency, GHG generation and fossil energy consumption. The model can examine a number of possible biomass utilization scenarios in any area. Moreover, it has been highlighted that use phases are design and vision decisions. More specifically, the model can show the various design phases and alternative scenarios for decision-making.

Using DEMBC for the designed biomass town areas, the current situation (103 areas) and the plans (71 areas) for biomass utilization were analysed and evaluated. A "utilization valuation model" by area was created using the data obtained from the available biomass town designs. Using the model, the environmental impacts and energy and economic efficiency were analysed based on the conversion technology. The conversion technology and the resource classification were arranged in a matrix, and change was analysed in terms of environmental impacts and of energy and economic efficiency. Also shown was the relationship between the introduction of the biomass utilization system and the characteristics of a region, such as biomass potential, population, area and financial situation. As a result it is possible to create an evaluation standard for planning biomass utilization.

Acknowledgements

This section is based on a FY 2007/2008 research project with a Waste Treatment Research Grant (K1916, K2060) from the Ministry of the Environment. Sincere thanks are owed to Professor Malcolm J. M. Cooper and Professor Francisco P. Fellizar Jr (Ritsumeikan Asia Pacific University) for their helpful comments.

REFERENCES

Forestry Machine Association (2003) "Shinrin Baiomasu no Shushu Yusou Kosuto no Teigen ni tsuite".

Ministry of Agriculture, Forestry and Fisheries, Agriculture, Forestry and Fisheries Research Council (2006) *Ine de Ushi wo Sodateru – Shiryou Ine ni yoru Kokusan-Gyu-Seisan*, Agriculture Research and Development Report, No. 15, pp. 1–17.

Morimoto Hidetsugu, Doi Kazuyuki, Hoshino Satoshi, Yuyama Yoshito, Kuki Yasuaki (2009) "Development of Integrated Assessment Model for Biomass Utilisation and Application of the Model: Case Study of 38 Municipalities Which Released the Biomass Town Design", *Journal of Rural Planning Association* 27 (Special Issue): 317–322.

Systemization Sub-team, Bio-recycle Project, ed. (2006) "Design and Evaluation of Biomass Use System".

Yuyama Yoshito (2004) "Developing Strategy on Diagnosis Model for Biomass Resources Circulative Use", *Journal of the Japanese Society of Irrigation, Drainage and Reclamation Engineering* 72(12): 1037–1040.

Yuyama Yoshito, Ikumura Takashi, Ohara Akihiko, Kobayashi Hisashi and Nakamura Masato (2006) "Evaluation of Performance and Costs of Various Biomass Conversion Technologies", Technical Report of the National Institute for Rural Engineering, No. 204: 61–103.

4-2

Analysis of energy, food, fertilizer and feed: Self-sufficiency potentials

Nobuyuki Tsuji and Toshiki Sato

4-2-1 Introduction

The overall food self-sufficiency rate in Japan is currently under 40 per cent, although in Hokkaido Prefecture it is almost 200 per cent (Ministry of Agriculture, Forestry and Fisheries, 2009). Out of the 47 prefectures in Japan, 5 have food self-sufficiency rates greater than 100 per cent (Hokkaido, 195 per cent; Akita, 174 per cent; Yamagata, 132 per cent; Aomori, 118 per cent; and Iwate 105 per cent) (Ministry of Agriculture, Forestry and Fisheries, 2009). Japan's energy self-sufficiency rate is only 4 per cent (19 per cent if nuclear power generation is included) (Agency for Natural Resources and Energy, 2008). Most potash and phosphate supplies are imported and dairy farming in Japan is heavily reliant on imported livestock feed. Farmers consume diesel while using farm machinery, chemical fertilizers and insecticides are made from fossil fuels, and transportation of livestock feed from foreign countries by ship consumes fuel. In addition, the storage of harvested crop products requires huge amounts of electricity. As such, agriculture and dairy farming are highly vulnerable to oil price rises. Current agricultural practices may be defined as "converting oil into food" (Saito, 2009). The energy profit ratio, defined as (output energy / invested energy), involved in producing rice was estimated at 2.5 in the 1960s, but by the 1990s the profit ratio had fallen to 0.7 (Sato, 2005). This shift shows that modern rice production, with its large paddy field sizes, has become extremely dependent on large inputs of fossil energy.

Designing our future: Local perspectives on bioproduction, ecosystems and humanity,
Osaki, Braimoh and Nakagami (eds),
United Nations University Press, 2011, ISBN 978-92-808-1183-4

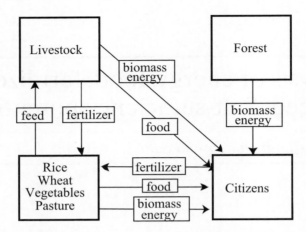

Figure 4.2.1 Potential energy and material flows in Furano.

Biomass is currently attracting considerable attention as a renewable energy source as humanity finally begins to confront the issue of fossil fuel depletion (Sato et al., 2010). There are many potential sources of biomass energy: biogas from livestock excrement or organic waste thrown away by humans, organic waste from agriculture (rice straw, rice hulls and wheat straw), woody biomass pellets, and so on. In general, most types of biomass energy have lower output densities than their fossil fuel counterparts. This characteristic confers several benefits, namely that biomass is renewable and that any waste product could be used as fertilizer. Therefore, sustainable agriculture should include a combination of systems for recycling agricultural products between humans and livestock. Agriculture provides food to humans and feed to livestock; livestock provide biomass energy for biogas; manure produced by cattle also provides fertilizer for agriculture; organic waste produced by citizens is used as biogas and fertilizer (see Figure 4.2.1). This web-like relationship is important to both energy and material flows. Here, four self-sufficiency rates are considered simultaneously: food, fertilizer, feed and energy for the city of Furano in northern Japan.

4-2-2 Material and energy flows in agriculture and dairy farming in Furano City

The city of Furano has a population of about 25,000 and encompasses a total area of 600 km² (of which 420 km² is forested). The total agricultural output, valued at JPY 18.84 billion in 2008, comprised 70 per cent vegetables (onions, carrots, water melons, melons, sweetcorn, and so forth), 11 per cent livestock, 5 per cent wheat and 4 per cent rice (Furano Agriculture and Forestry Department, 2008). This output required 6,070 hectares (ha), or about half of all land (10,397 ha) (Furano Branch of Kamikawa Agricultural Extension Centre, 2007).

Based on the current energy and material flows in Furano (Furano Branch of Kamikawa Agricultural Extension Centre, 2008), it is estimated that 70 tons of nitrogen (t-N) and 7,573 GJ of energy were applied to the paddy fields (776 ha) to produce rice (4,190 tons), straw (4,316 tons) and hulls (1,396 tons) and yielded JPY 272 million as income. A quarter of the straw was converted to feed for milch cows (standard usage rate in Japan) (NEDO, 2009). This feed contains 37.6 per cent total digestible nutrients (TDN) (standard usage rate in Japan) (NEDO, 2009); and 63 per cent (standard usage rate in Japan) of hulls were converted to 4.4 t-N (NEDO, 2009). After the farmers harvested their wheat, they used 66 per cent (standard usage rate in Japan) of the wheat straw as fertilizer (NEDO, 2009). Considerable potential exists to obtain energy from currently unused organic materials. There are 2,051 cows in Furano (Furano Branch of Kamikawa Agricultural Extension Centre, 2007), and all cow excrement is used as fertilizer. It is possible to produce biogas from livestock excrement and organic waste, as well as to obtain fertilizer after the energy has been extracted.

4-2-3 Mathematical analyses

The following assumptions are made: willow trees are planted in the field to derive thermal energy; this thermal energy is converted into fuel and/or electricity energy without any loss. Food, energy, feed (TDN) and fertilizer (nitrogen) self-sufficiency rates depend on the areas of agricultural production and willow cultivation, and on the number of cows. The optimal area allocation and the number of cows, where food, energy, feed and fertilizer self-sufficiency rates all exceed 100 per cent at the same time, are then decided. Define $\mathbf{x} = (x_1, x_2, x_3, \ldots, x_8, x_9, x_{10})$ as the area/number vector, where: x_1 = paddy field area; x_2 = wheat field area; x_3 = pasture area; $\ldots$; x_8 = willow field area; x_9 = the number of cows; x_{10} = the population size (refer to Table 4.2.1). The vector, $\mathbf{c}$, is the food calorific coefficient vector, and $\mathbf{e_b}$ and $\mathbf{e_f}$ are the produced biomass energy and consumed fossil energy coefficients, respectively. The remaining vectors are shown in Table 4.2.1 (see also Sato et al., 2010). Forests are not considered in this mathematical analysis because, although forests provide biomass energy, the forest industry in Furano is stagnant. The problem (1) is formulated as follows.

Table 4.2.1 Basic units, current values in Furano and definitions of vectors

No.		Vector	Paddy rice 1	Wheat 2	Pasture 3	Dent corn 4	Onion 5	Sweet corn 6	Carrots 7	Willow 8	Dairy cows 9	Population 10
Calorific value [Gcal/unit]		c	19.20^a	13.2^a	0^a	0^a	17.6^a	11.0^a	12.9^a	0	5.6^a	0
Biomass energy [GJ/unit]		e_b	49.80^b	15.3^b	0	0	0	0	0	140.0^c	4.7^b	0.2^b
Energy consumption [GJ/unit]		e_f	9.80^d	6.9^d	3.8^d	7.0^d	103.0^d	26.0^d	10.0^d	8.4^c	5.1^e	26.9^b
Feed production [TDNton/unit]		f	0.52^f	0	4.4^f	9.6^f	0	0	0	0	0	0
Nitrogen production [kg/unit]		n_p	5.70^f	16.0^f	0	0	0	0	0	0	54.0^f	1.8^g
Nitrogen consumption [kg/unit]		n_c	90^h	100^h	40^h	140^h	120^h	120^h	120^h	0	0	0
Income [1000JPY/unit]		y	350^d	260^d	-70^d	190^d	1060^d	870^d	540^d	-50^c	363^d	0
Current value		x_c	776	1600	1327	269	1358	375	365	0	2051	25,076
Unit			ha	ha	ha	ha	ha	ha	ha	ha	head	capita

Sources:
[a] Ministry of Agriculture, Forestry and Fisheries (2009).
[b] NEDO (2009).
[c] Larsson et al. (2003).
[d] Furano Branch of Kamikawa Agricultural Extension Centre (2008).
[e] Statistics Department, Minister's Secretariat, Ministry of Agriculture, Forestry and Fisheries (2008).
[f] Agricultural, Forestry and Fishery Bio-Recycle Project "Systematization Sub-team" (2006).
[g] Personal communication with Furano Centre for Environmental Health (nitrogen concentration in organic waste is 2.5%).
[h] Agricultural Policy Planning Department, Hokkaido (2002).

Problem (1) – The primitive case

As a first step, willow is ignored, because willow is not currently planted in the Furano fields. Then, the elements of the vectors x_8, e_{b8}, e_{f8} and n_{c8} are constantly set to 0 in this case. The problem is formulated as follows:

$$\text{Maximize } (\mathbf{e_b} - \mathbf{e_f}, \mathbf{x})$$

subject to

$$\sum_{i=1}^{8} x_i = \sum_{i=1}^{8} x_{ci}, \ (\mathbf{c}, \mathbf{x}) \geq 0.9 x_{10}, \ (\mathbf{f}, \mathbf{x}) \geq 4.9 x_9, \ (\mathbf{n_p}, \mathbf{x}) \geq (\mathbf{n_c}, \mathbf{x}),$$
$$(\mathbf{y}, \mathbf{x}) \geq r(\mathbf{y}, \mathbf{x_c}), \ x_{10} = 25076$$

where $(\mathbf{x}, \mathbf{y})$ is the inner product of vectors $\mathbf{x}$ and $\mathbf{y}$, that is, $\sum_i x_i y_i$. Here r is the ratio to current income, that is, $(\mathbf{y}, \mathbf{x_c})$ = JPY 3,259 million, where $\mathbf{y}$ and $\mathbf{x_c}$ are, respectively, the income and current area vectors in Furano (see Table 4.2.1). The objective function means maximizing net energy under the following constraint conditions:
1. The sum of the total field area is constant.
2. The food self-sufficiency rate (colorific base) is more than 100 per cent (0.9 Gcal/person/year – Ministry of Agriculture, Forestry and Fisheries, 2009).
3. The feed self-sufficiency rate is more than 100 per cent (4.9 TDN ton/head/year – Livestock Dictionary Editorial Committee, 1996).
4. The nitrogen self-sufficiency rate is more than 100 per cent.
5. Total income is greater than r (<1) times the current income even if the area allocation is changed.
6. The population is constant.

The above problem can be solved using a linear programming technique, because both the objective and the constraint functions are linear. Mathematica Ver. 6.0 was used (Wolfram Research, 2007). However, it was not possible to attain an ideal outcome in which all the self-sufficiency rates are equal to or exceed 100 per cent ($r = 1$) at the same time. Under the condition $0 \leq r < 0.9$, the result is shown in Figure 4.2.2.

Problem (2) – When willow is planted in the fields

The energy sufficiency rate is very low in the above case (Figure 4.2.2). Recently, willow has received a lot of attention as an energy source, and it may be planted in the field as an agricultural crop (Eriksson, 2008; Heller et al., 2003). In this case, x_8 will take a certain value, and e_{b8}, e_{f8}, and

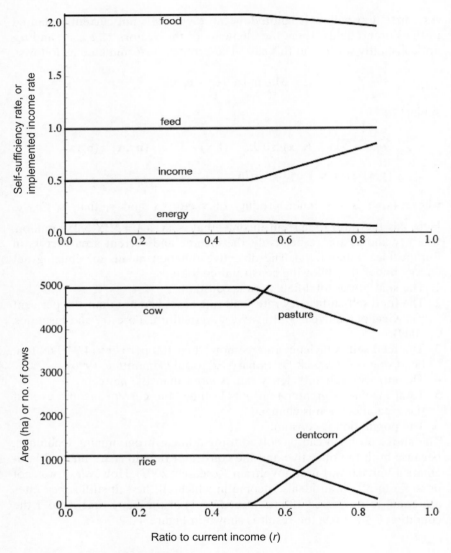

Figure 4.2.2 The case where willow is not planted.
Notes: The upper panel shows the maximum energy self-sufficient rate and the values of constraints when the ratio to current income (r) is changed. The lower panel shows the optimal allocation areas and the number of cows when r is changed.

c_{n8} will take the values shown in Table 4.2.1. The problem was reformulated as follows:

$$\text{Maximize } (\mathbf{e_b} - \mathbf{e_f}, \mathbf{x})$$

subject to

$$\sum_{i=1}^{8} x_i = \sum_{i=1}^{8} x_{ci}, \ (\mathbf{c}, \mathbf{x}) \geq 0.9 x_{10}, \ (\mathbf{f}, \mathbf{x}) \geq 4.9 x_9, \ (\mathbf{n_p}, \mathbf{x}) \geq (\mathbf{n_c}, \mathbf{x}),$$
$$(\mathbf{y} + \mathbf{y_e} + \mathbf{y_f} + \mathbf{y_n}, \mathbf{x}) \geq r(\mathbf{y}, \mathbf{x_c}), \ x_{10} = 25076,$$

where, $\mathbf{y_e}$, $\mathbf{y_f}$ and $\mathbf{y_n}$ are the benefit vectors newly defined as:

$$\mathbf{y_e} = 80[\text{JPY/L}] \frac{1000\mathbf{E}}{38.2[\text{MJ/L}]},$$

$$\mathbf{E} = \begin{cases} e_{bi} - e_{fi} & e_{bi} > e_{fi} \\ 0 & e_{bi} < e_{fi} \end{cases},$$

$$\mathbf{y_f} = 90000[\text{JPY/tonTDN}]\mathbf{f},$$

$$\mathbf{y_n} = 800[\text{JPY/KgN}]\mathbf{n_p},$$

where 80 [JPY/L] and 38.2 [MJ/L] are a price of and the energy obtained from duty-exempt diesel oil, respectively. In addition, a new vector, $\mathbf{E}$, is introduced as defined above. It is possible to get 49.8 GJ/ha of energy from rice straw and hull and 140 GJ/ha from willow. This biomass energy is equivalent to the earned benefit. It is assumed here that the produced renewable energy is measured as diesel, and that it has the same economic values as duty-exempt diesel. It is further assumed that the recycled TDN and nitrogen have the same values as TDN and nitrogen. The result (Figure 4.2.3) shows a larger energy self-sufficiency rate than in Problem (1). The ideal result, where all self-sufficiency rates are simultaneously 100 per cent, is realized at $r = 1$.

4-2-4 Discussion

Figure 4.2.2 shows that the energy self-sufficiency rate is very low if willow is not planted. In that case, energy plants and/or natural energy from outside the system are needed – wood pellets from the forest, for example. Figures 4.2.2 and 4.2.3, which represent the extreme cases, show the following:

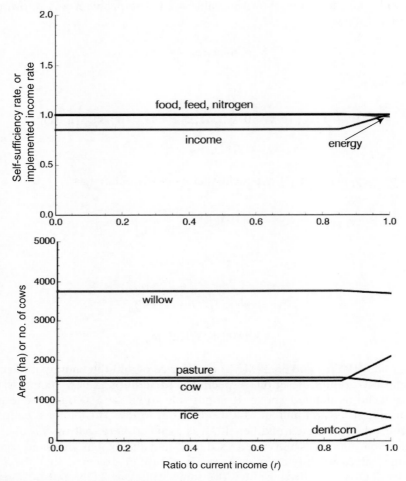

Figure 4.2.3 The case where willow is planted.
Notes: The upper panel shows the maximum energy self-sufficient rate and the values of constraints when the ratio to current income (r) is changed. The lower panel shows the optimal allocation areas and the number of cows when r is changed.

1. Not planting any onions, carrots or sweetcorn is optimal, in spite of the fact that onions have the highest economic value. Even if $r = 0$, which means that there are no income constraints, these vegetables are not allocated any areas. This is because these vegetables provide only economic benefits; they do not provide any energy, nitrogen or feed.

2. The number of cows increases as r increases. Cows can effectively increase income. As the number of cows increases, pasture and/or dent corn areas also increase.

Energy plants, cows and rice are the most important products in this system. Cows contribute significantly to nitrogen and food self-sufficiency rates and to income, whereas energy plant species contribute to energy self-sufficiency rates. Dent corn and pastures support cows. As such, a negative correlation we found between willow and cows via dent corn and pasture. Rice is an excellent product because it contributes to all self-sufficiency rates (food, energy, nitrogen and feed) and to income. Conversely, vegetables such as onions, carrots and sweetcorn are ignored despite having high economic values.

Limitations of the model

The following points have to be considered in the model:
1. Additional energy: no additional energy sources were considered when obtaining biomass energy from rice straw, hull, livestock excrement and willow.
2. Conversion rate: conversion losses were not considered when recycled nitrogen was obtained from organic waste.
3. Stationary state: it usually takes several years before biomass energy can be got from willow. Therefore, stationary states were not considered in this model.
4. Unlimited cows: a certain amount of land is needed to keep cows, and this area generally increases as the number of cows increases. Here it was assumed that an unlimited number of cows could be kept in a certain area. However, indirect constraints through feed were considered.
5. Changeability or usability of the field: usually, it is not easy to change from a vegetable field to a paddy field. As such, agricultural products depend on the soil conditions in the field, which may vary widely. Soil diversity was not factored into the model.
6. Over-evaluation: it is possible to save money as energy and fertilizer made from recycled biomass are used. It is assumed that these benefits will be achieved even if the sufficiency rate is over 100 per cent. Any amount exceeding 100 per cent will be sold outside the system.

The area allocation problem can be discussed from the viewpoints of many self-sufficiency rates, even if the previous points are not considered.

4-2-5 Conclusion

Four self-sufficiency rates were investigated using a mathematical model and a primitive assumption. It was possible to get an almost ideal outcome, that is, the four self-sufficiency rates were almost 100 per cent at the same time. This web-like relationship between compartments but also between agricultural products was the most important discovery. In addition, livestock were found to be central to the system, and rice was found to be an excellent agricultural product.

Acknowledgements

The authors would like to thank the City of Furano, especially Messrs Masaaki Hara, Hiroyuki Ueda, Katsuyuki Sekine and Ryuichi Kawasaki, for their kind assistance and cooperation in studying Furano. They would like also to thank the many people in Furano City who kindly agreed to be interviewed. They are grateful to Professors Mitsuru Osaki and Noriyuki Tanaka of the Centre for Sustainability Science (CENSUS), Hokkaido University, for many exciting discussions. This study was partly supported by the Ministry of Education, Culture, Sports, Science and Technology (MEXT) through Special Coordination Funds for Promoting Science and Technology, entitled "Sustainability Governance Project", and the Global Environment Research Fund (Hc-084) of the Ministry of the Environment, Japan.

REFERENCES

Agency for Natural Resources and Energy, Ministry of Economy, Trade and Industry (2008) *Energy White Paper 2008*, <http://www.enecho.meti.go.jp/topics/hakusho/2008/index.html> (in Japanese).

Agricultural, Forestry and Fishery Bio-Recycle Project "Systematization Subteam" (2006) "Design and Evaluation of Biomass Use System", Tsukuba, Japan (in Japanese).

Agricultural Policy Planning Department, Hokkaido (2002) "Fertilizer Application Guide", Hokkaido Agricultural Development and Extension Association, Sapporo, Hokkaido (in Japanese).

Eriksson, Lisa Naslund (2008) "Comparative Analysis of Forest Fuels in a Life Cycle Perspective with a Focus on Transport Systems", *Conservation and Recycling* 52: 1190–1197.

Furano Agriculture and Forestry Department (2008) "General Description of Furano Agriculture" (in Japanese).

Furano Branch of Kamikawa Agricultural Extension Centre (2007) "Furano Agriculture Data Book, 2006", Farming Promotion Group in Furano Region (in Japanese).

Furano Branch of Kamikawa Agricultural Extension Centre (2008) "New Agriculture Production Technique System in Furano Region, Ver. 2007" (in Japanese).

Heller, Martin C., Gregory A. Keoleian and Timothy A. Volk (2003) "Life Cycle Assessment of a Willow Bioenergy Cropping System", *Biomass and Bioenergy* 25: 147–165.
Larsson, S., C. Cuingnet, P. Clause, I. Jacobsson, P. Aronsson, K. Perttu, H. Rosenqvist, M. Dawson, F. Wilson, G. Mavrogianopoulus, D. Riddel-Black, A. Carlander, T. A. Stenstrom and K. Hasslgren (2003) *Short-rotation Willow Biomass Plantations Irrigated and Fertilized with Wastewaters*. Sustainable Urban Renewal and Wastewater Treatment, No. 37. Danish Environmental Protection Agency. Available at: <http://www2.mst.dk/udgiv/Publications/2003/87-7972-744-1/pdf/87-7972-746-8.pdf> (accessed 10 March 2010).
Livestock Dictionary Editorial Committee (1996) *Japanese Standard Animal Feeding Dictionary*. Tokyo: Youken-dou (in Japanese).
Ministry of Agriculture, Forestry and Fisheries (2009) "The Supply and Demand Table at 2006", <http://www.maff.go.jp/j/zyukyu/zikyu_ritu/pdf/ws.pdf> (in Japanese).
NEDO [New Energy and Industrial Technology Development Organization] (2009) "The Estimation of Biomass Endowments", <http://app1.infoc.nedo.go.jp> (in Japanese).
Saito, Yutaka (2009) "The Perspective on Sustainable Northern Biosphere", *Energy Shigen* 30(2): 39–43 (in Japanese).
Sato, Toshiki (2005) "The Actual State and Problems of Investing Energy in Rice Produce – On Investing Electrical Energy Analysis", *Bulletin of Hiroshima Perpetual University* 16(2): 15–26 (in Japanese).
Sato, T., N. Tsuji, N. Tanaka and M. Osaki (2010) "Food and Energy Self Sufficiency Potential Analysis Based on Biomass in Agriculture, Livestock, and Forest", *Journal of the Japanese Agricultural System Society* 26(1): 17–25 (in Japanese with English summary).
Statistics Department, Minister's Secretariat, Ministry of Agriculture, Forestry and Fisheries (2008) "Production Cost in Livestock", Association of Agriculture and Forestry Statistics, Tokyo, Japan (in Japanese).
Wolfram Research (2007) *Mathematica Version 6.0*. Champaign, IL: Wolfram Research, Inc.

4-3

Biogas plants in Hokkaido: Present situation and future prospects

Juzo Matsuda and Shiho Ishikawa

4-3-1 Introduction

Biogas plants are facilities that consist of a methane fermentation system to treat domestic livestock waste and an energy recovery system. These plants are intended to treat waste mainly in order to convert it into fertilizer and to produce biogas as an alternative fuel to oil. The use of biogas has attracted much attention owing to its ability to reduce carbon dioxide emissions, thus lightening the environmental load. The ability of biogas to decrease carbon dioxide emissions, despite the fact that it emits carbon dioxide itself, is based on the carbon-neutral concept. This means that terrestrial carbon is constantly transferred in a recurrent cycle: pasture grass—cattle—livestock waste—methane gas—carbon dioxide gas—pasture grass. This implies that, in order for biogas to remain carbon neutral, it is necessary to continuously reproduce biomass (pasture grass), which is its source.

Biogas plants have been constructed in Japan since 1998. However, they are simple treatment plants for domestic livestock waste and are ineffective as energy production facilities. It should therefore be noted that these plants merely provide a basic method of resolving the problems of domestic livestock waste, without truly addressing energy issues.

Biogas plant construction, having been established more than 10 years ago in Hokkaido, started to decline and, around 2004, trial biogas plants began to cease operations (see Figures 4.3.1 and 4.3.2). These plants were constructed at enormous cost and when it is time to replace their pumps, engines and so forth it is extremely uncertain whether the plant will

Designing our future: Local perspectives on bioproduction, ecosystems and humanity,
Osaki, Braimoh and Nakagami (eds),
United Nations University Press, 2011, ISBN 978-92-808-1183-4

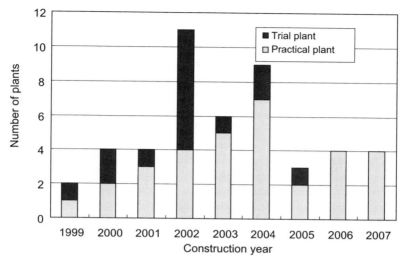

Figure 4.3.1 Number of biogas plants constructed by dairy farms in each year, 1999–2007.

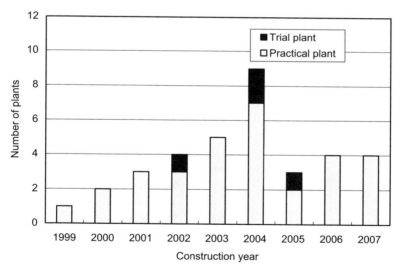

Figure 4.3.2 Number of biogas plants currently being operated by dairy farms, by year of construction.

continue operating or not. To permit the continued operation of these plants, it is now essential, from both economic and environmental perspectives, to create the legal conditions that will allow the sale of electricity and biogas generated by biogas plants under the Feed-in Tariff.

4-3-2 Biogas plants in Hokkaido

Figure 4.3.3 demonstrates the results of a multiple-choice interview survey conducted among 34 dairy farmers who had introduced biogas plants, as a way of measuring their degree of satisfaction with such plants. According to the figure, almost all farmers are satisfied with the digesters: specifically 16 were satisfied and 3 were not. Many farmers are very satisfied with gas holders, desulphurization systems, boilers and dehumidifiers. But it is clear that a majority of farmers are dissatisfied with the electricity generators (the co-generation system) and the digested slurry storage tanks. This is because of the frequent malfunctioning of the power generation engines, high management costs (including those for replacing oil), an inability to sell the electricity generated and the inadequate capacity of storage tanks. Under present design specifications, milch cattle produce discharges of about 60 kg/day. In reality, a high-lactating cow discharges around 70 kg/day. In addition, the waste water from the milking parlour passes through the digester to the storage tanks along with the digested slurry. As a result, the volume in the design specifications is in-

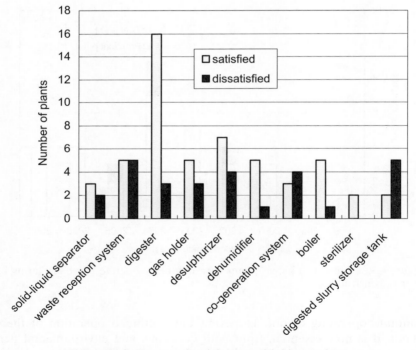

Figure 4.3.3 Degree of satisfaction with biogas plants.

sufficient, hence the farmers dislike the fact that, although they must retain the liquid during the six-month winter period, it is impossible to do so. Their dissatisfaction with biogas plants is directed at the equipment related to energy generation and at issues regarding the fermented fluids (digested slurry). In brief, the problem with biomass energy in Japan is the lack of outlets, that is, applications to consume or sell what is produced.

4-3-3 Biogas plant problems and improvement measures

Biogas plants were welcomed in Japan when they were introduced as a way to earn profits, as they had been in Germany. However, as the differences between Japan and Germany became apparent, the Japanese realized that these plants could not be viably profitable facilities.

The sale of electricity

Under the German system, although a subsidy is almost never provided to construct a plant – and if a subsidy is paid it is only a small amount – the price that electricity companies pay to purchase electricity generated by biogas at the outlet is set high for a period of 20 years under the Feed-in Tariff. In the case of a plant with a service life of 20 years, the investment can be recovered in about 7 years. The number of biogas plants has increased to about 4,000 in Germany because investors can achieve adequate economic returns.

In Japan, on the other hand, the minimum subsidies provided are about 50 per cent, and they are 75 per cent in Hokkaido. The Hokkaido Regional Development Bureau offers a 95 per cent subsidy towards biogas plant construction. Accordingly, a plant can be constructed with a low financial burden on farmers. However, there is no system allowing them to sell the electricity these facilities generate, and hence they are unable to earn any income from their plants. Moreover, the cost of operating a plant is extremely high, leading to a situation in which operations cannot be continued. This resulted in a fall in the number of plants from 47 to 35 over the 10-year period 1999–2008. Thus, no matter how much support is provided to unprofitable and unneeded facilities, biogas plants will clearly not survive economically without a subsidy to cover their outputs. However, there would be no problem if facilities that use electricity were installed on-site where there is a biogas plant. At one Hokkaido dairy farm, there are ice-cream, butter and cheese manufacturing factories near a biogas plant. Almost all of the electricity generated at the plant is consumed by these factories. As a result, the factories purchase far less power from electricity companies, ensuring that they are very economical. Unfortunately, such situations are extremely rare, even in Hokkaido.

The sale of biogas

Biogas is 60 per cent methane and 40 per cent carbon dioxide, with around 1,000 ppm equivalent of the corrosive gas hydrogen sulphide. Desulphurized biogas has a calorific value of approximately 23 MJ/m^3 and is 6A equivalent. If its carbon dioxide is removed, it becomes approximately 39 MJ/m^3 biomethane, which is 12A equivalent. It makes most economic sense when these gases are transported via a pipeline and sold. The biomethane can also be sold after being compressed and stored in tanks, but this process is subject to the extremely strict High Pressure Gas Safety Act, which makes it impossible to refine and compress more than 100 m^3 of biomethane per day. In addition, a lot of energy is employed in high compression; therefore, it is not suitable for ordinary farms.

When the Renewable Portfolio Standard (RPS) law was enacted and producers were obligated to add a specified percentage of biomethane to natural gas, the use of biogas expanded rapidly. In fact, in Germany, energy crops such as pasture grass or corn are methane fermented and the biomethane generated is transported to natural gas pipelines operated by several large cities.

On-site use and sale of digested slurry

Methane-fermented digested slurry is a high-efficiency organic fertilizer that can also be used as a soil improvement material. The rising price of chemical fertilizers and mixed feeds in 2008 greatly increased the value of digested slurry. Fertilizing with digested slurry has sharply cut the quantity of chemical fertilizer used by farmers and its application has increased yields of pasture grass, dent corn, wheat and other crops. In a number of cases reported from Hokkaido, it has even improved crop quality. Furthermore, some dairy farms have produced the same amount of milk as before, even though they have reduced the use of mixed feeds by using higher-quality pasture grass instead.

The decrease in the use of chemical fertilizers, combined with lowering the amount of mixed feed used, has had tremendous economic impacts. As explained above, the same can be said of on-site use of electricity generated. Farmers can even sell surplus digested slurry. The value of the fertilization constituents included in the digested slurry has been calculated to be more than 1,000 JPY/m^3. Yet, currently, it sells for several hundred JPY/m^3, which includes the slurry spreading charge by farmers who have biogas plants. In the future, if a suitable price is set, digested slurry could be a satisfactory source of earnings. Finally, the digested slurry has an extremely low odour, hence is inoffensive after application. Many livestock farmers want to construct biogas plants to reduce bad odours.

Unless the law is revised, it will be difficult to increase the income farmers earn by selling the generated electricity and biogas, as explained above. However, the use of digested slurry is progressing smoothly. Nonetheless, this degree of cost reduction and increase in income is not contributing to the spread of biogas plants. If the RPS is not revised and a feed-in tariff is not established, the extensive use of biogas plants seems unlikely in Japan.

4-3-4 The Japanese version of Green New Deal policies and biogas plants

The term "Green New Deal" policy has been frequently used in recent years. In the United States, President Obama's energy policies include energy self-sufficiency. Over the next decade, he has proposed an investment of USD 150 billion in renewable energies, as a part of the Green New Deal policies and energy conservation at public facilities, all of which are intended to create employment for several million people. Japan, too, has announced a bold policy of expanding the environmental business field by increasing its market scale from JPY 70 trillion to JPY 100 trillion per year, boosting its employment from 1.4 million to 2.2 million by 2015. Japanese policies promote activities that include the expansion of photovoltaic power production facilities and electric vehicles. In response to these initiatives, the Ministry of Economy, Trade and Industry has decided to introduce a Feed-in Tariff, increasing the price paid for power generated by photovoltaic power production from the present 24 JPY/kWh to 48 JPY/kWh.

What kinds of Green New Deal policies should be established in Japan? Japan's photovoltaic cell manufacturing technologies are the world's finest; hence it is necessary to increase their production. Japan must simultaneously turn its attention to biomass energy, which exists mainly in farming communities and outlying regions. It is in these regions where local plant manufacturing industry could be promoted to create employment.

In Hokkaido, 47 biogas plants have been constructed to treat livestock waste produced by milch cows, and approximately 80 such plants have been built nationwide. However, the operating costs of these plants are too high for many livestock farms to maintain. As a result, as stated above, only 35 plants are currently operating in Hokkaido. Many biogas plants perform co-generation, i.e. generating and selling electricity, and hence have been relying on earning profits. However, even though selling the electricity they produce would greatly lower their power costs, they cannot depend on the income generated because of the low selling price.

Take Hokkaido, for example. The selling price of electricity under the RPS is about 10.5 JPY/kWh during the day and 4.5 JPY/kWh for 10 hours at night. Under these conditions, the operating cost of co-generation is higher than the income based on the profits from selling electricity generated by biogas plants. Some farms actually discharge biogas into the atmosphere at night. If a 50 kW generator were operated for 10 hours at night and the electricity selling price were set at 24 JPY/kWh, which is the same as that now paid for photovoltaic electricity, the income generated would rise to approximately JPY 4.4 million per year, which would cover the operating management costs. In Germany, the electricity sold from biogas plants is around three times more expensive than the electricity that farmers buy; hence biogas plants are established to earn profits. As a result, the number of plants in operation has increased to about 4,000 during the past five years. It is imperative that ways are considered to encourage the introduction of more plants in Japan.

Under the leadership of the Ministry of Agriculture, Forestry and Fisheries, bioethanol production from agricultural products is on the rise in Japan. In Hokkaido, two factories producing 15,000 kL/year of bioethanol have been constructed. This project also faces extremely harsh conditions in light of falling oil prices and rising prices of rice, wheat and other agricultural products. In Japan, the prices of raw materials for energy production are high and yields are small; therefore costs are high, the energy profit ratio is low and the lifecycle assessment effects are small. The cultivation of crops to be used as raw materials for biofuel improves the environment by generating clean energy, cutting greenhouse gases and promoting agriculture, while simultaneously producing grain to safeguard against food shortages. In Japan, 1 million hectares of rice fields have been converted to other uses and 400,000 hectares of cultivable land, equal to the area of Saitama Prefecture, have been abandoned. If this 1 million hectares of land were restored to paddy fields and the rice used to manufacture biofuel, industry could be promoted and employment created.

However, a far superior energy balance and greater economic benefits would be obtained by manufacturing biogas through the process of methane fermentation of whole crop residues, including rice straw, wheat straw and whole crop silage, instead of using food grains to produce bioethanol. Technologies for dry methane fermentation plants that process such dried residues have already been developed. Moreover, regionally generated garbage, sludge and waste from food-processing plants are suitable biogas fermentation materials. As shown in Figure 4.3.3, the agricultural residue from farms may be supplied to livestock farms or directly to biogas plants. Both wet-type and dry-type biogas plants could be built and managed by local civil engineering or construction companies under the guidance of plant manufacturers. Because plants would likely be constructed at a number of places in farming regions, new industries and employment opportunities – far more than the manufacture of bioethanol would generate – would be created. It would be economically beneficial if underused farmland were used to cultivate rice or wheat, which is then used as a material for energy production, and if biogas plants were constructed that use agricultural residues as raw material. From an agricultural perspective, food self-sufficiency would improve, because preparations to deal with food shortage crises would be made. In addition, farmland would be conserved, thus improving the environment. Biogas plants will be able to accept and treat waste material, including garbage, livestock waste and sludge. It will also be possible to use digested slurry or compost generated as a residue from fermentation as a fertilizer. Agricultural products cultivated with the use of these materials will be supplied to

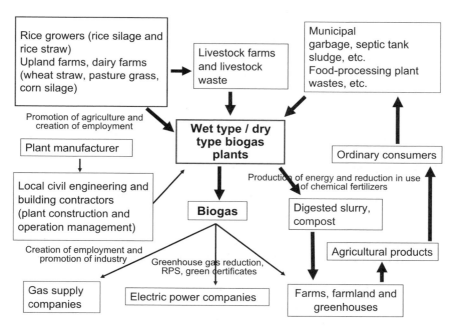

Figure 4.3.4 Achieving Green New Deal policies with biogas plants.

consumers and, completing the cycle, raw garbage and sludge will be used by biogas plants. The biogas produced could be either utilized directly or converted to energy. This electricity could be supplied to the region, thus reducing greenhouse gas emissions. Such measures would definitely move Japan in the direction of a low-carbon resource-circulating society.

In response to the setting of the selling price of electricity produced by photovoltaic generation at 48 JPY/kWh, the Feed-in Tariff will probably be extended to cover the sale of electricity produced by wind power and biomass. If this is not done, Japan's biomass energy industry will not develop. Furthermore, the revision of the RPS will probably affect not only electricity companies but also gas companies, by obligating them to use the biogas produced. When the outputs for energy from biogas plants are achieved, a biogas plant society in compliance with the Green New Deal policy will be established and a low-carbon society that originates in farming communities will no longer be a pipedream.

4-3-5 Summary

More than a decade has passed since the commencement of the full-scale introduction of biogas plants to deal with livestock waste in Hokkaido. Initially, both general contractors and foreign companies assumed that biogas plants would spread quickly. This assumption was based on the subsidies granted, ranging from a minimum of 50 per cent to 75 per cent in Hokkaido and even 95 per cent for Hokkaido Development Bureau projects. The dairy farms that introduced these plants were satisfied with the livestock waste treatment, but decided that the cost of management (including the cost of co-generation) was high. Moreover, farmers were aware that they could not sell the electricity generated or, even if they could, the price would be extremely low. As a result, some of the trial plants introduced on dairy farms began to shut down and were eventually removed. General contractors and foreign companies withdrew from the projects. Only about 35 of the more than 50 plants originally constructed are now in operation.

Recently, the soaring price of chemical fertilizers and mixed feeds has resulted in an extremely high value being put on digested slurry. The use of digested slurry has brought about an effective reduction in the use of chemical fertilizers and an increase in the yields and quality of agricultural products such as pasture grass, resulting in a decrease in the use of mixed feeds. These effects have had a significant economic impact. It is assumed that these results will halt the decline in the number of biogas plants.

Furthermore, the introduction of the methane fermentation method using agricultural residues, not only by livestock farmers but also by rice producers and dry field farmers, has improved energy efficiency, the environment and agriculture (food products) in farming regions. At the same time, use of this method will promote the expansion of industry and employment. Needless to say, however, policies that create the outlets for energy are required order to make this progress.

4-4
The bioenergy village in Germany: The Lighthouse Project for sustainable energy production in rural areas

Marianne Karpenstein-Machan and Peter Schmuck

4-4-1 The global situation and sustainability science

Looking at the brute fact that global carbon dioxide emissions have almost doubled since the UN Conference on Environment and Sustainable Development in Rio de Janeiro in 1992 – despite the ambitious *Agenda 21* document – one may ask whether fundamentally new societal approaches are needed to solve the global problems. One new strategy within science is "sustainability science", first formulated in 2001. Essentially five methodological principles of traditional science are replaced in this approach, because the solving of global problems – based on nonlinear, complex processes, and including long time lags between actions and their consequences – requires more than developing and testing hypotheses in laboratories. Instead of regarding scientific activities as value-free endeavours that work in a mainly monodisciplinary, analytical and linear way, with basic research and a strict division between research and application as the ideal, sustainability science (1) acts explicitly to support sustainable development, (2) works in synthetic and parallel ways, (3) takes an interdisciplinary approach, (4) is transdisciplinary and (5) combines research and application in action-oriented research, during which scientists initiate sustainability changes in a society and simultaneously perform research (Sheldon et al., 2000). The main reason for this new orientation within science is not that the traditional approach of science is wrong, but that it is too slow in solving actual global problems. If scientists want to contribute substantially to the sustainability revolution

---

Designing our future: Local perspectives on bioproduction, ecosystems and humanity,
Osaki, Braimoh and Nakagami (eds),
United Nations University Press, 2011, ISBN 978-92-808-1183-4

(McKenzie-Mohr, 2002), they have to act as citizens initiating changes toward sustainable development, in addition to their traditional role as scientific analysers.

In the Bioenergy Village Project, an example for sustainable energy production in rural areas, the methodological guiding principles of the "sustainability science" approach were applied in the following way. In 1999, an interdisciplinary group of scientists (sociologists, psychologists, political scientists, economists, agronomists and geologists) developed the vision of a bioenergy village and cooperated between 2000 and 2005 across disciplines with engineers and people from a suitable German village on the technical and social implementation of renewable energy plants. From 2000 until 2008, scientific research analysed the initial situation and the consequences of the ongoing process. Direct support for sustainable development was achieved: by diminishing the village's carbon dioxide emissions by about 3,600 tons/year and by transferring the model to many more villages in Germany with the same positive climate effects, the project contributed to counteracting further climate change (Karpenstein-Machan and Schmuck, 2007; Ruppert et al., 2008).

4-4-2 The technical concept of the bioenergy village Juehnde

In Juehnde, the technical concept of the village consists of three main components:

(1) Electricity and space heat are produced by burning biogas in a combined heat and power (CHP) generator with an electricity capacity of 680 kilowatts (kW). Liquid manure and field crops, cultivated on 250 hectares (ha) of arable land, are digested enzymatically by microorganisms under anaerobic conditions, and biogas is generated. The CHP converts biogas into roughly 35 per cent electricity and 50 per cent usable heat energy. Electricity is fed to the national electricity grid. Heat energy is used for the space heating and hot water demands of village households.

(2) In wintertime, additional heat energy is delivered by a central heating plant with a thermal capacity of 550 kW fired by locally produced wood chips.

(3) Heat energy is distributed as hot water into a 5.5 km-long hot water grid that connects the plant with the households. Heat transfer in the houses occurs through heat exchangers (which include a heat meter), which replace individual heating systems. Furthermore, to provide security in the event of a breakdown of the biomass plants and for periodic maintenance, a peak load boiler with a capacity of 1,600 kW fired by natural oil has been installed (see Figure 4.4.1).

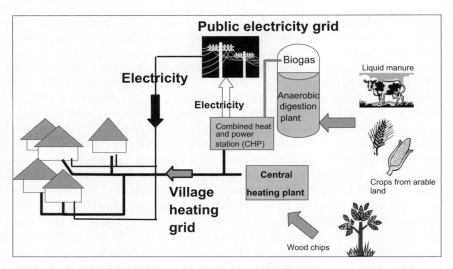

Figure 4.4.1 The technical concept of the bioenergy village Juehnde.

4-4-3 Guidelines to implement bioenergy projects in communities

Bioenergy can positively contribute to climate goals and rural development; however, if not implemented carefully and in cooperation with the people concerned, it could exacerbate the degradation of land, water bodies and ecosystems and increase greenhouse gas (GHG) emissions, leading to citizens' rejection of the initiative. Sustainable development will require agriculturally sustainable methods and high energy efficiency in the conversion and utilization of produced biomass energy. People should be involved in the development of renewable energy. The challenge lies in translating the opportunity into reality. In the Lighthouse Project "Bioenergy Village", a holistic approach was followed and ecological, economic, technical and social guidelines were formulated, the last two aspects being considered here together.

Ecological aspects and their realization in Juehnde

Guideline: Use local renewable energy sources (energy crops and organic residuals) to replace fossil and nuclear energy resources and to reduce GHG emissions

As a result of the change in the energy system to CO_2-neutral energy resources biomass, emissions of carbon dioxide were greatly reduced by 8.2

tons per capita per year (approximately 76 per cent; relating to heat clients). In total, 3,671 tons of CO_2 (t-CO_2) are saved every year in Juehnde (Ruppert et al., 2008). This is a remarkable contribution to climate protection.

Most of the necessary biomass to meet the electricity and heat requirements of the village comes from field crop production and grassland, grown at a maximum distance of 10 km from the energy plant. Manure biomass is delivered from four farms that are located less than 1 km from the energy plant. Woody biomass comes from surrounding forests. One can therefore conclude that the residents of Juehnde use their own community resources for energy production and thus avoid transporting these materials long distances as another way of reducing GHG emissions.

Guideline: Produce energy crops in a sustainable way, thus avoiding loss of diversity, degradation of land and pollution of groundwater

To optimize the ecological effects in energy crop rotations, an ecologically oriented cultivation system was developed at the University of Kassel (Karpenstein-Machan, 2001; Karpenstein-Machan, 2005; Scheffer and Stülpnagel, 1993). It is based on a diverse crop rotation, with several winter and summer crops. In moderate climates with a growing period of six months or more, two crops per year are feasible. This opens up numerous possibilities for enhancing the crop rotation with different species, reducing the input of pesticides and protecting the soil against erosion. In contrast to food production for human consumption, for a biogas plant the crops need to be harvested unripe, with a dry-matter content of about 30 per cent. At that growth stage (the milky stage of the grain), all parts of the crop can be used as a substrate for digestion. Responding to the demands of optimal fermentation, energy plants are harvested with a conventional fodder harvester and chopped into lengths of between 0.5 and 5 cm. Harvested chopped crops are directly transported to a clamp silo in order to make silage. This is done by compressing the harvested material with heavy tractors (see Plate 4.4.1).

After the implementation of the energy plants in Juehnde, farmers cultivated energy crops on 250 ha of land per year in a water protection area. The crop rotation has been enriched with new biomass crops such as winter triticale, winter rye, sunflowers, maize and a mixture of different crops for energy production. In comparison with conventional crops such as winter wheat for grain production, fewer pesticides (growth regulators, fungicides, insecticides) and less nitrogen are now used for biomass production (Karpenstein-Machan and Lootsma, 2009). The nitrogen balance for the energy crops was negative (N uptake was greater than N input through fertilizer). These changes in cultivation method minimize negative impacts in the water protection area, such as nitrate and pesticide

Plate 4.4.1 Photo of storage for harvested energy crops; in the background the biogas plant in Juehnde with two fermentation units (*Source*: © Marianne Karpenstein-Machan, 8 July 2007).

leaching and groundwater pollution. Because of the more varied energy crop rotation, the change in winter and summer crops and minimal tillage methods, soil erosion and losses in humus are avoided.

An energy balance was performed for energy crops. All supply chains for farm energy inputs, such as fuel for fieldwork and fuel for transportation, lubrication, machines, fertilizer, seeds, pesticides and silaging, were taken into account. Following these calculations, the energy input/output ratio was 1:19 for triticale and 1:18 for maize (Karpenstein-Machan and Lootsma, 2009). This ratio shows that the cultivation system is able to replace fossil energy with renewable energy on a remarkable scale.

Guideline: Use locally available animal manure and other organic residuals as energy sources to reduce competition for land and to curb GHG emissions.

The use of manure and other organic residuals for energy production has several fundamental advantages for energy projects: less arable land is needed for energy production, and therefore competition for land can be minimized; methane emissions can be reduced considerably; animal manure, especially from cows, improves the biological process in the biogas plant because manure contains methane bacteria.

In Juehnde, all available manure from animal production (10,000 m^3 per year from 700 cows and 1,300 pigs) is freshly processed in the biogas plant. Two times a week, the fresh manure is collected from the farms. In this way, methane losses from open manure storage tanks on the farms can be avoided. Calculations show that 309 t-CO_2-equivalent/year can be saved by processing the manure (Karpenstein-Machan and Lootsma, 2009; Ruppert et al., 2008). Furthermore, the use of manure reduces the inputs required from energy plants in Juehnde by 50 ha.

Guideline: Use digestate from energy production as a fertilizer on the fields to recycle nutrients and reduce inputs of mineral fertilizers

Mineral fertilizer use has been reduced considerably in Juehnde owing to the use of digestate. About 20,000 m^3 digestate, derived from energy crops and manure, is produced every year. That amount of digestate contains about 92 tons nitrogen, 16 tons phosphorus, 77 tons potassium and 10 tons magnesium, which are used as a liquid fertilizer on the fields around the village. Depending on the utilized area of arable land in Juehnde district, about 50 per cent of mineral fertilizer can be replaced by digestate. This approach could lead to big energy savings (about 100,000 litres oil equivalent = 280 t-CO_2/year) because the production of mineral fertilizer requires high inputs of fossil energy.

Guideline: In order to achieve high efficiency in the use of renewable resources, employ not only the electricity but also the heat energy from combined heat and power stations

Most biogas plants in Germany feed the electricity produced into the national electricity grid and do not use – or do not efficiently use – the additional heat energy. This waste of energy resources leads to a low efficiency factor for bioenergy. There should not be such a dissipation of resources in local renewable energy systems.

In Juehnde, the electricity is fed into the national grid and the heat energy is utilized to cover the heat demands of village households. The energy efficiency of the power station is 83 per cent (Karpenstein-Machan and Lootsma, 2009; Ruppert et al., 2008), which means that 83 per cent of the energy produced is utilized. About 150 households are connected to the hot water grid. As a result, individual oil and gas heaters have been replaced by heat energy from the hot water grid, which is derived from biomass. Therefore, the use of fossil oil for heating has been reduced by about 400,000 litres/year, and the use of electricity derived from fossil fuels has been reduced by about 2 million kWh in Juehnde. A surplus of 3 million kWh produced in the power station leads to a further replacement of fossil energy.

Economic aspects and their realization in Juehnde

Guideline: Produce at least the amount of electricity that is consumed in the village, in order to balance the plant's economy on a long-term basis by selling the electricity to the public provider

In Juehnde, twice as much electricity is produced as is consumed in the village. The high electricity production has a positive impact on the economy of the community-owned energy company. Calculation of the profitability of the bioenergy facilities of Juehnde shows that about 62 per cent of the turnover comes from sales of electricity (Ruppert et al., 2008). According to the feed-in regulations of the German Renewable Energy Act, the feed-in prices are paid for 20 years. Because the prices are fixed, the community-owned company is able to plan dependably for the long term.

Guideline: Provide at least 50 per cent of the heat energy demands of the village through biomass; use the heat and power stations to balance the economy by selling the heat to clients

The production of electricity balances the economy. However, the company will not normally be in the black from sales of electricity alone. The more heat that clients in the village require, the better the utilization of the so-called "lost heat" and the better the company's economic situation. Furthermore, the capital-intensive investment in the hot water grid is only economically sound if a majority of the households in the village decide to connect to the grid and meet their heat demands from the grid alone. For instance, in Juehnde, 75 per cent of the households participate in utilizing the community hot water grid. Calculation of the profitability of the bioenergy facilities of Juehnde in 2007 shows that the sales of heat energy to the clients requiring heat, together with the sales of electricity to the national grid, generated a profit of €159,000 (Ruppert et al., 2008). These results indicate success only two years after the establishment of the energy plants. In the event of rising inflation rates over the long term – resulting in higher costs – it will be possible to compensate biomass energy carriers, staff, insurance, maintenance costs and so forth by increasing heat energy prices. Currently, the heat energy price is fixed for five years.

Guideline: Heat clients and local farmers should be shareholders in the energy plants, with more than a 50 per cent share in order to benefit from the financial revenue

The big advantage is that the people of the village themselves can take part in the decision-making process related to all operations of the company, including prices and dividends, for example. They nominate individuals in whom they have confidence as the board of management and the supervisory board. Furthermore, every member of the community can nominate him- or herself to a board. With more than 50 per cent of the shares in the hands of the village residents, the people are independent of outside investors. Thus, the villagers benefit from the financial revenues of their energy company and the money circulates throughout the region, supporting the rural economy.

In Juehnde, the villagers decided to establish a cooperative with limited liability. In a cooperative, each member has one vote, independent of the total amount of the member's share. Members of the cooperative comprise the heat clients, farmers, other village residents and sponsors. The minimum value of shares required for membership was initially €1,500. The villagers hold 85 per cent of the cooperative's shares, so they can contribute to the development of their energy company in a democratic manner.

Guideline: When prices for fossil fuels rise, bioenergy villages that follow the Lighthouse Project Juehnde should run profitably without government grants

For the pilot project in Juehnde, the German government gave a one-time grant of 30 per cent of the total investment (€1.5 million). When the project was started by a team at the University of Göttingen, the price of oil was low (€0.35/litre), and alternative renewable energy systems could not be implemented on a sound economic basis without subsidies. Since then, oil prices have more than doubled. Together with the current version of the German Renewable Energy Act, which came into force in January 2009, the conditions for further bioenergy villages are now much better than in 2004, when Juehnde created a business plan for the energy plants. The feed-in rates have now been upgraded for smaller plants, and the use of animal manure for processing has been promoted. These measures should encourage a reduction in the use of crops and arable land and minimize competition for land. As the business plans of four other bioenergy villages in the district of Göttingen indicate – which are actually under construction – it is possible to establish bioenergy villages with almost no grants from government.

Social aspects and their realization in Juehnde

Guideline: Invite people in the community to participate in planning, implementing and investing in energy plants in order to gain a high rate of acceptance and commitment among the rural population

In democratic societies, there are no means to force people to establish a bioenergy village. Therefore, the participation of rural people in a bioenergy project has to be voluntary, based on conviction or invitation but not on coercion. Therefore, the challenge for the Lighthouse Project was primarily to create a social innovation, not a technical one. With the establishment of a bioenergy village, numerous changes occur for the residents. The villagers assume a unique social role as energy producers and distributors to meet the energy demands of their own village. Much state-of-the-art knowledge is required, and many new duties have to be distributed among the actors.

This process was organized by the university team, applying social success factors found in interviews with initiators of comparable successful community projects (Eigner and Schmuck, 2002). These factors include identifying potential initiators, making personal contact with the villagers, giving neutral scientific information to all people, behaving appropriately and respectfully toward detractors, creating a good rapport with the public media, using social occasions and existing networks to spread the concept, and visiting comparable energy plants. Eight working groups and a central planning group, composed mainly of village inhabitants, discussed all aspects of the very complex project during the preparation phase (2001–2003), thus ensuring that the people of the village would later consider the project as their own creation.

A psychological consequence of that broad engagement of many people in the village was people's deep satisfaction with the project. Residents in Juehnde today are proud of their project, and they feel less dependent on imports of fossil energy resources.

Guideline: Use "state of the art" biomass conversion technology (reliable techniques) to create trust and avoid technical problems and gaps in the supply of heat energy in the village

If a number of people have to be convinced to change to a new renewable energy system, they must be certain that the new technology is reliable; otherwise, they will not replace their individual gas or oil house heater and connect to the village heating grid. Technical problems can create gaps in the supply of heat energy and, for example, in extreme cases people cannot heat their rooms during the wintertime. To avoid such problems in Juehnde, three independent heat energy systems (combined heat and power station, central heating plant, peak load boiler) were established that can work either together or independently (see subsection 4-4-2). This energy "back-up system" makes it possible to deliver the necessary heat energy to the village inhabitants in accordance with their needs. This gives people the confidence to replace their individual heating systems, fuelled by fossil oil and gas, and obtain a more reliable energy system than they had before. Furthermore, they now have free space in the cellar and no more smells of fossil oil.

4-4-4 Transfer to other villages

The transfer of this model to other villages in Germany has worked well to date. The same social and technical installations as in the model village are under construction in four other villages (Reiffenhausen, Barlissen, Krebeck and Wollbrandshausen) in the district surrounding the university city of Göttingen. Across Germany, there are approximately 30 comparable villages with decentralized heat and electricity production based on locally available renewable resources, many of them directly inspired by the model village (e.g. Rai-Breitenbach, Lausheim).

Since the project was started in the village of Juehnde in 2001, there have been several thousand national and international visitors to Juehnde, among them politicians and scientists from the United States, China, Japan, Indonesia and numerous other countries. After the present authors visited Indonesia, the Indonesian government decided in 2006 to set up 1,000 bioenergy villages in that country. Implementing similar projects in other countries, however, requires careful adaptation of the social, economic and ecological sustainability guidelines to local conditions.

REFERENCES

Eigner, S. and P. Schmuck (2002) "Motivating Collective Action. Converting to Sustainable Energy Sources in a German Community", in P. Schmuck and W. Schultz (eds), *Psychology of Sustainable Development*. Norwell, MA: Kluwer Academic Publishers, pp. 241–260.

Karpenstein-Machan, M. (2001) "Sustainable Cultivation Concepts for Domestic Energy Production from Biomass", *Critical Reviews of Plant Science, Special Issue on Bioenergy* 20(1): 1–14.

Karpenstein-Machan, M. (2005) *Energy Crop Cultivation for Biogas Plant Operators*. Frankfurt: DLG (in German).

Karpenstein-Machan, M. and P. Schmuck (2007) "The Bioenergy Village. Ecological and Social Aspects in Implementation of a Sustainability Project", *Journal of Biobased Materials and Bioenergy* 1: 148–154.

Karpenstein-Machan, M. and A. Lootsma (2009) "Implementation of an Ecologically Sound Cultivation System for Energy Crops in the Bioenergy Village Juehnde" (in German), paper presented at the Annual Biogas conference, Hanover, 3–5 February. Published in the *Conference Proceedings* 18: 19–25.

McKenzie-Mohr, D. (2002) "The Next Revolution: Sustainability", in P. Schmuck and W. Schultz (eds), *Psychology of Sustainable Development*. Boston: Kluwer Academic Publishers, pp. 19–36.

Ruppert, H., S. Eigner-Thiel, W. Girschner, M. Karpenstein-Machan, F. Roland, V. Ruwisch, B. Sauer and P. Schmuck (2008) "The Bioenergy Village", unpublished research report, University of Göttingen.

Scheffer, K. and R. Stülpnagel (1993) "Wege und Chancen bei der Bereitstellung des CO_2-neutralen Energieträgers Biomasse – Grundgedanken zu einem Forschungskonzept", *Der Tropenlandwirt* 49: 147–161.

Sheldon, K., P. Schmuck and T. Kasser (2000) "Is Value Free Science Possible?", *American Psychologist* 55: 1152–1153.

4-5
Samsø: The Danish renewable energy island

Ryoh Nakakubo and Søren Hermansen

4-5-1 Introduction

Renewable energy in Denmark

Denmark is a small Scandinavian country with a land area of 43,000 km^2 and a population of about 5.5 million, and it is known for its high proportion of renewable energy for energy consumption. In the past 20 years, the proportion of total energy generated from renewable sources has increased from approximately 5 per cent to 17 per cent. In 2007, renewable energy production was calculated to be 130.2 petajoules (PJ), with wind power and biomass production accounting for 25.8 PJ and 90.5 PJ, respectively (see Figure 4.5.1). In biomass production, straw accounted for 18.3 PJ, wood 41.2 PJ, renewable waste 30.1 PJ and fish oil 0.8 PJ. In 2007, wind power accounted for 19.7 per cent of domestic electricity supply, a significant increase from only 1.9 per cent in 1990 (Danish Energy Agency, 2008). Therefore, the Danish government established a target of increasing the share of renewable energy to at least 30 per cent of total energy consumption by 2025 (Danish Energy Authority, 2007).

Historical background

The oil crisis in 1973 resulted in a radical change in people's attitude towards the use of resources. There was a great protest against the use of nuclear energy, and therefore a political as well as a societal will was

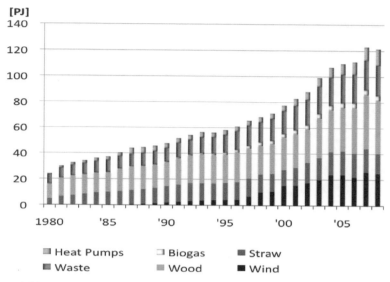

Figure 4.5.1 Production of renewable energy by type.
Source: Danish Energy Agency (2008).

formed to develop an energy sector based on renewable energy sources (Beuse et al., 2000). Wind energy and biomass-based district heating systems also attracted much attention in the 1980s (Odgaard, 2000). Following the Toronto Conference on the Changing Atmosphere in 1988 and the establishment of the Intergovernmental Panel on Climate Change, the Danish government initiated the preparation of a new and more comprehensive plan of action for the use of renewable energy, especially biomass and wind energy. This energy strategy, called "Energy 2000", aimed at constructing land-based wind turbines with a capacity of 1,500 MW by 2005 (Odgaard, 2000). In the spring of 1996, the Danish government endorsed a new energy strategy called "Energy 21", with a target of meeting 35 per cent of Danish energy consumption by renewable energy by 2030 (Odgaard, 2000).

Initiation of the renewable energy project in Samsø

In 1997, the Danish government announced a competition for the local municipalities of various islands to propose a plan for a complete change in the energy supply to renewable energy over a period of 10 years. The idea behind this competition was to focus on renewable energy and study the maximum extent to which renewable energy could be achieved in a well-defined area using the available technology with almost no generous

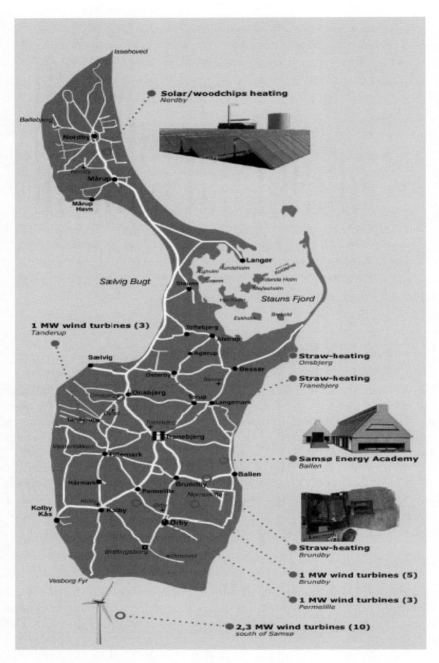

Plate 4.5.1 Map of Samsø (*Source*: Samsø Energy Academy).

grants. The criterion for choosing a winner was a realistic plan for obtaining 100 per cent renewable energy in 10 years, and involvement of various local actors was considered important (Jørgensen et al., 2007; Lunden, 2003). Samsø Island won the competition in October 1997, and several projects to make Samsø a 100 per cent renewable energy island began.

General information about Samsø

The location of Samsø is shown in Plate 4.5.1. It is 26 km long and 7 km wide at its maximum width, and has an area of 114 km^2. The population was 4,100 in 2008. Agriculture is the primary business sector, and many vegetables and fruits are exported from the island. Tourism is the second main business sector, with almost half a million tourists staying overnight each year.

Energy status

After the renewable energy project was initiated in Samsø in 1997, the energy sources for consumption on the island changed drastically. The share of total heat production produced by renewable energy increased from about 25 per cent in 1997–1999 to about 65 per cent in 2005. In 1997, only 5 per cent of the island's electricity was derived from wind turbines on the island and almost all of the island's electricity was imported from the mainland's coal-based power plants (Jakobsen, 2008). Wind turbines in Samsø now provide 100 per cent of its electricity.

To summarize, renewable energy, which accounted for only 13 per cent of Samsø's energy supply in 1997, increased to 99.6 per cent in 2005, and Samsø is now regarded as the Danish renewable energy island. Details of the renewable energy project are given in the following sections.

4-5-2 Renewable energy island project in Samsø

District heating systems

To increase renewable energy consumption for heating, four district heating stations were constructed in which straw and wood chips produced by local farmers are incinerated. The main reasons for connecting to district heating stations are the rational use of energy, relatively simpler functioning and maintenance compared with individual heating solutions, and savings on heating bills. When the project was initiated, existing homes could link to the district heating system on a voluntary basis. Only new buildings in areas with existing or planned district heating were obliged to connect to the district heating system. In Denmark, a high registration fee, typically about DKK 36,000 (1 DKK = €0.13 in 2009), is normally paid when the district heating system is established, as well as when connecting to existing district heating systems. In this project, the registration fee for district heating systems before constructing the plants was set at only DKK 80 to encourage locals to join the district heating systems. A consequence of using this model is high heating prices, because the repayment of investment costs is added to the price of the heat delivered. However, the price is still reasonable compared with the high costs of heating by oil or electricity.

Tranebjerg District Heating Station, which provides more than 90 per cent of Tranebjerg's heat supply, began operations in 1994 using straw for producing heat. This station is owned by NRGi, a local utility company, and the initial investment cost was DKK 26.3 million. The project was not financially supported by the government. Nordby District Heating Station, which provides about 80 per cent of the heating for buildings in the area, began operations in 2002, with 80 per cent of its heat supply from burning wood chips and 20 per cent from a 2,500 m^2 solar heating system. This station is also owned by NRGi, and the initial investment cost was DKK 20.5 million. This investment was partly subsidized with a grant of about DKK 9 million. Onsbjerg District Heating Station began operations in 2003, and supplies about 80 buildings with heat generated by burning straw. Kremmer Jensen ApS owns this plant and the initial investment was DKK 8.5 million. A grant of DKK 3 million was paid by the Danish Energy Authority. Brundby District Heating Station began operations in 2004 and supplies 232 buildings with heat generated by burning straw. This station is owned by consumers in a limited liability association. The capital costs were DKK 16.2 million, and the project received a grant of DKK 2.5 million from the Danish Energy Authority. In total, DKK 71.5 million was initially invested for the four district heating systems, DKK 14.6 million of which was partly subsidized.

Individual renewable systems

Approximately 1,200 homes in rural areas where houses are a long way from the district heating system are dependent on individual systems for space and water heating. Several energy campaigns, which included energy exhibitions and professional advice about energy solutions, were carried out in these areas to increase the use of individual renewable energy systems instead of electricity for heating. As a result of these campaigns, the renewable energy systems installed since 1998 include about 100 solar heating systems, 120 biomass-based burners and 35 heat pump systems. About 15 per cent of homes in rural areas are now supplied with energy entirely from renewable energy sources. Approximately DKK 15 million was invested in individual renewable energy production units, with a subsidy of about DKK 3 million by the Danish Energy Authority. These campaigns resulted in an increase in the share of total heat production generated by renewable energy to about 65 per cent in 2005 from about 25 per cent in 1997–1999.

Land-based wind turbines

To make Samsø self-sufficient in electricity, a project to erect land-based wind turbines was initiated in 1998 (Plate 4.5.2). There was no shortage of investment because farmers with potential wind turbine sites on their land were keen to invest in their own wind turbines. Several public meetings were held to make the project open to local residents and to generate positive interest in investing in the wind turbines in this project. A

Plate 4.5.2 Land-based wind turbines

collective ownership system was also introduced to facilitate implementation and secure broad public support. The idea of collective ownership was based on reserving shares for the general public. In 2000, 11 wind turbines of 1 MW were erected in three areas: Tanderup, Permelille and Brundby. About DKK 66 million was invested in these wind turbines. Of the 11 wind turbines, 2 are collectively owned (divided into 5,400 shares and owned by 450 investors) and the other 9 are individually owned by farmers. The land-based turbines produced 100 terajoules (TJ) in 2005, which was equal to Samsø's electricity consumption for that year.

Offshore wind turbines

There was no available technology to meet the energy needs of the transport sector with renewable energy. Thus, erecting offshore wind turbines was planned to compensate for energy consumption in the transport sector on the island. In 2002, 10 2.3 MW offshore wind turbines were constructed at a cost of DKK 250 million 3.5 km south of Samsø along the Paludan Flak reef (Plate 4.5.3). The municipality of Samsø owns five of these turbines and large investors own three of them. The other two are cooperatively owned by about 1,500 small shareholders. The collective ownership of the wind turbines led to considerable local influence over the project. All residents of Samsø were offered ownership. The offshore turbines cost DKK 10.4 million per MW, which was much higher than the cost of DKK 6 million per MW for land-based turbines. However, off-

Plate 4.5.3 Offshore wind turbines (*Source*: Samsø Energy Academy)

shore turbines can produce 3,500 MWh per MW installed capacity, whereas land-based turbines can generate only 2,300 MWh. The offshore turbines produced 285 TJ in 2005, which was more than sufficient to compensate for the energy consumption from the transportation sector, which was 210 TJ for that year.

4-5-3 Overview of the renewable energy project in Samsø

Renewable energy share of Samsø

Heat and electricity produced by renewable energy successfully increased after initiating the renewable energy island project. The proportion of total heat produced by renewable energy increased from about 25 per cent in 1997–1999 to about 65 per cent in 2005. As a result, the consumption of fuel oil for heating purposes decreased significantly from 133 TJ in 1997 to 74 TJ in 2005. In 1997, wind turbines on the island produced only 5 per cent of Samsø's electricity. Today, the amount of electricity produced by land-based and offshore wind turbines greatly exceeds the electricity consumption of the island. In 2005, 386 TJ was produced by the wind turbines and 286 TJ was exported from the island. Consequently, 99.6 per cent self-sufficiency in renewable energy was achieved in Samsø with an average investment of about DKK 100,000 per capita and public subsidies of about DKK 7,000–8,000 per capita.

The change in energy resources on the island is shown in Figure 4.5.2. Fuel oil consumption decreased by 15 per cent from 1997 to 2005, mainly because of an approximately 50 per cent decrease in oil consumption for heating purposes. An increase in the consumption of renewable energy from sources such as straw, wood pellets and solar heat was achieved by the four district heating systems and individual renewable energy systems after initiating the renewable energy project. The amount of electricity imported decreased drastically after 2001, and imported electricity was replaced by electricity produced by the land-based and offshore wind turbines.

Environment

Emissions of carbon dioxide (CO_2), nitrogen oxides (NOx), sulphur dioxide (SO_2) and airborne particles were significantly reduced by the energy transitions achieved by the renewable energy project. These reductions were mainly due to electricity production from wind turbines. In 1997, emissions of CO_2, NOx and SO_2 per Samsø resident were 11 tons, 80 kg and 20 kg, respectively. In 2005, these emissions had decreased to −4 tons,

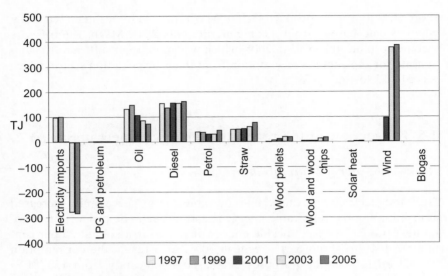

Figure 4.5.2 The change in energy resources on Samsø.
Source: Samsø Energy Academy.

−2 kg and −1 kg, respectively, by replacing electricity based on fossil fuel with wind-based electricity.

Tourism

Samsø is a popular tourist destination because of its beautiful scenery (there are about 500,000 overnight bookings annually). Gamboa and Munda (2007) indicated that factors in local conflicts and opposition to the installation of wind turbines are the visual impact and fear of the effects on tourism. This was also the case in Samsø. However, for Samsø, a new type of tourism – ecotourism – was developed by the renewable energy project. Politicians, business executives, grassroots organizations, media representatives and students began visiting Samsø from all over the world to learn about the renewable energy island (Jørgensen et al., 2007). In 2001, about 1,000 people with a professional background in renewable energy visited Samsø.

4-5-4 Factors contributing to the success of Samsø's renewable energy project

Several factors contributed to making Samsø a 100 per cent renewable energy island: subsidies, local involvement, and the characteristics of Samsø as an island.

Subsidies

Direct public subsidies for the district heating systems, offshore wind turbines and private renewable energy systems accounted for DKK 30 million. It is believed that these subsidies contributed greatly to the increase in the proportion of renewable energy consumption in Samsø. Furthermore, a major economic incentive for renewable energy takes the form of support for electricity production (Law No. 944 of 27 December 1991), which is the only scheme providing a subsidy for running costs in Denmark (Odgaard, 2000). Under this law, a subsidy of DKK 0.27 per kWh was provided until 1999. Because of rapid technological developments, this subsidy proved to be enormous. Thus, from 2000, green certificates with a value of DKK 0.10–0.27 per kWh, which is set by the market, gradually replaced the fixed electricity production subsidy for all new renewable energy plants and existing wind turbines (Odgaard, 2000). The production costs for offshore and land-based electricity account for DKK 0.35–0.38 and DKK 0.30 per kWh, respectively. This is more than the production cost of coal-based electricity from newly built power plants, which is DKK 0.25 per kWh (Odgaard, 2000). Therefore, the running cost subsidy was essential for the erection of wind turbines in Samsø.

Local involvement

Many studies have reported strong opposition to local wind energy projects, although there is good public support for wind energy (Bell et al., 2005; Gross, 2007; Thele, 2008; Wolsink, 2005). The commonest explanation for local opposition is the not-in-my-backyard (NIMBY) syndrome. In Denmark, however, local ownership through cooperatives seems to diminish local opposition by compensating for the disadvantages (Hvelplund, 2005). About 80 per cent of the installed wind energy capacity in Denmark is owned by local individuals and cooperatives (Krohn, 2002; Loring, 2006). In Samsø, local involvement also seemed to play a major role in obtaining local acceptance of the renewable energy project.

During the 10-year period in which the renewable energy project was implemented in Samsø, many public meetings were held, which were open to all local residents (Jørgensen et al., 2007). Thus, residents had a lot of opportunities to participate in the discussion and planning phases, and the developers took their opinions into account (Jakobsen, 2008). In addition, the opportunity to invest in cooperatives resolved the debate regarding wind turbines, because the possibility of local investment and ownership gave the citizens of Samsø an opportunity to understand the presence of wind turbines (Jørgensen et al., 2007). The municipality of Samsø owns five of the offshore turbines. This indicates that all citizens own wind turbines and can earn money. Hermansen, leader of the Samsø Energy Academy, stated: "If the project was owned by a big private company, I think people would have been angry and have had a bad feeling. I believe that local participation and local ownership are completely decisive for the success of the project" (Jakobsen, 2008).

The characteristics of Samsø as an island

The characteristics of Samsø Island seemed to be the factors contributing to the success of the renewable energy project. First, Samsø's energy consumption of 130 MJ/capita is 19 per cent lower than that of Denmark. Second, Samsø's land resources of 2.8 ha/capita, which can be used to produce biomass, are more than three times larger than those of Denmark. Samsø's wind resources per inhabitant are also relatively high. These conditions in Samsø seem to have made it less difficult to achieve 100 per cent self-sufficiency with renewable energy. In addition, the fact that Samsø's local economy had been declining for many years before the renewable energy project was initiated seemed to motivate islanders to make it a renewable energy island, which would create good jobs and other opportunities.

4-5-5 Future challenges

The main purpose of the renewable energy project in Samsø was to make it a 100 per cent renewable energy island. The renewable energy supply met 99.6 per cent of demand in 2005 and therefore the objective of the project was fulfilled. Furthermore, implementing the project resulted in the development of ecotourism on the island. Today, Samsø is known as the Danish renewable energy island.

Although this renewable energy project was a success, two challenges to the achievement of a sustainable society were apparent. The targets of the project included a 25 per cent reduction in demand for heat and a 15 per cent reduction in electricity consumption for heating purposes. Several campaigns were conducted to achieve these targets, and approximately DKK 15 million was invested for energy conservation measures such as extra home insulation and window replacement. However, total heat consumption in Samsø in 2005 was 10 per cent higher than that in 1997, despite a 5 per cent decrease in the island's population. In addition, total electricity consumption did not change between 1997 and 2005. The trend to build larger houses and the increasing use of electrical products seemed to nullify conservation measures. This indicates the difficulty in changing human habits regarding energy savings.

In addition, it should be noted that, even though Samsø produces much more electricity annually than it consumes, it still needs backup from conventional power plants to act as a buffer against the instability of wind-based electricity. Thus, Samsø still relies on non-renewable energies to some extent. Therefore, further challenges, such as linking up with pumped-storage hydroelectricity or hydrogen plants, will need to be faced in order to achieve a completely sustainable society.

However, the renewable energy project in Samsø was a fruitful demonstration of sustainability, which focused on combining technology, economy, environment and social management. The importance of local ownership and involvement throughout the planning process should be highlighted as essential factors in the success of the renewable energy project in Samsø.

REFERENCES

Bell, D., T. Gray and C. Haggett (2005) "The 'Social Gap' in Wind Farm Siting Decisions: Explanations and Policy Responses", *Environmental Politics* 14(4): 460–477.
Beuse, E., J. Boldt, P. Maegaard, N. I. Meyer, J. Windeleff and I. Østergaard (2000) "Vedvarende Energi i Danmark. En Krønike om 25 Opvækstår 1975–2000", Organisasjonen for Vedvarende Energi.
Danish Energy Agency (2008) "Energy Statistics 2007", available at: <http://ens.dk/en-US/Sider/forside.aspx> (accessed 11 March 2010).
Danish Energy Authority (2007) "A Visionary Danish Energy Policy 2025", available at: <http://ens.dk/en-US/Sider/forside.aspx> (accessed 11 March 2010).
Gamboa, G. and G. Munda (2007) "The Problem of Wind Farm Location: A Social Multi-criteria Evaluation Framework", *Energy Policy* 35: 1564–1583.
Gross, C. (2007) "Community Perspectives of Wind Energy in Australia: The Application of a Justice and Community Fairness Framework to Increase Social Acceptance", *Energy Policy* 35: 2727–2736.
Hvelplund, F. (2005) "Renewable Energy Policy in Denmark", in Danyel Reiche (ed.), *Handbook of Renewable Energies in the European Union*. Frankfurt am Main: Peter Lang Publishing.
Jakobsen, I. (2008) "The Road to Renewables: A Case Study of Wind Energy, Local Ownership and Social Acceptance at Samsø", Master's thesis, Faculty of Social Science, University of Oslo.
Jørgensen, P. J., S. Hermansen, A. Johnsen and J. P. Nielsen (2007) "Samsø a Renewable Energy-Island: 10 Years of Development and Evaluation", PlanEnergi and Samsø Energy Academy.
Krohn, S. (2002) "Danish Wind Turbines: An Industrial Success Story", Danish Wind Industry Association.
Loring, J. M. (2006) "Wind Energy Planning in England, Wales and Denmark: Factors Influencing Project Success", *Energy Policy* 35: 2648–2660.

Lunden, M., ed. (2003) "The Whisper of Wings", Elverkongensdatter.
Odgaard, O. (2000) "Renewable Energy in Denmark", Danish Energy Agency, Copenhagen. Available at: <http://www.agores.org/Publications/EnR/Denmark%20REPolicy2000%20update.pdf>.
Thele, F. (2008) "Vindkraft i Motvind – kontroversen om Havsul-prosjektene", in J. Hanson and O. Wicken (eds), *Rik på Natur: Innovasjon i en Ressursbasert Kunnskapsøkonomi*. Bergen: Fagbokforlaget Vigmostad & Bjørke AS.
Wolsink, M. (2005) "Wind Power Implementation: The Nature of Public Attitudes: Equity and Fairness Instead of 'Backyard Motives'", *Renewable and Sustainable Energy Reviews* 11: 1188–1207.

4-6

Self-sustaining models in India: Biofuels, eco-cities, eco-villages and urban agriculture for a low-carbon future

B. Mohan Kumar

India, home to over 1 billion people, has a population growth rate of 1.93 per cent per annum, well above the global average (Census of India, 2001). The population has nearly tripled in the past 50 years, from 361 million in 1951 to 1.027 billion in 2001, during which rapid strides were made in agriculture and industry. The country's economy has also been growing rapidly, with real growth rates of gross domestic product remaining consistently more than 5 per cent in the past decade. Despite the recent global economic downturn, the Indian economy grew by 8.7 per cent between 2007 and 2008 at 1999–2000 prices (MOF, 2007/8). Rising atmospheric CO_2 levels, however, are a major drawback of the ever-increasing fossil fuel consumption associated with India's rapid economic growth. According to the *World Energy Outlook* (IEA, 2002), India is already the world's fifth-largest emitter – at 982 Tg CO_2 (Tg = teragram = 10^{12} g) – and its per capita CO_2 emissions are projected to increase to 1.6 Mg (megagram = 10^6 g) by 2030. There have been a few attempts to utilize renewable energy resources (biofuels and biomass) and to develop sustainable urban and rural environments (eco-cities and eco-villages) in India. This section gives an overview of these developments, but will exclude exploratory efforts to tap solar, wind and wave energy, which are beyond the scope of this review.

4-6-1 Biofuels

In the quest for a low-carbon future, substituting fossil energy and cutting down on the use of resources by living parsimoniously and self-sufficiently

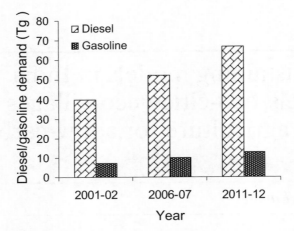

Figure 4.6.1 Potential demand for diesel and gasoline in India.
Source: Planning Commission (2003).

are paramount. Although bioenergy can make significant contributions to the world's growing needs for clean energy (Turkenburg et al., 2000), this is as yet an emerging concept in India. The Government of India (GOI), however, regards biofuels as a feasible option for augmenting future fuel supply (PSA, 2006). The long-term expectation is that India will be a producer of ethanol and a net importer of biodiesel. To promote the utilization of biofuels in the fuel mix, GOI on 30 September 2003 launched a 5 per cent ethanol doping programme for petrol in nine states (Andhra Pradesh, Goa, Gujarat, Haryana, Karnataka, Maharashtra, Punjab, Tamil Nadu and Uttar Pradesh) and four union territories (Daman and Diu, Dadra and Nagar Haveli, Chandigarh and Pondicherry). The National Mission on Biodiesel, covering an area of 400,000 hectares (ha) (Planning Commission, 2003), is another major initiative to find a renewable alternative for increasing fuel consumption (Figure 4.6.1).

On 11 September 2008, GOI also issued a National Policy on Biofuels, which, among other things, proposes to set up a National Biofuel Coordination Committee and Biofuel Steering Committee for biofuel development (Ministry of New & Renewable Energy, 2008). The salient features of this policy are as follows: an indicative target of 20 per cent blending of bioethanol and biodiesel by 2017; augmenting indigenous production of non-edible oilseeds on waste/degraded/marginal land; establishing biodiesel plantations on wasteland; prescribing a minimum support price for biodiesel oilseeds and a minimum purchase price for bioethanol; bringing biodiesel and bioethanol under the ambit of "Declared Goods" to ease their transportation within and across states; and a "tax holiday" on biofuels.

As part of national efforts to promote a biofuel mix, Indian Railways has evaluated various biodiesel blends (5, 10 and 20 per cent) for running diesel locomotives. Results not only indicate reduced emissions but also suggest energy security through low consumption of fossil fuels. Indications are that even partial replacement of the approximately 2 billion litres of diesel fuel per annum (approximately 5,000 diesel locomotives) by biodiesel would substantially reduce fossil fuel consumption (PSA, 2006). Indian Railways is also in a unique position to vertically integrate all activities relating to production, processing and utilization of biodiesel. For example, apart from utilizing biodiesel for running locomotives, it has earmarked spare land (approximately 700 ha) for producing oil crops such as *Jatropha curcas* (jatropha), is collecting surplus oilseeds from the state forest departments and is setting up a 400 Mg per year transesterification plant (PSA, 2006).

4-6-2 Energy farming in India

Energy farming or "petro-farming" is a sensible option to ameliorate the chronic energy deficiency that India experiences. Tree-borne oilseed crops (TBOs) are particularly important in this respect. A wide spectrum of hydrocarbon-yielding plants such as *J. curcas, Pongamia pinnata, P. glabra, Hevea braziliensis, Madhuca indica, M. longifolia, Calophyllum inophyllum, Salvadora persica* and *S. oleoides* is available (PSA, 2006; Kalita, 2008). The predominantly tropical climate and extensive land resources – with over 55 million ha of degraded land (Table 4.6.1) – are additional factors that favour TBO production in India. Table 4.6.2 illustrates the land requirements under different scenarios of biodiesel blending. Many TBOs can tolerate abiotic stresses, implying their suitability for degraded sites, including moderately saline soils, where other crops seldom perform.

Interest in using jatropha as a feedstock for biodiesel production has been growing around the world in recent years (Achten et al., 2008) and there is a special focus on this crop in India (PSA, 2006). Predisposing factors in this respect are the relatively high total oil content of jatropha seeds (~32 per cent), with 97.6 per cent neutral lipids, 0.95 per cent glycolipids and 1.45 per cent phospholipids (Rao et al., 2009), and a favourable fatty-acid composition (e.g. triglyceride alkyl chains of the oil predominantly contain palmitic, oleic and linoleic acids; Tapanes et al., 2008) with low phospholipid content, which aid its processing into biodiesel. In addition, the press cake derived from the oil extraction process is a good fertilizer material and the organic wastes can be digested to produce biogas (CH_4). These attributes have persuaded investors,

Table 4.6.1 Various categories of wasteland in India, 2003

Category	Wasteland (million ha)	Per cent of total geographical area covered
Gullied and/or ravinous land	1.90	0.60
Land with or without scrub	18.79	5.94
Waterlogged and marshy land	0.97	0.31
Land affected by salinity/alkalinity – coastal/inland	1.20	0.38
Shifting cultivation area	1.88	0.59
Underutilized/degraded notified forest land	12.66	4.00
Degraded pastures/grazing land	1.93	0.61
Degraded land under plantation	0.21	0.07
Sands: inland/coastal	3.40	1.07
Mining/industrial wasteland	0.20	0.06
Barren rocky/stony waste/sheet rock area	5.77	1.82
Steep sloping area	0.91	0.29
Snow-covered and/or glacial area	5.43	1.72
Total wasteland area	55.27	17.45

Source: DOLR (2003).

Table 4.6.2 Potential demand for biodiesel in India at different blending rates and related land requirements

Year	Biodiesel @ 5%		Biodiesel @ 10%		Biodiesel @ 20%	
	Demand (Tg)	Land (million ha)	Demand (Tg)	Land (million ha)	Demand (Tg)	Land (million ha)
2005/6	2.48	2.07	4.96	4.14	9.91	8.28
2006/7	2.62	2.19	5.23	4.38	10.47	8.76
2011/12	3.35	2.79	6.69	5.58	13.38	11.19

Source: Planning Commission (2003).

policymakers and clean development mechanism (CDM) project developers to consider it as a substitute for fossil fuels and as a mechanism for emission abatement. It has been estimated that 10 million ha of jatropha plantations could generate 7.5 Tg of fuel annually, which would provide year-round employment for 5 million people (PSA, 2006). However, large variations in dry seed yield (1,172 to 6,700 kg ha^{-1} yr^{-1}; Achten et al., 2008), owing to biotic and abiotic factors, is a major constraint in expanding jatropha cultivation.

Indigenous species such as *sal* (*Shorea robusta*), *neem* (*Azadirachta indica*), *undi* (*Calophyllum inophyllum*), *mahua* (*Madhuca indica*), *karanja* (*Pongamia pinnata*), *Michelia champaca* and *Garcinia indica* yield suitable-grade oil on an economic scale. Annual production of such oilseeds in India is more than 20 Tg, with *mahua* alone accounting for 181 Gg (gigagram = 10^9 g; Kaul et al., 2003). Some of these seeds have a high oil content: *M. champaca* and *G. indica* yield 45.0 per cent and 45.5 per cent oil, respectively. Fatty-acid composition, iodine value and cetane number also indicate their suitability for use as biodiesel (Hosamani et al., 2009). However, such indigenous TBOs have not been adequately exploited in India (Ghadge and Raheman, 2005). Exotic species such as *Copaifera langsdorfii*, *C. multijuga* (yielding 40–60 L of oil per tree in a single year) and *Euphorbia* spp. (*E. lathyris* and *E. antisyphilitica*; "gasoline trees") too hold promise. But, despite their high bio-crude potential, they are as yet under-exploited in India.

As TBOs and other hydrocarbon-yielding plants evolve further as substitutes for fossil fuel, attempts are under way to improve their agronomic yields, develop small-scale extraction plants, evaluate chemical composition and evolve possible ways of modifying the biosynthetic routes to produce more desirable end products (Kalita, 2008). The resultant higher productivity and profitability may engineer shifts in land-use patterns favouring energy crops. However, this biofuel route to CO_2 emission reduction is not always risk free, especially in populous countries such as India where extensive replacement of food crops by energy crops may adversely affect food availability, access, stability and utilization (FAO, 2008).

4-6-3 Bioethanol

Interest in bioethanol (that is, ethanol produced using biomass for feedstock) is also rising in India (e.g. 5 per cent blending in petrol). The country's projected demand and availability of ethanol (2001–2017) for petrol blending are shown in Table 4.6.3. Ethanol availability is currently sufficient to meet demand, even up to 10 per cent (11th Five-Year Plan target, 2007–2012). Most of the ethanol in India is extracted from molasses produced in the sugar-making process (PSA, 2006). Being the second-largest producer of sugarcane (*Saccharum officinarum*) after Brazil, this is not surprising (Singh, 2000).

Lignocellulosic biomass (crop residues and "purpose-grown" woody biomass) is another potential feedstock for bioethanol production, albeit not adequately exploited in India. Hydrolysis of cellulosic and lignocellulosic biomass feedstocks can open the way towards low-cost and efficient

Table 4.6.3 Ethanol demand and availability for gasoline blending in India

Year	Ethanol demand 5% blending (million L)	Ethanol availability (million L)
2001/2	416.14	537
2006/7	592.72	2309
2011/12	756.35	2054
2016/17	965.30	1754

Source: Planning Commission (2003).
Note: Ethanol availability for blending is the difference between total production and utilization for industrial and portable uses.

production of ethanol. The total crop residue production during 1996/7 in India was estimated to be 626 Tg of air-dry weight (Ravindranath et al., 2005) – rice, wheat, sugarcane and cotton accounting for approximately 66 per cent. Municipal solid waste material (MSW) is yet another source of residues. Most Indian cities generate considerable quantities of waste (0.1–0.62 kg/capita/day; Kumar et al., 2009). However, inadequate human resources, financial resources, implements and machinery required for effectively handling MSW hamper its utilization. The conversion of cellulosic and lignocellulosic biomass feedstocks is also more difficult than the conversion of sugar and starch. This is because lignocellulosic materials require pretreatment by mechanical and physical means (e.g. steam) to clean and size the biomass and to destroy its cell structure to make it accessible to further chemical or biological treatment.

4-6-4 Eco-cities, eco-villages and biomass towns to promote materials and energy circulation

Evolution of the eco-city concept in India

In many countries, national and local governments have initiated efforts to reduce the carbon footprint of the built environment through recycling of urban wastes. The "biomass town", a concept focusing on the centralized utilization of waste biomass in cities, towns and villages in Japan (MAFF, 2006), is a prominent example of this. "Eco-city" is another urban development model that aims at complete recycling of urban wastes. Such programmes generally advocate balancing the economic, environmental and social dimensions of development, and aim at improving environments at local and regional scales.

Although well known elsewhere in the world, eco-cities are a relatively new concept in India, albeit sustainability and frugal lifestyles are ingrained in traditional Indian culture (Kumar, 2008). However, certain initiatives, consistent with international efforts, were made in India for a sustainable "urban humanosphere" (the ecological and social environment in which people live). The Zoning Atlas Initiative of the Central Pollution Control Board (CPCB) launched in 2001 for the siting of industries (CPCB, 2001) and the Zero Garbage Town (ZGT) concept implemented in Namakkal municipality in Tamil Nadu (NRI-India, n.d.) are forerunners of this. In particular, the Namakkal project, formulated by the National Productivity Council, aims to create an eco-friendly municipality incorporating futuristic waste management concepts. Central to the project are poverty alleviation and the empowerment of women through the recovery, recycling and re-use of wastes, besides capacity-building in the area of green productivity.

Drawing on the experiences of ZGT and similar other initiatives, an eco-city programme (ECP) was launched by CPCB during the 10th Five-Year Plan (2002/3 to 2006/7) with technical support from the German Technical Cooperation. It was conceptualized to improve the environment of small- and medium-sized towns, and to achieve sustainable development through a comprehensive urban improvement system employing practical, innovative and non-conventional solutions and delivering visible environmental improvement (CPCB, 2005). During the first phase, the programme was operationalized in six towns: Kottayam–Kumarakom (Kerala), Ujjain (Madha Pradesh), Vrindavan (Uttar Pradesh), Tirupati (Andhra Pradesh), Puri (Orissa) and Thanjavur (Tamil Nadu). The project was implemented by CPCB in partnership with the respective State Pollution Control Boards and municipalities, on a 50:50 cost-sharing basis.

Surjan and Shaw (2008) evaluated the ECP at Puri and found that there were distinct improvements in public and private goods ("amenity" and "productivity"), greater availability of investment subsidies, the coming into effect of recycling-oriented legislation, access to technological resources from the private sector, and widespread recognition of the urgency of acting on environmental issues. There were also certain shortcomings; for example, spatial development led by the master plan did not receive recognition in the political and social arena nor was it fully implemented. Also, most decisions relating to land use, housing and infrastructure were ad hoc and subject to political interference. Despite such pitfalls, ECP is regarded as one step forward in the transformation of a messy urban waste disposal programme to a more organized sustainable development initiative. Efforts are also being made to replicate it in other cities (e.g. the Taj Eco-City Project, Uttar Pradesh, and the pilot phase Eco-Industrial Estate projects at Vishakhapatnam, Nacharam and Mallapur in Andhra Pradesh; CPCB, 2005).

Eco-villages

Whereas the eco-city programme focuses on established towns and cities, eco-villages are *de novo* entities that integrate the ecological, economic, social, cultural and spiritual sustainability of urban or rural areas. Harmony with nature in human interactions and social structure and preventing environmental deterioration are fundamental philosophies of an eco-village. Basic to it is cross-cultural adaptation of diverse people living and working together with a common vision of building a more sustainable future, as well as developing and testing different models of economy, energy and cooperative living. Simply put, an eco-village is an agglomeration (rural or urban) aiming for self-sufficiency, and where the priority is to place humans and the environment at the centre of its interests. To achieve this, the focus is on aspects such as permaculture (designing human settlements and agricultural systems that mimic nature), ecological building, green production, renewable energy and self-sufficient agriculture (Siracusa et al., 2008).

Auroville, Puducherry: the "City of Dawn"

Auroville, established on 28 February 1968, is a well-known eco-village in India.[1] It is situated on the Coromandel Coast of south India, and draws inspiration from the vision and work of the renowned seer and spiritual visionary Sri Aurobindo. Auroville is a unit of the Global Ecovillage Network, which is an apex confederation of people and communities dedicated to living "sustainable-plus" lives.[2] Auroville is envisaged as a city for up to 50,000 inhabitants from around the world, and today it has 2,000 inhabitants from some 30 countries. They live in about 100 clusters of varying sizes, separated by village and temple lands and surrounded by Tamil villages.

In 1988, GOI passed the Auroville Foundation Act to safeguard the development according to its charter. The principal activities of the Foundation are afforestation, organic agriculture, educational research, healthcare, village development, town planning, groundwater management, cultural activities and community services. The wasteland reclamation programmes implemented at Auroville, in which more than 1,000 ha of degraded land have been "re-greened" by planting over 2 million multipurpose tree saplings, were particularly successful and have attracted national and international acclaim. Soil and water conservation measures such as contour risers and mini-check dams have significantly enhanced the life-support potential of this bioregion. Auroville also aims to revive local health traditions and ancient Indian medical systems such as *Ayurveda* and *Siddha*. For this purpose, an ethno-medicinal forest (to conserve medicinal plants), an outreach nursery (for propagation and distribution) and a Bio-Resources Centre (for education, training and research in the use of locally available medicinal plants in primary healthcare) have been established. Health education, preventive care and treatment of diseases, empowerment of women, and providing education to village children, as well as encouraging the growth of community spirit and a sense of self-confidence through social initiatives, micro-projects and awareness campaigns, are also being focused on at Auroville.

4-6-5 Urban organic agriculture to aid organic recycling in eco-cities

Urban agriculture integrates various practices such as soil-less cultures (hydroponics), substrate cultures, container gardens and home gardens. Two promising special forms of urban and peri-urban agriculture are allotment gardens (land parcels allocated to individuals or households for personal use through government or private enterprises, and organized into self-governing associations) and community gardens (maintained by a group of individuals or households who produce agricultural goods collectively for their own consumption). Rooftop gardening is also becoming increasingly popular in many densely populated cities of India. Apart from providing nutritious food to the community, such gardens offer an array of ecological benefits (Drescher et al., 2006) such as clean air, organic recycling, ameliorating microclimates and conserving urban water resources. These gardens would help in the successful implementation of the eco-city's integrated solid waste management programme. Segregated biodegradable wastes from households can be delivered to the allotment gardens where they are converted into compost. However, there is a need for mainstreaming such concepts into overall city planning and development under ECP.

4-6-6 Conclusions

Triggered by favourable national policy and growing concerns about fossil energy sources, a new bioenergy industry is emerging in India. Within the bioenergy industry, bioethanol and biodiesel take an important position, whereas agricultural residues and woody biomass are underexploited. In recent years, the solid biomass sector has shown specific adaptations. The production of bioenergy from specialized bioenergy crops, however, should be limited to surplus land or land unsuitable for agriculture. Exploratory efforts under the eco-city programme have shown promising results, and urban and peri-urban agriculture, if promoted, may further augment organic waste utilization.

Notes

1. See the Auroville website at: <http://www.auroville.org/> (accessed 12 March 2010).
2. See the Global Ecovillage Network website at: <http://gen.ecovillage.org/about/index.html> (accessed 12 March 2010).

REFERENCES

Achten, W. M. J., L. Verchot, Y. J. Franken, E. Mathijs, V. P. Singh, R. Aerts and B. Muys (2008) "Jatropha Bio-diesel Production and Use", *Biomass and Bioenergy* 32: 1063–1084.

Census of India (2001) "2001 Census", available at: <http://www.censusindia.net/> (accessed 12 March 2010).

CPCB [Central Pollution Control Board] (2001) "Highlight 2001: Environmental Planning and Mapping", <http://www.cpcb.nic.in/oldwebsite/Highlights/Highlights01/ch-13.html> (accessed 12 March 2010).

CPCB [Central Pollution Control Board] (2005) "Eco-City Program", *Annual Report 2004–2005*, New Delhi. Available at: <http://www.cpcb.nic.in/oldwebsite/Eco-city%20Program/Eco_Industrial_Estates.html> (accessed 12 March 2010).

DOLR [Department of Land Resources] (2003) "Category-wise Wastelands of India", Ministry of Rural Development, Government of India. Available at: <http://dolr.nic.in/fwastecatg.htm> (accessed 12 March 2010).

Drescher, A. W., R. J. Holmer and D. L. Iaquinta (2006) "Urban Homegardens and Allotment Gardens for Sustainable Livelihoods: Management Strategies and Institutional Environments", in B. M. Kumar and P. K. R. Nair (eds), *Tropical Homegardens: A Time-Tested Example of Sustainable Agroforestry*. Dordrecht: Springer Science, pp. 317–338.

FAO [Food and Agriculture Organization of the United Nations] (2008) "Bioenergy, Food Security and Sustainability – Towards an International Framework", High-Level Conference on World Food Security: The Challenges of Climate Change and Bioenergy, Rome, 3–5 June 2008, HLC/08/INF/3. Available at: <http://www.fao.org/fileadmin/user_upload/foodclimate/HLCdocs/HLC08-inf-3-E.pdf> (accessed 12 March 2010).

Ghadge, Shashikant V. and Hifjur Raheman (2005) "Biodiesel Production from Mahua (*Madhuca indica*) Oil Having High Free Fatty Acids", *Biomass and Bioenergy* 28: 601–605.

Hosamani, K. M., V. B. Hiremath and R. S. Keri (2009) "Renewable Energy Sources from *Michelia champaca* and *Garcinia indica* Seed Oils: A Rich Source of Oil", *Biomass and Bioenergy* 33: 267–270.

IEA [International Energy Agency] (2002) *World Energy Outlook 2002*, 2nd edn. Paris: International Energy Agency.

Kalita, Dipul (2008) "Hydrocarbon Plant – New Source of Energy for Future", *Renewable and Sustainable Energy Reviews* 12: 455–471.

Kaul, S., A. Kumar, A. K. Bhatnagar, H. B. Goyal and A. K. Gupta (2003) "Biodiesel: A Clean and Sustainable Fuel for Future", Scientific Strategies for Production of Non-Edible Vegetable Oils for Use as Biofuels, All-India Seminar on National Policy on Non-Edible Oils as Biofuels. Sustainable Transformation of Rural Areas (SuTRA), Indian Institute of Science in Bangalore, unpublished.

Kumar, B. M. (2008) "Forestry in Ancient India: Some Evidences on Productive and Protective Aspects", *Asian Agri-History* 12(4): 299–306.

Kumar, S., J. K. Bhattacharyya, A. N. Vaidya, Tapan Chakrabarti, Sukumar Devotta and A. B. Akolkar (2009) "Assessment of the Status of Municipal Solid Waste Management in Metro Cities, State Capitals, Class I Cities, and Class II Towns in India: An Insight", *Waste Management* 29: 883–895.

MAFF [Ministry of Agriculture, Forestry and Fisheries, Japan] (2006) "Biomass Nippon Strategy", Tokyo (in Japanese). Available at: <http://www.maff.go.jp/j/biomass/pdf/h18_senryaku.pdf> (accessed 12 March 2010).

Ministry of New & Renewable Energy, Government of India (2008) "National Policy on Biofuels", New Delhi. Available at: <http://www.indiaenvironmentportal.org.in/files/biofuel-policy_0.pdf> (accessed 6 April 2010).

MOF [Ministry of Finance, Government of India] (2007/8) *Economic Survey 2007–2008*. New Delhi: Oxford University Press, India. Available at: <http://indiabudget.nic.in/es2007-08/esmain.htm> (accessed 12 March 2010).

NRI-India (n.d.) "Namakkal Municipality – A Zero Garbage City", <http://www.3rkh.net/3rkh/files/02%20NAMAKKAL%20MUNICIPALITY.pdf> (accessed 12 March 2010).

Planning Commission (2003) *Report of the Committee on Development of Biofuel*. New Delhi: Planning Commission, Government of India.

PSA [Principal Scientific Adviser to the Government of India] (2006) "Report of the Working Group on R&D for the Energy Sector for the Formulation of the Eleventh Five-Year Plan (2007–2012)", New Delhi.

Rao, Kotte S., Pradosh P. Chakrabarti, B. V. S. K. Rao and R. B. N. Prasad (2009) "Phospholipid Composition of *Jatropha curcas*: Seed Lipids", *Journal of American Oil Chemists Society* 86: 197–200.

Ravindranath, N. H., H. I. Somashekara, M. S. Nagaraja, P. Sudha, G. Sangeetha, S. C. Bhattacharya and P. Abdul Salam (2005) "Assessment of Sustainable Nonplantation Biomass Resources Potential for Energy in India", *Biomass and Bioenergy* 29: 178–190.

Singh, S. K. (2000) *India Sugar Annual Report 2000*, GAIN Report #IN0019. US Embassy, New Delhi, India.

Siracusa, Giuseppe, Angela D. La Rosa and Paolo Palma E. La Mola (2008) "New Frontiers for Sustainability: Energy Evaluation of an Eco-village", *Development and Sustainability* 10: 845–855.

Surjan, Akhilesh K. and Rajib Shaw (2008) "Eco-city to 'Disaster-Resilient Eco-Community': A Concerted Approach in the Coastal City of Puri, India", *Sustainability Science* 3: 249–265.

Tapanes, Neyda C. O., Donato A. Gomes Aranda, Jose W. De Mesquita Carneiro and Octavio A. Ceva Antunes (2008) "Transesterification of *Jatropha curcas* Oil Glycerides: Theoretical and Experimental Studies of Biodiesel Reaction", *Fuel* 87: 2286–2295.

Turkenburg, Wim C., Jos Beurskens, André Faaij, Peter Fraenkel, Ingvar Fridleifsson, Erik Lysen, David Mills, Jose R. Moreira, Lars J. Nilsson, Anton Schaap and Wim C. Sinke (2000) "Renewable Energy Technologies", in *World Energy Assessment of the United Nations, UNDP, UNDESA/WEC*. New York: United Nations Development Programme, pp. 219–272.

5
Self-sustaining local and regional societies

5-1

Forest biomass for regional energy supply in Austria

Yoshiki Yamagata, Florian Kraxner and Kentaro Aoki

5-1-1 Introduction and background

Forest bioenergy to reduce greenhouse gas emissions

Energy from forest biomass, commonly described as near carbon neutral, can be turned into heat, biofuels and various chemical materials. Renewable energy from biomass will play an essential role in reducing the carbon intensity of energy and decoupling energy use from CO_2 emissions (IEA, 2007). The wood energy sector will be strongly influenced and supported by energy policies aimed at mitigating climate change and diversifying the national energy portfolio to enhance energy security (Olsson et al., 2009). Experiences from the Austrian bioenergy sector, which is largely based on forest biomass, might serve as examples of regional sustainability.

Background on the forest environment in Austria

Austria is a country with abundant wood resources and vast experience in forest management. Of the total land area, 3.96 million hectares (ha), or about 47 per cent, is forested, and the country has an estimated growing stock of 1.1 billion solid m^3 (BFW, 2002). The area of protected forest is 20 per cent. Meanwhile, almost the entire managed forest area is third-party-certified by the Programme for the Endorsement of Forest Certification Schemes and partly by Forest Stewardship Certification,

---

Designing our future: Local perspectives on bioproduction, ecosystems and humanity, Osaki, Braimoh and Nakagami (eds),
United Nations University Press, 2011, ISBN 978-92-808-1183-4

guaranteeing sustainable forest management through the labelling of Austrian wood products (Kraxner and Rametsteiner, 2005).

5-1-2 Overview of major forest bioenergy utilization in Austria

Historical overview of bioenergy development in Austria

Austria is one of the pioneers of modern forest biomass utilization, with a history of market development for energy and fuelwood resources reaching back many centuries. As a forest-rich country, fuelwood has always played a key role in industry, especially salt mining and steel production. The first sustainability regulations for forestry date back several centuries to when massive deforestation for mining and steel-making activities resulted in catastrophic soil erosion, mud slides, flooding and avalanches. In parallel with industrial development, the domestic use of fuelwood became increasingly sophisticated. Iron stoves replaced tile stoves to minimize wood fuel inputs, maximize heat output and increase user convenience (e.g. requiring refills only once a day).

In the early 1950s, forestry-related industries such as large sawmill companies started large-scale renewable heat production using biomass derived mainly from forestry and sawmill residues (Nemestothy, 2006). In the mid-1980s, heating projects of various scales began to supply biomass energy to rural villages and towns. These projects usually began by supplying public buildings such as schools, hospitals, nursing homes or town halls, and institutions such as village centres (Weiss and Rametsteiner, 2005).

In the second step of expanding district heating, increasing numbers of private homes are being connected to extended heating grids for comfort reasons, and because heating with wood is seen as traditional in Austria. The installation of decentralized district heating plants has both co- and cross-benefits, such as steering local economies by paying for thinning products and other previously valueless forest residues, and job and value creation in rural areas. This development helped make vital forest-thinning measures economically feasible for local, small-scale forest owners, which contributed to improving forest quality and health.

Further technological advances in wood pellet heating systems have led to increased biomass utilization in private households since 1995. Biomass-fired combined heat and power (CHP) plant projects for both district heating and electricity production have been introduced throughout Austria since 2003.

Renewable energy use in Austria

Renewable energy meets 25 per cent of Austria's total primary energy demand, of which 59 per cent is derived from biomass, 36 per cent from hydropower, 3 per cent from solar thermal and heat pump power and 2 per cent from wind and photovoltaic energy (2007). The biomass itself – 213 PJ (petajoule = 10^{15}) – is composed of fuelwood (30 per cent), wood chips from sawmill residues/bark (27 per cent), burnable organic waste (14 per cent), black liquor from paper production (12 per cent), wood chips from forest residues such as thinnings and bark (6 per cent), pellets (4 per cent) and other resources such as biofuels, biogas and energy crops (7 per cent). The share of renewable energy in gross electricity consumption was 54.8 per cent in 2007, which placed Austria first among the 27 European Union (EU) countries.

Innovation of biomass heating plants in Austria

The development of biomass district heating took off in Austria during the 1990s, when the country achieved significant increases in biomass heat production. The number of newly established biomass district heating plants has risen constantly since 1980 (see Figure 5.1.1). By 2008, there were over 7,100 biomass heating plants with capacities ranging from 101 kW to 1 MW (megawatt), and about 1,000 biomass systems with capacities larger than 1 MW across Austria (Furtner and Haneder, 2008). In the same year, the cumulative heat capacity of biomass heating plants above 0.1 MW reached around 4.7 GW (gigawatts).

With respect to CHP plants, there were 20 woody biomass systems (electricity production of more than 0.4 MW) in 2002. These biomass-fired CHP plants are used by private pulp and paper industries (using black liquor and bark) and wood-processing companies (AEA, 1999). As shown in Figure 5.1.2, the number of biomass CHP plants has increased remarkably. By 2008, over 100 biomass CHP plants generated about 348 MW of renewable electricity and almost 2 GW of heat (Nemestothy, 2008).

Pellet home-heating systems in Austria

Estimates indicate that more than 90 per cent of all pellet boilers installed in Austria are of domestic origin. Around 14 pellet boiler manufacturers produced about 11,100 units for the domestic market in 2008. The total production of pellet stoves totalled over 50,000 units, of which around 80 per cent were exported during 2008. About 65,000 households

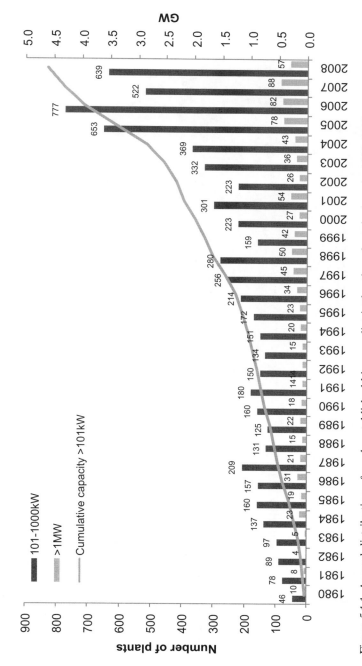

Figure 5.1.1 Annual distribution of newly established biomass district heating plants in Austria, 1980–2008.
Source: Authors' compilation; Furtner and Haneder (2008).
Note: Figure shows woodchip-based heating plants only (including residential buildings, micro-grid systems, district heating plants and industrial plants).

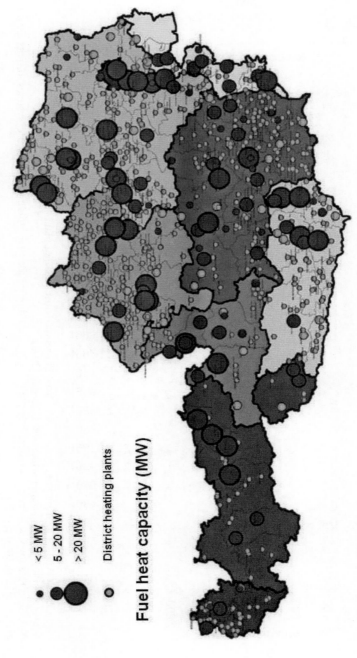

Figure 5.1.2 Location and capacity of biomass district heating and CHP plants in Austria, 2008.
Source: modified from Nemestothy (2008).

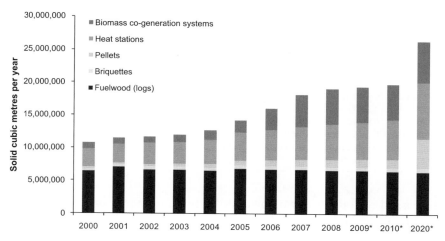

Figure 5.1.3 Sector-wise distribution of wood demand for energy production in Austria.
Source: Authors' compilations; Ministry of Life (2009).
Note: * forecast for 2009, 2010 and 2020.

and industrial companies currently use pellets to heat their homes, offices and workshops.

Wood resources and their allocation

Figure 5.1.3 clearly shows that the share of raw timber used for biomass district heating has been constantly increasing over the last 10 years. Projections even indicate an increase of up to 100 per cent for this particular sector, a common development in the pellet and CHP sectors, based on the annual results seen over the past decade. Conventional biomass on the other hand will stagnate during the same period and will not be able to keep up with the increases in other sectors, based on predictions for coming years. It seems as if other modern forest-based feedstocks will increasingly replace fuelwood by 2020.

5-1-3 Selected examples of district heating systems of different capacities in Austria

This subsection briefly describes three typical Austrian biomass district heating systems operating at varying scales of 0.5 MW to 20 MW and grid lengths of 70 m to 70 km. This analysis is part of a broader study aimed at identifying socioeconomic drivers as well as factors that hinder

Table 5.1.1 Summary of system facility for micro-grid heating: Private farmer's initiative, province of Lower Austria (as of 2008)

1 × woodchip heat boiler (Hargassner GmbH): 0.55 MW
1 × 15 loose m^3 of indoor woodchip storage besides the boiler
Feedstock consumption: 130 loose m^3/year
Grid length: 70 m, three households in direct vicinity
Established: 2003
Initial investment costs: €43,000
Annual CO_2 emission reductions: 22 tons
Feedstock: woodchips from forest residues, corn core, *Miscanthus* briquettes
Operators: 1 person (farmer, part time)

successful regional forest biomass utilization for energy supply (heat and electricity generation).

Micro-grid heating, private farmer's initiative, province of Lower Austria (Table 5.1.1; Plate 5.1.1)

The farmer primarily intended to reduce fossil fuel consumption of a conventional boiler by utilizing previously discarded biomass residues from his cropland and forest. Following the successful installation of the heating device and a one-year test period, another neighbour wanted to join the grid the following year. After connecting two neighbouring households, the system reached its technical limit. Trialling different feedstocks led to maintenance problems with the heating device. The farmer built the system by himself without subsidies.

Medium-scale district heating, farmers' association, province of Styria (Table 5.1.2; Plate 5.1.2)

In 1990, 18 members of a local farmers' association began seeking ways to utilize low-quality timber. Small forest owners in the region faced difficulties selling low-quality timber on the market, especially after wind-throw calamities, owing to the sudden oversupply of timber. One solution was to establish an association for biomass heat supply. Initially the association planned only to supply heat to a public school and buildings surrounding the biomass plant. It faced difficulties in convincing local homeowners to connect to the plant's district heating grid because of fairly low oil prices (€0.29/litre) and concerns over the long-term viability of the project. However, the town supported the initiative because it wanted to use renewable energy to heat public buildings. At that time, two regional political issues arose: the town wanted to avoid an expensive obligatory connection to a natural gas pipeline (30 km from town);

Plate 5.1.1 Photo of a former barn used as additional storage facility for wood chips (left). Photo of the biomass boiler (centre). Photo of the indoor woodchip storage beside the boiler (right). (*Source*: © K. Aoki and F. Kraxner)

Table 5.1.2 Summary of system facility for medium-scale district heating: Farmers' association, province of Styria (as of 2008)

2 × woodchip heat boiler (Urbas GmbH): 1.8 + 1.2 MW = 3 MW
1 × back-up oil boiler
1 × 35,000 m^3 of woodchip storage
3 × 400 litre water tank
Feedstock consumption: 50 m^3/day (on average), 18,000 m^3/year
Grid length: 7 km, 320 households and public buildings
Established: 1996
Initial investment: €2.9 million
Annual CO_2 emission reductions: 70 tons
Feedstock: woodchips from forest residues and low-quality timber, bark
Operators: 3 people (farmers, part time)

and the mayor joined the farmers' association in an attempt to promote the Climate Alliance (Klimabündnis).

Large-scale district heating with CHP, town supply, province of Tirol (Table 5.1.3; Plate 5.1.3)

The key motivation for proposing a municipal biomass power plant in the town (population 13,000) was to curb the severe air pollution that occurred in the valley in winter owing to temperature inversions. Further incentives were that the town needed to improve its heat supply infrastructure anyway, and that the local government wanted to utilize regional renewable energy sources (with no gas connection, the town was completely dependent on other fossil fuels for its energy supply). Abundant forests in the region ensured that forest biomass (i.e. wood chips, sawdust, bark) was available as a domestic renewable energy resource. Moreover, the town has many sunny days (about 2,000 sunshine hours annually) so sunlight could also be utilized as an alternative energy resource.

Driven by the active engagement of motivated local people, the town joined the Climate Alliance in 1998. In April 2000, the town kicked off the biomass heat supply project, together with two experienced and financially strong project partners: the provincial energy suppliers of Tirol and Styria. As a result, the largest biomass CHP plant in Europe (as of 2000) based on the organic Rankine cycle system, with a final grid length of 65 km (2006), was constructed. In 2006, about 97 per cent of the required energy was generated from biomass, and about 3 per cent was derived from fuel oil and the photovoltaic solar panels installed on the plant's roof. The biomass plant substitutes for about 9,650,000 litres of heating oil annually. The biomass power plant has had a significant economic impact on the forest sector. Including the purchase of fuelwood,

Plate 5.1.2 Photo of open-air storage place for chipped wood in front of the heat plant (left). Photo of low-quality timber storage before chipping (centre left). Photo of a heat boiler (centre right). Photo of a wide-bucket excavator used for loading the wood chips (right). (*Source:* © K. Aoki and F. Kraxner)

Table 5.1.3 Summary of system facility for large-scale district heating with CHP: Province of Tyrol (as of 2008)

3 × woodchip boiler (Kohlbach GmbH): 7 + 6 + 8.7 MW = 21.7 MW
2 × extra-light fuel oil back-up boiler: 11 + 11 MW = 22 MW
1 × thermal solar collector panel (630 m^2): 0.35 MW
Electric capacity from biomass ORC-CHP system: 2.5 MW (Turboden)
Feedstock consumption: 800 m^3/day (winter), 20 m^3/h, 80,000 m^3 annually
Grid length: 65 km (together with a second biomass plant)
Established: 2000
Initial investment: €24.77 million
Annual CO_2 emission reductions: 6,000 tons
Feedstock: woodchip, sawdust and bark
Operators: 6 (permanent staff)

the local energy agency invested about €1.2 million annually in the domestic timber industry and promoted the regional added-value chain. Since the establishment of the biomass power plant, the total amount of wood harvested in the Eastern Tirol district has increased by 21 per cent. Local farmers (small-scale forest owners) and forest industries deliver the forest biomass. The majority of the feedstock comes from within a 50 km radius.

5-1-4 Conclusions

Entrepreneurship, innovation and the creation of local value

Biomass energy activities can bring positive political and socioeconomic impacts to local communities (carbon-neutral region, sustainable forest management, job creation, increase in community revenue, creation of regional value, etc.). Regenerative heat and power production using regional forest biomass creates multiple regional values and can induce new markets, such as for low-quality timber and wood residues (e.g. sawdust and bark in the wood industry), which consequently steers vital forest management actions such as thinning. In addition, renewable energy supply systems create new jobs and businesses in the feedstock (e.g. forest biomass delivery, chipping firms) and energy (e.g. grid connection, operation, maintenance) sectors, thereby stimulating local economies. Some regions have successfully created a job market for additional services, such as eco- and bioenergy tourism, with the regions receiving large numbers of visitors to see their innovative activities (Madlener and Koller, 2007).

The qualitative analysis carried out in this study has focused on the evolution of biomass plant establishment in Austria, and why, how and in

Plate 5.1.3 Photo of the CHP biomass plant (left). Photo of the town on a clear winter day (centre left). Photo of the town during inversion of weather conditions (centre right). Photo of a wide-bucket excavator under the solar panel roof of the biomass plant (right). (*Sources*: <http://www.stadtwaerme-lienz.at/>; <http://www.brg-lienz.tsn.at/>, © K. Aoki and F. Kraxner)

what socioeconomic circumstances new biomass plant projects can successfully be launched in order to utilize local forest resources to supply renewable, and hence carbon-neutral, bioenergy on a regional basis.

Based on this analysis, the following set of essentials can be learned from the Austrian experience:

(1) *Policies*: Decisionmakers and policymakers need to provide the right incentives and create an environment enabling the initiation of local biomass projects by placing a high priority on renewable energy sources, to combat climate change and ensure future energy security. Such policies need to be combined with those addressing sustainable rural development, the environment and local societies.

(2) *Policy tools*: A decisive instrument for the support of regional bioenergy development is the provision of special subsidies. In Austria, these policy tools, initially intended to subsidize local economies and the energy autarchy, have recently evolved into climate policy tools aimed at tackling Austrian CO_2 emission targets. Many subsidies exist for establishing decentralized biomass plants in combination with district heating grids (local, regional, national and EU funds). On the other hand, financial support also exists for those willing to change their present (fossil fuel based) heating system into forest biomass (wood pellets, wood chips or fuelwood) systems or to connect their homes to one of the numerous public heating grids. Further economic incentives include, *inter alia*, cheap credit.

(3) *Knowledge and public opinion*: The public's knowledge of forest, forestry and bioenergy as well as of local history and traditions has turned out to be crucial for the successful establishment and maintenance of forest-based bioenergy projects. Additionally, it is extremely important for local initiators to be aware of public opinion in project areas in order to steer and support targeted educational measures and other events to disseminate knowledge.

(4) *Sustainable forest management and forest certification*: Effective forest management taking into consideration sustainability criteria and consequent forest management certification turned out to be a positive incentive when conducting thinning measures and harvesting activities for biomass use.

(5) *Synergetic approach*: A local forest-based bioenergy project might be the perfect basis for adopting a synergetic approach to multiple issues such as the climate change problem by sequestering carbon in standing forest biomass and long-lasting forest products, by using wood as a healthy and renewable construction material under the green building aspect or by substituting fossil-based energy with forest-based bioenergy. Addressing rural development issues by combining sustainable forest management and its certification with environmental protection and the (local) forestry industry, in addition to

closing the loop by linking recreational issues and eco-tourism by creating positive public opinion about planned or existing bioenergy projects, can be an important part of such synergies.

REFERENCES

AEA [Austrian Energy Agency] (1999) "Biomass-Fired Combined Heat and Power (CHP) Plants", <http://www.energyagency.at/fileadmin/aea/pdf/energietechnologien/englisch/projekt-biomasse-kraft-waerme-en.pdf> (accessed 12 March 2010).

BFW [Bundesforschungs- und Ausbildungszentrum für Wald] (2002) "Österreichische Waldinventur 2000–2002" [Austrian Forest Inventory 2000–2002], Federal Forest Office, Vienna (in German). Available at: <http://bfw.ac.at/700/700.html> (accessed 12 March 2010).

Furtner, K. and H. Haneder (2008) "Biomasse – Heizungserhebung 2008", NÖ Landes-Landwirtschaftskammer, Abteilung Betriebswirtschaft und Technik.

IEA [International Energy Agency] (2007) *World Energy Outlook 2007 Edition – China and India Insights*. Paris: OECD/IEA.

Kraxner, F. and E. Rametsteiner (2005) "Western Europe Certifies 50%, and North America 30%, of Their Forests: Certified Forest Products Markets, 2004–2005", Chapter 9 in UNECE/FAO, *Forest Products Annual Market Review, 2004–2005*. Timber Bulletin vol. 58, ECE/TIM/BULL/2005/3. New York and Geneva: United Nations. Available at: <http://www.unece.org/timber/docs/fpama/2005/2005_fpamr.pdf> (accessed 27 April 2010).

Madlener, R. and M. Koller (2007) "Economic and CO_2 Mitigation Impacts of Promoting Biomass Heating Systems: An Input–Output Study for Vorarlberg, Austria", *Energy Policy* 35: 6021–6035.

Ministry of Life (2009) *Facts and Figures 2009*. Vienna: Federal Ministry of Agriculture, Forestry, Environment and Water Management. Available at: <http://gpool.lfrz.at/gpoolexport/media/file/Facts_and_Figures_2009.pdf> (accessed 12 March 2010).

Nemestothy, K. (2006) "Case: Business Opportunities in Wood Energy Production", presentation on 8 September, Austrian Energy Agency.

Nemestothy, K. (2008) "Wood Fuel Availability and Security", Austrian Chamber of Agriculture, presentation at Wood Energy Solutions Conference, 4 June, Koli, North Karelia, Finland.

Olsson, O., B. Hillring, J. Vinterbäck, W. Mabee, A. Wahl, K. Skog, H. Spelter and R. Hartkamp (2009) "Continued Growth Expected for Wood Energy Despite Turbulence of the Economic Crisis: Wood Energy Markets, 2008–2009", Chapter 9 in UNECE/FAO, *Forest Products Annual Market Review, 2008–2009*. Geneva Timber and Forest Study Paper 24, ECE/TIM/SP/24. New York and Geneva: United Nations.

Weiss, G. and E. Rametsteiner (2005) "The Role of Innovation Systems in Non-Timber Forest Products and Services Development in Central Europe", *Economic Studies* 6(1): 23–36.

5-2

Transition initiatives: A grassroots movement to prepare local communities for life after peak oil

Noriyuki Tanaka

5-2-1 Introduction

There are both optimistic and pessimistic views on the availability of cheap and sufficient fossil fuels for the maintenance of our present lifestyle. Fossil fuels will definitely run out in the future, and how long we can rely on them is still debatable. Here, the progression of a distinctive movement started by a pessimistic school teacher from a very small town in rural England – Totnes in Devon – to a worldwide regional fossil fuel free movement is described.

The movement has spread rapidly in Europe, the United States, Canada, Australia, New Zealand and even Japan. One of the driving forces for this movement is the enthusiasm of Rob Hopkins, founder of Transition initiatives, but the real driving force has been the systematic change in attitudes at a community level towards creating a low-carbon society by providing concrete strategies and tools. The final outcome aims to establish regional autonomy in terms not only of energy but also of culture, economy and social systems as a whole for the revitalization of rural communities.

5-2-2 Beginning of the movement

According to Dr Richard Heinberg of the Post Carbon Institute in the United States, Transition initiatives started in England, in Totnes, Devon,

---

Designing our future: Local perspectives on bioproduction, ecosystems and humanity, Osaki, Braimoh and Nakagami (eds),
United Nations University Press, 2011, ISBN 978-92-808-1183-4

where Dr Rob Hopkins, a permaculture teacher and the founder of this movement, held the first meeting of enthusiastic people packed into a small community hall in 2006. Soon after, in 2007, BBC radio announced this as "one of the most dynamic and important social movements of the 21st century". This movement's rationale is that the era of oil energy has peaked and will soon be over. Based on this, the movement aims to initiate locally based action to provide society with renewable energy and resources. The initiatives quickly developed into a franchise system and became the Transition Network movement. This initiative is based on a step-by-step procedure called "12 key steps" and criteria to obtain the designation of an official "Transition Town" (TT). A TT is given exclusive access to different kinds of Transition Training from the Transition Network, such as:
- *Training for Transition* – a two-day fundamentals course in setting up and running successful Transition Initiatives
- *Train the trainers* to build a team of people who can deliver the two-day Training for Transition course
- *Talk Training* for delivering public talks on Transition to raise awareness

A designated town is also obliged to attend certain periodic conferences and to maintain a prescribed level of activities.

5-2-3 Step-by-step approach for empowering people and community

The Transition Network has established criteria for obtaining the official designation of TT. The criteria are posted on the Transition Network website, and any town can apply by using a standard form to provide information to check its compliance with the criteria. The 16 criteria are as follows:[1]

1. an understanding of peak oil and climate change as twin drivers (to be written into constitution or governing documents)
2. a group of 4–5 people willing to step into leadership roles ...
3. at least two people from the core team willing to attend an initial two day training course ...
4. a potentially strong connection to the local council
5. an initial understanding of the 12 steps to becoming a TT
6. a commitment to ask for help when needed
7. a commitment to regularly update your Transition Initiative web presence ...
8. a commitment to make periodic contributions to the blogs on Transition Network ...

9. a commitment, once you're into the Transition, for your group to give at least two presentations to other communities in your vicinity ...
10. a commitment to network with other TTs
11. a commitment to work cooperatively with neighbouring TTs
12. minimal conflicts of interests in the core team
13. a commitment to work with the Transition Network re grant applications for funding from national grant giving bodies ...
14. a commitment to strive for inclusivity across your entire initiative. We're aware that we need to strengthen this point in response to concerns about extreme political groups becoming involved in transition initiatives. One way of doing this is for your core group to explicitly state their support for the UN Declaration of Human Rights (General Assembly resolution 217 A (III) of 10 December 1948). You could add this to your constitution (when finalised) so that extreme political groups that have discrimination as a key value cannot participate in the decision-making bodies within your transition initiative. There may be more elegant ways of handling this requirement, and there's a group within the network looking at how that might be done.
15. a recognition that although your entire county or district may need to go through transition, the first place for you to start is in your local community. It may be that eventually the number of transitioning communities in your area warrant some central group to help provide local support, but this will emerge over time, rather than be imposed. (This point was inserted in response to the several instances of people rushing off to transition their entire county/region rather than their local community). Further criteria apply to initiating/coordinating hubs – these can be discussed person to person.
16. and finally, we recommend that at least one person on the core team should have attended a permaculture design course.

Based on the information submitted to the Transition Network office, the Network team decides on the designation of the applicant town. As of July 2009, 192 cities in 11 nations had been designated as TTs. The number of cities by country is given in Table 5.2.1. The UK has 121 cities, followed by the United States with 35, and Australia with 18. Each city has four to five people who are core Transition activists. Core members execute the 12 key steps provided by the *Transition Initiatives Primer* (Brangwyn and Hopkins, 2008). All TT activities can be found on the Transition Network's website.[2]

5-2-4 Manual for Transition

The Transition Network has created a manual for the Transition movement, outlining the 12 key steps for Transition Initiatives (given in Table 5.2.2). Sequential instructions are given and training programmes for

Table 5.2.1 Designated Transition Towns (as of July 2009)

Countries	Number of towns
UK	121
United States	35
Australia	18
New Zealand	7
Canada	5
Chile	1
Finland	1
Germany	1
Italy	1
Japan	1
Netherlands	1

members of TTs are periodically held in England and elsewhere. The *Transition Initiatives Primer* (Brangwyn and Hopkins, 2008) has now been translated from English into nine languages, including German, Japanese, Spanish and Italian. These are available on the Internet. In 2008, Rob Hopkins (Hopkins, 2008) published a comprehensive book entitled *The Transition Handbook*. This landmark book provides detailed nuts and bolts, practical knowledge, skills and encouragement to all kinds of readers such as students, teachers, activists and, of course, actual practitioners in TTs all over the world.

5-2-5 The Transition movement in Japan

Three cities, Fujino, Hayama and Koganei, are self-nominated, and one of them, Fujino, was officially designated in 2008. At present, 11 core members have been actively involved in this movement in Japan and have their own web page.[3]

5-2-6 Epilogue

In the twenty-first century, the dependence of modern society on oil is at a dangerous level. It is quite clear that modern society, in both developed and developing countries, is subject to unprecedented vulnerability in terms of energy supplies. Recently, uncontrolled investment in the international oil market created a huge price surge. Every sector in society was affected and endangered by this. The Transition movement is surely a great way to increase the resilience of towns. The main message of the initiatives, "Be prepared for peak oil", should be adopted throughout the world.

Table 5.2.2 The 12 steps of Transition

Step	Explanation
#1. Set up a steering group and design its demise from the onset	This stage puts a core team in place to drive the project forward during the initial phases. We recommend that you form your Steering Group with the aim of getting through stages 2–5, and agree that once a minimum of four sub-groups (see #5) are formed, the Steering Group disbands and reforms with a person from each of those groups. This requires a degree of humility, but is very important in order to put the success of the project above the individuals involved. Ultimately your Steering Group should consist of one representative from each subgroup.
#2. Awareness raising	This stage will identify your key allies, build crucial networks and prepare the community in general for the launch of your Transition initiative. For an effective Energy Descent Action plan to evolve, its participants have to understand the potential effects of both Peak Oil and Climate Change – the former demanding a drive to increase community resilience, the latter a reduction in carbon footprint. Screenings of key movies (Inconvenient Truth, End of Suburbia, Crude Awakening, Power of Community) along with a panel of "experts" to answer questions at the end of each, are very effective. (See ["Movies for raising awareness", *Transition Initiatives Primer*] for the lowdown on all the movies – where to get them, trailers, what the licensing regulations are, doomster rating vs solution rating) Talks by experts in their field of Climate Change, Peak Oil and community solutions can be very inspiring. Articles in local papers, interviews on local radio, presentations to existing groups, including schools, are also a part of the toolkit to get people aware of the issues and start thinking of solutions.
#3. Lay the foundations	This stage is about networking with existing groups and activists, making clear to them that the Transition Initiative is designed to incorporate their previous efforts and future inputs by looking at the future in a new way. Acknowledge and honour the work they do. and stress that they have a vital role to play. Give them a concise and accessible overview of Peak Oil, what it means, how it relates to Climate Change, how it might affect the community in question, and the key challenges it presents. Set out your thinking about how a Transition Initiative

Table 5.2.2 (cont.)

Step	Explanation
	might be able to act as a catalyst for getting the community to explore solutions and to begin thinking about grassroots mitigation strategies.
#4. Organise a Great Unleashing	This stage creates a memorable milestone to mark the project's "coming of age", moves it right into the community at large, builds a momentum to propel your initiative forward for the next period of its work and celebrates your community's desire to take action.
	In terms of timing, we estimate that 6 months to a year after your first "awareness raising" movie screening is about right.
	The **Official Unleashing of Transition Town Totnes** was held in September 2006, preceded by about 10 months of talks, film screenings and events.
	Regarding contents, your Unleashing will need to bring people up to speed on Peak Oil and Climate Change, but in a spirit of "we can do something about this" rather than doom and gloom.
	One item of content that we've seen work very well is a presentation on the practical and psychological barriers to personal change – after all, this is all about what we do as individuals.
	It needn't be just talks, it could include music, food, opera, break dancing, whatever you feel best reflects your community's intention to embark on this collective adventure.
#5. Form working groups	Part of the process of developing an Energy Descent Action Plan is tapping into the collective genius of the community. Crucial to this is to set up a number of smaller groups to focus on specific aspects of the process. Each of these groups will develop their own ways of working and their own activities, but will all fall under the umbrella of the project as a whole.
	Ideally, working groups are needed for all aspects of life that are required by your community to sustain itself and thrive. Examples of these are: food, waste, energy, education, youth, economics, transport, water, local government.
	Each of these working groups is looking at their area and trying to determine the best ways of building community resilience and reducing the carbon footprint. Their solutions will form the backbone of the Energy Descent Action Plan.

Table 5.2.2 (cont.)

Step	Explanation
#6. Use Open Space	We've found Open Space Technology to be a highly effective approach to running meetings for Transition Initiatives. In theory it ought not to work. A large group of people comes together to explore a particular topic or issue, with no agenda, no timetable, no obvious coordinator and no minute takers. However, we have run separate Open Spaces for Food, Energy, Housing, Economics and the Psychology of Change. By the end of each meeting, everyone has said what they needed to, extensive notes had been taken and typed up, lots of networking had taken place, and a huge number of ideas had been identified and visions set out. The essential reading on Open Space is Harrison Owen's *Open Space Technology: A User's Guide*, and you will also find Peggy Holman and Tom Devane's *The Change Handbook: Group Methods for Shaping the Future* an invaluable reference on the wider range of such tools.
#7. Develop visible practical manifestations of the project	It is essential that you avoid any sense that your project is just a talking shop where people sit around and draw up wish lists. Your project needs, from an early stage, to begin to create practical, high visibility manifestations in your community. These will significantly enhance people's perceptions of the project and also their willingness to participate. There's a difficult balance to achieve here during these early stages. You need to demonstrate visible progress, without embarking on projects that will ultimately have no place on the Energy Descent Action Plan. In Transition Town Totnes, the Food group launched a project called "Totnes – the Nut Tree Capital of Britain" which aims to get as much infrastructure of edible nut bearing trees into the town as possible. With the help of the Mayor, we recently planted some trees in the centre of town, and made it a high profile event.
#8. Facilitate the Great Reskilling	If we are to respond to Peak Oil and Climate Change by moving to a lower energy future and relocalising our communities, then we'll need many of the skills that our grandparents took for granted. One of the most useful things a Transition Initiative can do is to reverse the "great deskilling" of the last 40 years by offering training in a range of some of these skills.

Table 5.2.2 (cont.)

Step	Explanation
	Research among the older members of our communities is instructive – after all, they lived before the throwaway society took hold and they understand what a lower energy society might look like. Some examples of courses are: repairing, cooking, cycle maintenance, natural building, loft insulation, dyeing, herbal walks, gardening, basic home energy efficiency, making sour doughs, practical food growing (the list is endless).
	Your Great Reskilling programme will give people a powerful realisation of their own ability to solve problems, to achieve practical results and to work cooperatively alongside other people. They'll also appreciate that learning can truly be fun.
#9. Build a Bridge to Local Government	Whatever the degree of groundswell your Transition Initiative manages to generate, however many practical projects you've initiated and however wonderful your Energy Descent Plan is, you will not progress too far unless you have cultivated a positive and productive relationship with your local authority. Whether it is planning issues, funding or providing connections, you need them on board. Contrary to your expectations, you may well find that you are pushing against an open door.
	We are exploring how we might draft up an Energy Descent Action Plan for Totnes in a format similar to the current Community Development Plan. Perhaps, one day, council planners will be sitting at a table with two documents in front of them – a conventional Community Plan and a beautifully presented Energy Descent Action Plan. It's sometime in 2008 on the day when oil prices first break the $100 a barrel ceiling. The planners look from one document to the other and conclude that only the Energy Descent Action Plan actually addresses the challenges facing them. And as that document moves centre stage, the community plan slides gently into the bin (we can dream!).
#10. Honour the elders	For those of us born in the 1960s when the cheap oil party was in full swing, it is very hard to picture a life with less oil. Every year of my life (the oil crises of the 70s excepted) has been underpinned by more energy than the previous years.
	In order to rebuild that picture of a lower energy society, we have to engage with those who directly remember the transition to the age of Cheap Oil, especially the period between 1930 and 1960.

Table 5.2.2 (cont.)

Step	Explanation
	While you clearly want to avoid any sense that what you are advocating is "going back" or "returning" to some dim distant past, there is much to be learnt from how things were done, what the invisible connections between the different elements of society were and how daily life was supported. Finding out all of this can be deeply illuminating, and can lead to our feeling much more connected to the place we are developing our Transition Initiatives.
#11. Let it go where it wants to go …	Although you may start out developing your Transition Initiative with a clear idea of where it will go, it will inevitably go elsewhere. If you try and hold onto a rigid vision, it will begin to sap your energy and appear to stall. Your role is not to come up with all the answers, but to act as a catalyst for the community to design their own transition.
	If you keep your focus on the key design criteria – building community resilience and reducing the carbon footprint – you'll watch as the collective genius of the community enables a feasible, practicable and highly inventive solution to emerge.
#12. Create an Energy Descent Plan	Each working group will have been focusing on practical actions to increase community resilience and reduce the carbon footprint.
	Combined, these actions form the Energy Descent Action Plan. That's where the collective genius of the community has designed its own future to take account of the potential threats from Peak Oil and Climate Change.
	The process of building the EDAP is not a trivial task. It's evolving as we figure out what works and what doesn't.

Source: Brangwyn and Hopkins (2008: 24–28).

Notes

1. "Criteria for becoming an 'official' Transition Initiative", <http://www.transitionnetwork.org/community/support/becoming-official>.
2. See <http://www.transitionnetwork.org/>.
3. See <http://www.transition-japan.net/>.

REFERENCES

Brangwyn, Ben and Rob Hopkins (2008) *Transition Initiatives Primer – Becoming a Transition Town, City, District, Village, Community or Even Island. Version 26.* The Transition Network. Available at: <http://www.transitionnetwork.org/sites/default/files/TransitionInitiativesPrimer%283%29.pdf> (accessed 12 March 2010).

Hopkins, Rob (2008) *The Transition Handbook: From Oil Dependency to Local Resilience.* Totnes: Green Books.

5-3
Rebuilding *satoyama* landscapes and human–nature relationships

Kazuhiko Takeuchi

The author began working on *satoyama* landscape issues in 1999, in a joint research project entitled "Building a Nature Conservation Strategy for *Satoyama* Landscapes". At that time, *satoyama* landscapes were defined as land-use mosaics of rural villages, comprising *satoyama* woodlands, grasslands, dry fields, paddy fields, human settlements, artificial ponds, rivers and irrigation channels. The *satoyama* featured in the study were limited to woodlands and grasslands used for firewood and charcoal, and to agricultural woodlands and grasslands. It was considered that, to conserve *satoyama* landscapes, it was important to maintain natural environmental features such as landform, soils, vegetation and animal communities as well as to consider several other matters, such as the establishment of legal and economic frameworks for conservation and involving a wide range of stakeholders. The question of how to reconstruct in modern society the important role that *satoyama* landscapes once played as sources of natural resources and energy was also pondered.

Recently, the environmental administration in Japan has begun to view *satoyama* landscapes as significant. The Second National Biodiversity Strategy of Japan, enacted in 2002, highlighted three biodiversity crises. In addition to the loss of natural environments as a result of development and the disturbance of ecosystems caused by the arrival of exotic species, the deterioration of natural environments that occurs when people cease to care for them was also identified as problematic. When human-maintained natural environments, such as *satoyama* landscapes, cease to be used as sources of natural resources and energy, ecological

Designing our future: Local perspectives on bioproduction, ecosystems and humanity,
Osaki, Braimoh and Nakagami (eds),
United Nations University Press, 2011, ISBN 978-92-808-1183-4

succession takes over, and the precious flora and fauna that once inhabited these environments face the threat of extinction. In fact, it has been demonstrated that most of the currently endangered flora and fauna would be protected through the sustainable maintenance of *satoyama* landscapes.

In June 2007, the Japanese Cabinet approved "Becoming a Leading Environmental Nation Strategy in the 21st Century: Japan's Strategy for a Sustainable Society". This aims to create a sustainable society by integrating the notions of a low-carbon society, a resource-circulating society and a society in harmony with nature. Within this strategy, the *Satoyama* Initiative was proposed as a new endeavour aimed at making Japan a country in which society and nature coexist harmoniously. It is a genuine attempt to rebuild and maintain within a modern society the traditional symbiotic human–nature relationships that were once seen in *satoyama* landscapes, and it has expanded the perspective of the environmental administration to encompass natural and semi-natural environments throughout Japan. Simultaneously, it will also strengthen links between the fields of agriculture, forestry and fishing.

5-3-1 Rediscovered *satoyama*

Although the term "*satoyama*" was used in the Edo period, a forest ecologist called Tsunahide Shidei was responsible for the word becoming entrenched in Japanese society. Professor Shidei (2000) stated, "In order to express the concept of agricultural woodlands in a way that was easy to understand, I decided to turn the word '*Yamazato*' (village in a mountainous area) back-to-front and call it '*Satoyama*' (mountain in a village area)." Although he made this assertion in the 1960s, it is only recently that *satoyama* has become a widely used expression in the Japanese lexicon.

Why has this word become so popular? One major impetus has been the advance of large-scale developments such as residential and resort areas. When land is developed, natural environments disappear. Until that happened, people took the presence of *satoyama* for granted and did not fully appreciate their worth. It was only after these landscapes were lost as a result of development that people began to understand their value for the first time. It was probably then that the word *satoyama*, which evokes feelings of nostalgia, entered people's consciousness. Afterwards, it spread throughout society as the longing grew for natural environments in close proximity. Implied in the word *satoyama* is the recollection of an archetypal hometown, and thus the desire to revive the positive relationship that people once had with nature.

At the same time, ecologists have also begun to suggest that *satoyama* landscapes are important because they nurture an abundant biodiversity. Although the climate of south-west Japan is ideal for dense evergreen forests, open deciduous woodlands have long been maintained there, a phenomenon that can only be attributed to humans actively maintaining *satoyama*. Ephemeral plant species of the spring, such as the Japanese dogtooth violet, have persisted there since the last ice age. Even though such flora originated from warm climatic environments, they exhibit phases similar to deciduous trees that grow in cold climatic environments. With the fuel and fertilizer revolutions, the human management of such flora declined and the resultant vegetation succession has progressed to such an extent that relict species such as these are facing what is known as a biodiversity crisis brought on by under-use.

Thus, from the perspective that nature in *satoyama* landscapes ought to be maintained into the future, simply protecting *satoyama* landscapes from development and leaving them well alone is not enough. Rather, attempts must be made to construct new human–nature relationships, even if they differ from the traditional models, by involving humans in appropriate environment management. If this happens, for the first time ever it will be possible to maintain *satoyama* in the true sense of the word.

An example of appropriate human management of a *satoyama* landscape being resumed is the restoration by Mr Koichi Tagoku and his associates of the Zushi-Onoji Historical and Natural Environmental Conservation Area located in Machida City, Tokyo. They were the original landholders but sold the land to the Tokyo Metropolitan Government, which designated it a Historical and Natural Environmental Conservation Area. Once in the hands of the Tokyo metropolis, no *satoyama* landscape maintenance was ever conducted, which was considered problematic. The original landowners thus petitioned the metropolis to allow them to form a neighbourhood association for managing the *satoyama* landscape. They logged the trees in the *satoyama* woodland in order to stimulate regrowth. They also regenerated wetland rice paddies in the valley where cultivation had been abandoned. Through such proactive management actions, the semi-natural landscape recovered (see Plate 5.3.1).

The reality, however, is that in most parts of Japan the cultivation of *satoyama* woodlands and rice fields is steadily being abandoned, leading to the deterioration of *satoyama* landscapes. Such a situation cannot be described as *satoyama* landscape conservation by way of appropriate human–nature relationships. A key issue is how to resume human management in such areas while simultaneously maintaining biodiversity. Ultimately, the only way to maintain *satoyama* landscape biodiversity is by rebuilding human–nature relationships (including economic activities).

Plate 5.3.1 Restoration of paddy field in Machida City, Tokyo (Upper: abandoned; lower: restored. Photos taken by Mr Koichi Tagoku)

Incidentally, the expression "*satoyama* landscape" may seem limited to land uses such as woodlands and rice paddies, but one extremely important *satoyama* that has been in dramatic decline since the Meiji period is grasslands. About a century ago, grasslands occupied approximately 10 per cent of Japan's total land area. Today, however, they account for only 1 per cent or less. This is because, in the past, hay was in demand as thatch for roofing material and as livestock feed. That all changed when the number of thatched roofs declined significantly and most livestock feed came to be imported from abroad. As a result, the number of semi-natural grasslands in Japan has drastically decreased, as have the unique fauna that once inhabited and bred in these areas. At the foot of Mount Aso in Kyushu, cattle grazing and burning has maintained a unique grassland landscape called *Kusasenri* (vast grasslands). If *Kusasenri* is to be revived, the problem of how to link landscape maintenance with regional economic activities must be overcome.

5-3-2 Rebuilding dynamic human–nature relationships and *satoyama* landscapes

As described above, human utilization of *satoyama* has waned since the fuel and fertilizer revolutions of the 1960s and, as a consequence, human–nature relationships have become more tenuous and the biodiversity that these semi-natural landscapes supported has declined. However, looking back into the past, examples can be found of mountainous areas denuded owing to the overexploitation of *satoyama*. Professor Seiroku Honda, who taught Forestry at Tokyo Imperial University, lamented such environmental deterioration in "Akamatsu Bokoku-ron" (Thesis on the ruin of the nation of red pines) (Honda, 1900). Thus, we should realize that positive human–nature relationships were revived from states of environmental devastation owing to afforestation and the stimulation of ecological succession. After the fuel and fertilizer revolutions of the 1960s, *satoyama* regressed and declined. Thus, we should understand that *satoyama* are dynamic landscapes.

One major feature of *satoyama* is their shared utilization as "commons", a fact that is significant if one considers the future of *satoyama*. Accordingly, any re-evaluation of *satoyama* will need to be undertaken alongside a re-evaluation of the commons. It is important to propose new commons appropriate for a modern society. *Satoyama* commons should consist of a large number of entities, including farmers and foresters, new members, urban residents, administrative bodies and non-profit organizations, establishing joint management associations that act as single entities.

Regional management bodies must be established to handle the total management of the commons, and systems must be developed so that, in addition to agriculture, forestry and fishing, these entities will also be responsible for the management of manufacturing, processing and distribution activities. The author has proposed public regional management companies in the past (Takeuchi, 2001); however, it is essential that organizations be established that will form the management entities for the various areas, and that systems be developed that will enable the people who work there to be paid wages, to have vacations on public holidays and do their jobs in a secure manner.

Another major feature of *satoyama* is that in the past there were systems for recycling biomass resources in place. These systems must be reinstituted. The notion of regional resource-circulating zones was proposed in the Basic Plan for Establishing a Resource-Circulating Society, which was amended in March 2009. *Satoyama* landscape revival will also entail the formation of new regional resource-circulating zones. It is important that the question of how to recycle the biomass resources of *satoyama* landscapes is paid due consideration.

Figure 5.3.1 illustrates *satoyama* landscape reconstruction possibilities in terms of energy consumption in a case study of Saku District in

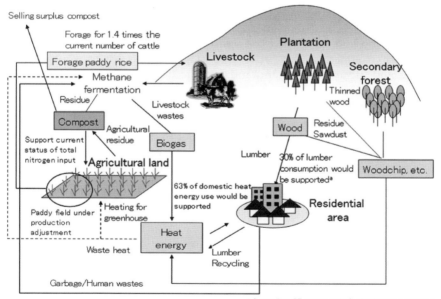

Figure 5.3.1 Biomass recycling potential in Saku City, Nagano Prefecture.
Source: Harashina and Takeuchi (2004).

Nagano Prefecture (Harashina and Takeuchi, 2004). By assessing the biomass recycling potential of such *satoyama* landscapes, it will be possible to consider the outcomes of "produce locally, consume locally" resource and energy consumption systems. In this case study, the community was fully self-sufficient in terms of timber. On the other hand, in areas where the cultivation of rice fields had been abandoned, the community was able to enhance its level of food self-sufficiency by producing rice for livestock feed, the production of which does not require a large labour force, and also to supply approximately 20 per cent of its own energy needs through the use of wood chips and pellets. Furthermore, in addition to being able to recycle the city's raw garbage and human waste as organic fertilizer, a degree of energy recovery was also possible. It is important that such biological resource recycling systems be discussed at a community level.

Regional infrastructure that enables elderly people to lead rich and healthy lives will be important as the Japanese population ages. The creation of a society in which productivity is not so high but life is sustainable, and in which people are able to enjoy life while utilizing the resources of their region, is also consistent with future visions of Japanese society. Dr Masami Iriki, former director of the Yamanashi Institute of Environmental Sciences, visualizes future ageing societies as "Eco-Longevity Societies", combining an ecological society with a society oriented to long life. It is hoped that, in addition to living longer, people will also lead healthy and happy lives. Against such a backdrop, *satoyama* landscape environments may prove important places in terms of human health and lifestyle fulfilment, and become a great asset to twenty-first-century Japan.

If *satoyama* landscapes are revived as attractive places, they will be a source of pride to Japan on the world stage, and a potential reason for international visitors from China and South Korea to visit Japan. If *satoyama* landscapes can be translated into opportunities for international exchange, this may also lead to economic growth in rural areas.

5-3-3 Disseminating the *satoyama* landscape theory to Asia

Needless to say, although *satoyama* landscapes are a Japanese concept, many similar landscape types may be observed across monsoon Asia. Plate 5.3.2 shows a terraced rice paddy landscape on the island of Bali in Indonesia. In monsoon Asia, trees and shrubs are still precious biomass resources in rural villages, and land is farmed intensively. Nevertheless, modernization is beginning to destroy human–nature relationships in rural communities there too.

Plate 5.3.2 Rice terrace in Bali, Indonesia

It might be possible to construct larger regional recycling zones by observing a larger spatial unit – a "watershed" – and reviewing the human–nature relationships in that area. By doing so, it might be possible to develop a "produce locally, consume locally" system in which biological wastes produced in cities are returned to rural villages and, simultaneously, the agricultural and forestry products produced by those rural villages are used in the cities. In effect, this would lead to the development of sustainable societies in which traditional and modern communities are in harmony with each another. It will be possible to rebuild the human–nature relationships that people in modern societies hope for through new methods of engineering and economics and the formation of new zones – systems that differ from those used in the past.

Agroforestry, a concept similar to that of the *satoyama* landscape, is widely known around the world. Plantations that have a high degree of production efficiency (usually entailing the formation of one-layer plant communities) can place a significant burden on the environment. Compared with this production method, agroforestry is relatively inefficient; however, it aims to balance human resource utilization with ecosystem conservation by replacing artificial ecosystems with multi-layered natural ecosystems.

A joint study with Professor Mohan Kumar of the Kerala Agricultural University in India compared agroforestry in India with *satoyama* landscapes in Japan (Kumar and Takeuchi, 2009). The authors observed that the cases were similar to one another: in both landscapes, hilly and upland areas were dissected by valleys and rice paddies extended throughout the valleys. The cases also resembled one another in that multi-layered plant communities were used as the local residents sought sustainability. In the future, they would like to develop similar comparative studies in other regions and to investigate the validity of the *satoyama* concept as a vehicle for building new human–nature relationships. At such a juncture, the discussion should cover the rediscovery of traditional knowledge and its utilization in a modern context.

At the Tenth Conference of the Parties to the Convention on Biological Diversity (COP 10) in October 2010, long-term strategies for conserving global biodiversity will be discussed. For the time being, the strategies will look ahead as far as 2020; in the long term, however, the goal will be 2050. Within that strategy, it is anticipated that the Japanese notion of the *satoyama* landscape will be disseminated throughout the world as a dynamic, sustainable model. Based on understanding knowledge and traditions from all over the world in relation to landscapes (other than *satoyama* landscapes) that are also concerned with the issues of humans coexisting with nature, the sustainable utilization of natural resources and the social systems that support them, various aspects of knowledge will be contributed towards the development of sustainability, and the main focus will continue to be rural villages in developing countries in Asia. It is hoped that a common strategy will be enthusiastically disseminated to the rest of the world.

BIBLIOGRAPHY

Harashina, Koji and Kazuhiko Takeuchi (2004) "Study on Regional Bioresource Utilization and Recycling System in a Case of Saku City, Nagano Prefecture, Central Japan", *Journal of the Japanese Institute of Landscape Architecture* 67: 741–744 (in Japanese with English abstract).

Honda, Seiroku (1900) "Wagakoku no Chiryoku no Suijaku to Akamatsu", *Toyo Gakugei Zasshi* 230: 465–469 (in Japanese).

Kumar, Mohan B. and Kazuhiko Takeuchi (2009) "Agroforestry in the Western Ghats of Peninsular India and the Satoyama Landscapes of Japan: A Comparison of Two Sustainable Land Use Systems", *Sustainability Science* 4(2): 215–232.

Shidei, Tsunahide (2000) "Miscellaneous Notes of Farm Forest", *Bulletin of Kansai Organization for Nature Conservation* 22(1): 71–77 (in Japanese).

Takeuchi, Kazuhiko (2001) "Nature Conservation Strategies for the 'SATOYAMA' and 'SATOCHI', Habitats for Secondary Nature in Japan", *Global Environmental Research* 5(2): 193–198.

Takeuchi, Kazuhiko, Atsushi Tsunekawa and Izumi Washitani, eds (2001) *Satoyama no Kankyo gaku*. Tokyo: University of Tokyo Press (in Japanese).

Takeuchi, Kazuhiko, Robert D. Brown, Izumi Washitani, Atsushi Tsunekawa and Makoto Yokohari, eds (2002) *Satoyama: The Traditional Rural Landscape of Japan*. Tokyo: Springer Japan.

5-4
Local commons in a global context

Makoto Inoue

Accumulated academic knowledge of the management of common-pool resources (CPRs) seems an effective way of bridging the local and global perspectives, even though efforts to deal with this issue are still inadequate. This section will demonstrate the fundamental concept related to the design of local commons in a global context, which contributes to the realization of self-sustaining local and regional societies. First, the history and the present situation in East Kalimantan are described by applying the dynamic framework of cultural ecosystems. Within this context, the importance of constructing "collaborative governance", or *kyouchi*, is demonstrated. Then the prototype design guidelines, or *kyouchi* principles, are outlined, including graduated membership, the commitment principle, and the fair distribution of benefits. The concept of collaborative governance has a theoretical affinity with deliberative democracy, except for the commitment principle, which is highly innovative.

5-4-1 Commons in transition

The local vs. the global: Conflict

"Think globally, act locally." Many people interested in "global" environmental problems share the common sense expressed by this phrase. This might imply that "local" praxis is important and recognized only when it supports the "global" environment or universal values. When "global" interests conflict with "local" ones, most people believe the former should have priority over the latter.

"Think locally, act globally." This phrase paradoxically focuses on the "local", because it states that action must contribute to conservation of the "global" environment if we take action while thinking seriously about the "local". It indicates that the solution of "local" environmental problems is indispensable for the solution of "global" ones, and that "local" values should not be inferior to "global" ones. The thinking, however, does not seem to assume a situation in which the interests clash with each other.

In reality, "global" environmental problems are deeply connected with "local" ones. It is not possible to imagine that the "local" environment of a place would be preserved in a pristine condition if the "global" environment were utterly destroyed. Similarly, the degradation of a particular "local" environment might have negative effects on the "global" environment, by shortening the march towards "global" environmental collapse. Efforts focused on the "global" environmental problem may be called the *satellite view approach*, and efforts focused on the "local" environmental problem may be called the *human view approach*. Both approaches are needed to tackle the problem.

However, not every scholar can make full use of both approaches. Moreover, it is not easy to discover and elaborate a new in-between approach, or a *bird's eye view approach*. This is why attempts have been made to bridge the central *human view approach* and the secondary *satellite view approach*. Accumulated knowledge of CPR management provides a foothold in this endeavour.

CPR management in globalization

CPRs, such as forests, wild animals, rivers, coastlines and seas, are characterized by low excludability and high subtractability. Thus it is difficult to apply the management institution for private goods, characterized as high excludability and high subtractability, as well as public goods, characterized as low excludability and low subtractability. This feature of CPRs has inspired social scientists to tackle the issue.

Ostrom (1990) indicated the conditions for the institutions in which the people of local communities practise collective management of CPRs in a sustainable way. The importance of "nested enterprises", which emphasize the relationship of the local commons with a wider unit, is stated as one of the eight design principles (Ostrom, 1990, 2005). Other scholars (Berkes, 2002; Stern et al., 2002) have also pointed out the significance of these issues in the face of economic globalization and political democratization.

The target of research on CPR management, however, has focused on "local" praxis. The theory of CPR management has not sufficiently dealt with the issue of bridging the *human view approach* and the *satellite view approach*. This section illustrates the fundamental concept as it relates to the design of local commons in the global context, which contributes to the realization of self-sustaining local and regional societies.

5-4-2 Dynamism of the cultural ecosystem: The reality in East Kalimantan

The concept of a cultural ecosystem

The concept of a cultural ecosystem (Kawakita, 1989) seems useful for understanding the relationship between human society and the natural environment. A cultural ecosystem consists of three components: the subject or "human society", an object or "natural environment", and an interface that couples human society and the natural environment – "culture". The influence of the subject on the object is termed the "human force"; the influence of the object on the subject is termed the "environmental force".

Kawakita (1989) structured culture into four aspects as follows:
- Technical aspects have a tangible and direct relation to the natural environment (Kakeya, 1994).
- Biological aspects consist of "economic aspects", which are necessary for the metabolism of material and energy, and "welfare aspects", which represent medical care and public welfare.
- Organizational aspects have an indirect relation with the natural environment, and consist of family, kinship, community, administrative agencies, other organizations and social norms. The organization of human society influences the relationship between society and the environment.
- Metaphysical aspects include a sense of values and views of the world, and deeply encroach on human society.

In general, technical aspects and biological aspects are integrated into a "material culture"; organizational aspects are interpreted as part of an "institutional culture"; metaphysical aspects are called a "mental culture".

The concept of a cultural ecosystem is certainly an appropriate framework for understanding the dynamism of "cultural change" in a particular location, because the mutual influences among these three components can be analysed using the devices of human force and environmental force. In most cases observed, however, "acculturation" – defined as cultural change caused by the impact of external factors, such as infiltration of the monetary economy and national policy – seems more important in understanding the human–nature interface.

The dynamic cultural ecosystem approach

Inoue (1995) proposed a "dynamic framework of cultural ecosystems" to add external factors to Kawakita's cultural ecosystem and to express broader nested social spheres in the framework. The approach was derived from the experience of a three-year fieldwork project in East Kalimantan, Indonesia. The specific cultural ecosystem of a Kenyah village or a quadrant (Figure 5.4.1) is located in a "social circle" of the Kenyah people that occupies a part of the "small world" of the Malay in the "inner world" of Southeast Asia (Yano, 1990). The Kenyah social circle is

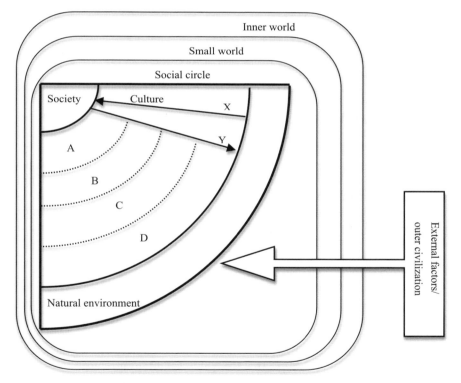

Figure 5.4.1 Dynamic framework of cultural ecosystems.
Source: Modified from Inoue (1995).
Notes: A: Mental culture (values, etc.); B: Institutional culture (organization, etc.); C: Material culture (economy, welfare); D: Material culture (technology); X: Environmental forces; Y: Human forces.

not stationary but rather an elastic political sphere. External factors are regarded as "outer civilization".

Concrete steps to utilize the framework in order to understand the human–nature interface in the "dynamic cultural ecosystem approach" are shown below:
- to outline the natural environment or landscape observed;
- to study the material culture (for example, technology, mentioned above), the economy and people's welfare;
- to study the institutional culture – for example, social organization, human networks and written/tacit rules;
- to study the mental culture behind tangible phenomena;
- to evaluate the direct impact of external factors on a specific aspect of the cultural ecosystem;
- to evaluate the indirect influence on other aspects of the cultural ecosystem by examining human forces and environmental forces.

Brief description of the Kenyah Dayak

Proto-Malayan ethnic groups on Borneo Island are generically called "Dayak". The Dayak people are classified into five groups linguistically, one of which is the "Kayan-Kenyah group", which includes the Kenyah people.

The native land of the East Kalimantan Kenyah people is called Apau Kayan and is located on the central plateau of Borneo Island. At the start of the 1950s, the Kenyah people began to move from the Apau Kayan to the Mahakam and Kayan basins. The inhabitants of most Kenyah villages are self-sufficient in staple foodstuffs, but trade with other regions is vital for obtaining other necessities of life. Six villages were selected as research sites: two villages in Apau Kayan, three on the middle reaches of the Mahakam River and one in the vicinity of Samarinda, the provincial capital.

The recurrent swidden system and its changes

The successive changes the Kenyah people had made to the swidden system were verified (Inoue and Lahjie, 1990) and contrasted with the other excellent comparative studies conducted on some different ethnic groups (Colfer, 1983; Hadi et al., 1985; Kartawinata et al., 1984; Vayda, 1981; Vayda et al., 1985). Although the Kenyah's swidden agriculture is quite restrained from an ecological point of view, its fragility has seen it transformed from traditional to non-traditional swidden agriculture.

As the Kenyah people migrated downstream, their livelihoods were incorporated into a market economy, which led to changes in their lifestyle and social structure, such as work organization, customary land tenure systems and mutual aid systems in daily life (Inoue and Lahjie, 1990). In the course of these changes, the previously sustainable swidden agriculture system was also transformed into a less sustainable form. Only the rotation system is summarized here (Inoue, 2000).

Almost all the swiddens are shifted every year, and the main annual crop is upland rice, often mixed with maize, cucumber and eggplant. Traditionally, the Kenyah people have classified the secondary growth after harvesting rice as a succession stage. The age range of each category is not fixed because the recovery rate of the vegetation depends on such site conditions as soil fertility. For example, vegetation can be deemed *jekau* (or secondary forest with few weeds) within 7 years on some plots, but in over 10 years on others. In the most remote village in Apau Kayan, the people do not fell the forest again until the establishment of *jekau lataq* (or old secondary forest), where the average tree diameter is the same as a human waist.

However, in the villages on the middle reaches of the Mahakam River, people revert to swiddens even if the secondary vegetation is still shrub-sized or *jue dumit*. Therefore, the rotation system in these villages could have a negative influence on the environment. Moreover, in the village near Samarinda, the people have commenced commercial pepper cultivation on swiddens where upland rice was formerly harvested.

To understand the current circumstances, the categorization of sustainable resource utilization (Inoue, 1995, 1998) shown below is useful:
- "Haphazard sustainable use" is defined as the mode of utilization whereby unconscious action brings about sustainable use of resources.
- "Incidental sustainable use" is defined as the mode of utilization whereby conscious actions for other purposes achieve sustainable use.
- "Intended sustainable use" is accomplished where sustainable resource management is achieved as intended. Then some regulations are incorporated into customary law. For example, there are regulations concerning tools, seasons and species in utilizing *iriai-rin* (or communal forests) in Japan.

The Kenyah did not incorporate "intended sustainable use" into their swidden system but instead achieved "incidental sustainable use", because they intended to reduce the labour required for weeding. This is the reason their swidden system was fragile in terms of sustainability.

Historical view of the cultural ecosystem in Apau Kayan

Using the results of comparative studies among six villages, the "dynamic cultural ecosystem approach" was applied to replicate the historical changes in the human–nature interface in Apau Kayan.

Prior to the colonial period, the Kenyah resided in Apau Datah, a highland area where primary forests, secondary forests, bush, grassland and swiddens were scattered throughout their settlements (Inoue, 1995). Livelihoods consisted of producing upland rice, cereals, root crops and vegetables in their territory, where recurrent long-cycle swidden agriculture was practised. They also depended on forests for non-timber forest products such as fruits, nuts, medicines, birds, monkey, deer, wild boar and fish.

The Kenyah were incorporated into the Southeast Asian trade network, trading forest-procured products with Chinese and Malay merchants for ironware and jars. This material culture was sustained by the institutional culture, in which hereditary social classes characterized the political system. The upper classes monopolized the mercantile trade to accumulate wealth, whereas commoners survived under the patronage of the aristocrats. It is important to note that land had no function as a source of wealth because it belonged to the community and each person had only usufruct rights. The mental culture was characterized by a folk religion in which people practised fortune-telling according to birds, conducted healing rituals and were shaped by taboos. In their thinking, many spirits dwelled in the forests.

The Kenyah were not aware of sustainable natural resource use and management, which led to creeping degradation of the natural environment owing to population increase and shortened fallow periods. Contrary to general expectations, human forces dominated environmental forces. Eventually, after a conflict with the Iban people in Sarawak, the Kenyah migrated from Apau Datah into Apau Kayan in the first half of the nineteenth century.

The Dutch government expanded its influence over Apau Kayan during the colonial period of the early twentieth century. The colonial administration affected the mental culture but not the material or institutional cultures in Apau Kayan (Inoue, 1995). A Protestant mission started work in Apau Kayan in the 1930s, and Protestantism gradually supplanted the forest spirits of the folk religion. This situation generated a kind of anomie in the mental culture, and foreshadowed the subsequent rapid changes in the cultural ecosystems as a result of the influence of human forces. On the other hand, the habit of *peselai*, or journeying to Sarawak or the Kalimantan lowlands for temporary work, promoted the gradual infiltration of a more developed material culture into Apau Kayan and allowed the Kenyah to enjoy a more civilized lifestyle. Long-term changes in the material culture may have caused environmental forces to influence the mental culture over time.

By the end of the Second World War, the human population was straining the limited carrying capacity of the Apau Kayan lands. This led to changes in the land tenure system, from usufruct rights to lasting claims in which a person or family who slashed and burnt vegetation to make swidden were repeatedly guaranteed the use of the same plot of land for swidden.

After the Republic of Indonesia's independence, the cultural ecosystem in Apau Kayan underwent rapid changes. The social class system formally faded away as Christian egalitarianism acted on the institutional culture. People developed an awareness of individualism as a result of selling forest products to merchants. Finally, a mass population outflow from Apau Kayan took place as the Kenyah were drawn to the outer civilization. As a result, the fallow period of swidden agriculture in Apau Kayan was extended (representing a change in the material culture), and the significance of customary forest conservation, or *tana' ulen*, in which swidden agriculture was prohibited, decreased.

The function of local communities weakened because administrative officers and the police took over several traditional roles (representing a change in the institutional culture). The government tried to invest in infrastructure such as roads, public halls, clinics, airports, schools, generators and water pipes, as well as to provide subsidies for the necessities of life, including salt, sugar, kerosene and detergent (representing a change in the material culture). In the 1990s, the population in Apau Kayan began to increase again.

Threat and opportunity

At the end of 2005, the Indonesian Minister of Agriculture revealed details of a government plan to develop the world's largest oil palm plantation along the border between Kalimantan and Malaysia, with Apau Kayan situated in the target area. The government intended to develop 3 million hectares of new oil palm plantations to meet domestic and global biofuel demand. Prior to the news release, the Indonesian President and the Chamber of Commerce and Industry had met with the Chinese government and private companies to conclude Memorandums of Understanding in order to finance the USD 567 million project (Wakker, 2006).

Fears that the project would threaten the customary rights over forest resources held by the indigenous Dayak people, including the Kenyah, as well as the forest ecosystem biodiversity led to a sustained lobbying campaign by civil society, the Indonesian media and foreign diplomats. As a result, the government's National Development Planning Agency changed the width of the border zone from 5–10 km to 100 km. The Minister of Forestry has stated that his ministry will not release protected forests in the border zone for oil palm plantations.

The government also stated that it would prioritize plantation development in "abandoned areas", although President Susilo Bambang

Yudhoyono continues to support the overall border development programme in Kalimantan. The definition of "abandoned areas", however, is controversial, because most of the swidden areas, especially bush and forest fallows, could be regarded as "abandoned areas" by the government.

The people might be able to tackle these threats and even transform them into opportunities if they act appropriately in the near future. The government of Indonesia launched a decentralization policy in 1999 that gave considerable powers of local autonomy to both the district (*kabupaten*) and city (*kota*) levels, which were formerly subordinate to the province (*propinsi*). In West Kutai district in the province of East Kalimantan, the governor approved the organization of a multi-stakeholder working group (Inoue, 2003). The group drafted an ordinance for district forest management that was approved by the local parliament in November 2002. The group also drafted an ordinance for community forestry; this was approved in June 2003.

These efforts were remarkable in two respects. First, systems for forming consensus-building mechanisms realized through multi-stakeholder approaches, based on the participation of local people in natural forest regions where valuable forests still remain, are extremely rare and innovative in Asia and the Pacific. Second, these systems can be regarded as a brand-new form of "nested forest governance" in which coordination and complementarity are established between governance at the provincial level and the district level.

However, one cannot take an optimistic view on the issue of the development of oil palm plantations for the time being. In many villages in East Kalimantan, the villagers are split into supporters or opponents of the oil palm companies. Some people, however, have started to demonstrate unprecedented decision-making by creating their own pilot test for industrialization via a combination of small-scale rubber plantations, rattan gardens and orchards, standing against the large-scale oil palm plantations. Such phenomena imply the emergence of new modes of institutional culture, in which the local people have room for autonomous decision-making, owing to the decentralization policy or external factors.

5-4-3 Collaborative governance: Bridging the commons to larger societies

Glocalization strategy

When designing the local commons in a global context, the concept of "glocalization", in which closure and openness as well as inherent values and universal values are adjusted, seems effective. Under this strategy, "collaborative governance" of forests might be constructed (Inoue, 2004).

This type of governance, which is organized through collaboration among stakeholders with diverse interests in local forest use and management, incorporates various steps such as appraisal, planning, implementation, monitoring and evaluation.

In the field, however, it was impossible to satisfactorily accommodate every stakeholder. Even though equal participation by all stakeholders should be ensured, the voices of the people residing in forest regions – usually minorities with less political power – might not be ultimately reflected in government policies.

In order to avoid any deterioration in local autonomy, it is essential for all the stakeholders to consent to the "principle of involvement" (Inoue, 2003), which recognizes the rights of stakeholders to speak and make decisions in a capacity that corresponds to their degree of involvement in and commitment to forest use and management. Under this principle, the local people who frequently visit and take care of the forest might be expected to have more power over the decision-making process. In this way, various stakeholders are able to agree on the legitimacy of the opinions of outsiders as well as of the local people.

Collaborative governance based on the "principle of involvement", however, may not be established if the local people stick fast to the kind of narrow-minded localism that is completely exclusive of outsiders. Thus "open-minded localism" is required, in which the local people aim to open their resources and the environment to outsiders.

Prototype design guidelines for collaborative governance of forests

The sphere of collaborative governance is not identical to the administrative area and scale. It may be formed within a local community, beyond the communities and local government or even beyond the nation. The sphere of collaborative governance (or *kyouchi*) looks like a mandala, in which other spheres overlap with some parts.

In order to facilitate the glocalization strategy, Inoue (2009) has proposed some prototype design guidelines for the collaborative governance of forests. These guidelines, or *kyouchi* principles, are derived from and evolved out of the design principles for CPRs (Ostrom, 1990, 2005; McKean, 1999; Stern et al., 2002), in which the importance of linkage with outside organizations and nested enterprises was pointed out but not further evolved. Three of the nine guidelines are of Inoue's own creation and are quite important for the realization of collaborative governance of the commons.
- **Design guideline 3** (graduated membership): based on "open-minded localism", some of the local people act as "core members". They have the strongest authority, and cooperate with other members who have relatively weaker authority.

- **Design guideline 4** (the commitment principle): the "principle of involvement" (Inoue, 2004) was a concept to embrace the authority to make decisions in the arena and to have a voice in the forum. The "commitment principle" here refers clearly to the authority to make decisions in the arena. Decision-making is not equal, but should be fair and just.
- **Design guideline 5** (the fair distribution of benefits): benefit distribution is not necessarily equal, but is fair in accordance with cost burdens.

5-4-4 Gaining approval of the larger society

The participants and players in the collaborative governance system are assumed to be those interested in specific issues. Their interests are often temporary and fickle. That is why reliable core members are indispensable for continuous activities. Moreover, both the members and their activities should gain the approval of a majority of the society. Once they acquire legitimacy in the broader society, collaborative governance will mature into a robust system in which the main actors of the local commons take action powerfully in collaboration with outsiders.

Gaining the approval of the larger society is connected to the concept of "deliberative democracy". Often called "discursive democracy", this is a system of political decisions that relies on citizen deliberation to formulate sound policies. In deliberative democracy, legitimate law-making can arise only through public deliberation by the people. The voices of the participants should be equal in a deliberative democracy (Cohen, 1997). This principle is important in order for the participants to speak out freely regardless of their social status, and to be bound only by the results of deliberation. In contrast, Guideline 4 (the commitment principle) avoids the influence of social status, because in the real world it seems impossible for participants to speak out freely regardless of their social status. Theoretical investigation to demonstrate the correctness of this concept is required as the next step.

REFERENCES

Berkes, F. (2002) "Cross-scale Institutional Linkages: Perspective from the Bottom up", in E. Ostrom et al. (eds), *The Drama of the Commons*. Washington, DC: National Academy Press, pp. 293–321.

Cohen, J. (1997) "Deliberation and Democratic Legitimacy", in J. Bohman and W. Rehg (eds), *Deliberative Democracy: Essays on Reason and Politics*. London: MIT Press, pp. 67–91.

Colfer, C. J. P. (1983) "Change and Indigenous Agroforestry in East Kalimantan", *Borneo Research Bulletin* 15(1): 3–21 and 15(2): 69–87.

Hadi, S., S. Hadi and R. Hidayat (1985) "Swidden Cultivation in East Kalimantan", in *Swidden Cultivation in Asia*, UNESCO Regional Office for Education in Asia and the Pacific, pp. 74–149.

Inoue, Makoto (1995) *Swidden Agriculture and Tropical Forest: Changes in Traditional Swidden Systems in Kalimantan*. Tokyo: Kobundo (in Japanese).

Inoue, Makoto (1998) "Evaluation of Local Resource Management Systems as the Premise for Introducing Participatory Forest Management", *Journal of Forest Economics* 44(3): 15–22.

Inoue, Makoto (2000) "Mechanism of Changes in the Kenyah's Swidden System: Explanation in Terms of Agricultural Intensification Theory", in Edi Guharidja, Mansur Fatawi, Maman Sutisna, Tokunori Mori and Seiichi Ohta (eds), *Rainforest Ecosystems of East Kalimantan: El Nino, Drought, Fire, and Human Impacts*. Tokyo: Springer-Verlag, pp. 167–184.

Inoue, Makoto (2003) "Diverse Management of Indonesian Forests: A New Governor Gives Locals a Greater Say in Their Resources", *International Herald Tribune and the Asahi Shimbun*, 4 April.

Inoue, Makoto (2004) *In Search of the Principle of Commons*. Iwanami-shoten (in Japanese).

Inoue, Makoto (2009) "Design Guidelines for Collaborative Governance (*kyouchi*) of Natural Resources", in T. Murota (ed.), *Local Commons in Globalized Era*. Kyoto: Minerva-shobou, pp. 3–25 (in Japanese).

Inoue, Makoto and Abubakar M. Lahjie (1990) "Dynamics of Swidden Agriculture in East Kalimantan", *Agroforestry Systems* 12(3): 269–284.

Kakeya, Makoto (1994) "Diverse Aspects of Environmental Socialization", in Makoto Kakeya (ed.), *Environmental Socialization*. Tokyo: Yuzankaku, pp. 3–15 (in Japanese).

Kartawinata, K. et al. (1984) "The Impact of Development on Interactions between People and Forests in East Kalimantan: A Comparison of Two Areas of Kenyah Dayak Settlement", *The Environmentalist* 4, Supplement 7: 87–95.

Kawakita, Jiro (1989) "Environment and Culture", in Takeshi Kawamura and Hideshige Takahara (eds), *Environmental Science II: Human Society*. Tokyo: Asakura Shoten, pp. 1–33 (in Japanese).

McKean, M. A. (1999) "Common Property: What Is It, What Is It Good for, and What Makes It Work?", in C. Gibson, M. A. McKean and E. Ostrom (eds), *Forest Resources and Institutions*. Rome: FAO.

Ostrom, E. (1990) *Governing the Commons*. Cambridge: Cambridge University Press.

Ostrom, E. (2005) *Understanding Institutional Diversity*. Princeton, NJ: Princeton University Press.

Stern, P. C., T. Dietz, N. Dolsak, E. Ostrom and S. Stonich (2002) "Knowledge and Questions after 15 years of Research", in E. Ostrom, T. Dietz, N. Dolsak, P. C. Stern, S. Stonich and E. U. Weber (eds), *The Drama of the Commons*. Washington, DC: National Academy Press, pp. 445–489.

Vayda, A. P. (1981) "Research in East Kalimantan on Interaction between People and Forests: A Preliminary Report", *Borneo Research Bulletin* 13(1): 3–15.

Vayda, A. P., et al. (1985) *Final Report: Shifting Cultivation and Patch Dynamics in an Upland Forest in East Kalimantan, Indonesia*. A joint project of the United States and Indonesian Man and the Biosphere (MAB) program, Rutgers University.

Wakker, Eric (2006) *The Kalimantan Border Oil Palm Mega-Project*. Commissioned by Milieudefensie/Friends of the Earth Netherlands and the Swedish Society for Nature Conservation (SSNC), AID Environment.

Yano, Toru (1990) "Seeking the Image of Region", in T. Yano (ed.), *Methodology of Southeast Asian studies*. Tokyo: Kobundo, pp. 1–30 (in Japanese).

5-5
Risk and resource management

Hiroyuki Matsuda

5-5-1 The theory of maximum sustainable ecosystem services

Ecosystems offer a variety of benefits (World Resources Institute, 2005). Harvests from agriculture, forestry and fisheries are just a small part of ecosystem services (Costanza et al., 1997). Ecosystem services include support services such as soil formation, photosynthesis and nutrient cycling; provisioning services such as food, water, timber and fibre; regulating services that affect climate, floods, disease, waste and water quality; and cultural services that provide recreational, aesthetic and spiritual benefits (World Resources Institute, 2005). The yield of fisheries comes under provisioning services. The existence of bio-resources may maintain these services.

There are some criticisms of the theory of maximum sustainable yield (MSY) (Matsuda and Abrams, 2008). The MSY fishing policy does not reflect uncertainty in stock estimates (measurement errors) or in the relationship between the spawning stock and recruitment (process uncertainties). The MSY policy also ignores the complexity of ecosystem processes because the theory uses single stock dynamic models (Matsuda and Katsukawa, 2002). Hiroyuki Matsuda and Peter Abrams (2006) analysed the maximum sustainable yield from entire food webs with each species being fished independently. They concluded that the MSY policy does not guarantee the coexistence of species and they proposed the concept of "constrained MSY", which maximizes the sustainable yield under which all species could coexist. Matsuda et al. (2008) incorporated ecosystem services into an optimal fisheries policy, and they called the optimal policy that maximizes total ecosystem services the maximum sustainable ecosystem service (MSES). Ecosystem services other than fisheries yield probably depend on the standing biomass, whereas the fisheries yield depends on the catch amount. The standing biomass probably monotonically decreases as the fishing effort increases, whereas the fisheries yield is a unimodal function of the fishing effort.

Suppose the fisheries yield, $Y(C)$, from a single target species is an increasing function of the catch amount, C; the cost of fisheries, $k(E)$, is an increasing function of the fishing effort, E; and the ecosystem service other than fisheries yield, $S(N)$, is an increasing function of the standing biomass, N. The total ecosystem service, V, is written as a function of N and C.

$$V(N, C) = Y(C) - k(E) + S(N). \tag{1}$$

Hereafter, $S(N)$ is simply the utility of the standing biomass. The catch amount, C, is an increasing function of both the fishing effort, E, and the stock biomass, N.

The following fish stock dynamics are assumed:

$$dN/dt = f(N)N - C(E,N), \tag{2}$$

where t is an arbitrary time unit and $f(N)$ is the per capita reproduction rate. $f(N)$ is a decreasing function of N owing to the density effect, and $f(N)N$ is usually a unimodal function of N. In these equations, process uncertainty in $f(N)$ and implementation errors in $C(E,N)$ are ignored.

The equilibrium point denoted by N^* that satisfies $dN/dt = 0$ in equation (2) depends on E. In general, the stable equilibrium point decreases as the fishing effort increases. This is intuitively understandable. The utility of the standing biomass, $S(N)$, is probably a convex or a sigmoid curve because the utility probably saturates when the stock is sufficiently abundant. In the first step, the optimal fishing effort is obtained that maximizes the total ecosystem service $V(N,C)$ given by equation (1) at the equilibrium, denoted by V^*. Since N^* and C^* are functions of E, the total ecosystem service at the equilibrium is a function of E.

Consider the optimal fishing effort that maximizes V^*. Matsuda and Abrams (2008: 737–744) called V^* the maximum sustainable ecosystem service (MSES) and denoted the fishing effort for MSES by E_{MSES}. If the utility of the standing biomass $S(N)$ is ignored, as is usually assumed in classical fisheries theory, V^* is a unimodal function of E and the optimal fishing effort, denoted by E_{MEY}, is well known by the maximum economic yield (MEY). If the cost of fisheries is negligible ($k(E) = 0$), E_{MEY} becomes the effort at the maximum sustainable yield (MSY), denoted by E_{MSY}.

If $S(N)$ is an increasing function of N, the E_{MSES} is always smaller than the E_{MSY} and E_{MEY}. However, the distance between E_{MSES} and E_{MEY} depends on the magnitude and curvature of $S(N)$. The E_{MSES} is smaller when the derivative of the utility of the standing resource with respect to the stock is of a larger magnitude (dS/dN is larger). If $|dS^*/dE| > dY^*/dE - dk/dE$ at $E = 0$, dV^*/dE is negative at $E = 0$ and a fishing ban can be optimal.

The fishing effort for actual fisheries is often considered to be larger than E_{MSY}, and it is unsustainable. Classical fisheries science recommends MSY or MEY. Recently, marine ecologists have recommended a no-take zone for ecosystem-based management (Pikitch et al., 2004). The fishing effort for a maximum sustainable ecosystem service is between the fishing ban and E_{MEY} (Figure 5.5.1). A paradigm shift is needed to encourage environment-friendly fisheries (Castilla and Defeo, 2005).

Adaptive management predicts and monitors changes in the ecosystem, and subsequently reviews and adjusts the management and use of natural resources. Such predictions and monitoring are best accompanied by feedback controls, such as the verification of hypotheses based on

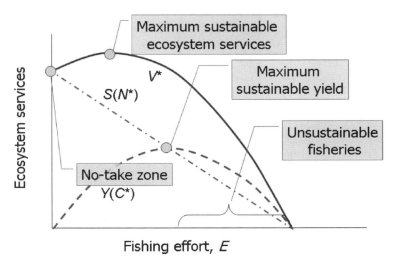

Figure 5.5.1 Schematic relationship between overfishing, maximum sustainable yield, maximum sustainable ecosystem services and no-take zone.
Notes: The dotted line, broken curve and bold curve represent the utility of the standing biomass ($S(N^*)$), fisheries yield ($Y(C^*)$) and total ecosystem services (V^*), respectively.

the results of monitoring in order to review and modify management activities.

5-5-2 The magnitude of fisheries' impact on marine ecosystems

The history of fisheries is characterized by overfishing (Hannesson, 1996). The theory of MSY takes into account the long-term yield from a living marine resource. This theory explicitly assumes a negative relationship between yield and standing stock quantities. In addition, marine ecosystems are characterized by uncertainty, dynamic properties and complexity. However, the classical MSY theory does not include any of these factors. Thus, it is not surprising that the MSY theory and its derivatives have not worked for fisheries management (Matsuda and Abrams, 2008).

Instead of a theory of single species management, an ecosystem approach has been popular. Matsuda and Abrams (2008) made 11 recommendations that could both increase the food resources derived from fish and reduce the chances of overexploitation or extinction: (1) catch fish at lower trophic levels; (2) do not use fish as fish meal; (3) reduce discards before and after landings; (4) establish food markets for temporally fluctuating fish at lower trophic levels; (5) improve the food-processing technology used on small pelagic fishes; (6) switch the target fish to correspond to the temporally dominant species; (7) conserve immature fish, especially when the species is at a low stock level; (8) develop technologies for selective fishing; (9) conserve both fish and fisheries; (10) say goodbye to traditional MSY theory; and (11) monitor not only the target stock level but also any other indicator of the "entire" ecosystem.

There are many warnings about threats to marine ecosystems. Pauly et al. (2002) calculated the mean trophic level (MTL) of fisheries catch and showed that the MTL decreased from about 3.5 in 1950 to about 3.3 in 1990. This conclusion implies overfishing because the fish being harvested are increasingly coming from the less valuable lower trophic levels as populations of higher trophic level species are depleted. The Convention on Biological Diversity uses the mean trophic level of marine fisheries catch as an indicator of marine ecosystem integrity and ecosystem goods and services (Secretariat of the Convention on Biological Diversity, 2006). Pauly et al. (2002) called the decline of the MTL "fishing down". Ransom Myers and Boris Worm (2003) argued that the biomass of top predators, including tuna, has been reduced by 90 per cent relative to levels prior to the onset of industrial fishing. In addition, Jean-Jacques Maguire et al. (2006) noted that about three-quarters of stocks are either fully exploited or overexploited.

Despite these warnings, there is some criticism of these arguments. Hampton et al. (2005) claim that the magnitude of tuna stock decline estimated by Myers and Worm (2003) is an overestimation. Although the southern bluefin tuna (*Thunnus maccoi*, SBT) is listed on the "critically endangered" list by the International Union for Conservation of Nature (IUCN), the extinction risk of SBT is without doubt smaller than that of the blue whale, which is listed as "endangered". It is very unlikely that SBT will become extinct within the next 50 years (Matsuda et al., 1998), although it will be difficult to achieve the target of the Commission for the Conservation of Southern Bluefin Tuna of recovering spawning stock biomass to the 1980 level by 2020 (Mori et al., 2001).

The MTL of global marine landings has not shown a monotonic decline, but rather has fluctuated from decade to decade (see Figure 5.5.6 of World Resources Institute, 2005). The global MTL was low around 1970 and the 1980s, but catches of Peruvian anchovy and Japanese sardine were both large. The theory of "fishing down" is based on the assumption that the major target species is a high-price and higher trophic level fish. This assumption is not true everywhere. The proportion of low-value fish in total fish consumption is about 80 per cent in developing countries, but it is about 10 per cent in developed countries (Delgado et al., 2003). In Japan, the MTL was 3.6 in 1960, 3.1 in 1990 and 3.6 in 2000. Therefore, Japanese fisheries are characterized by a higher MTL than the world average, and their MTL has not shown a long-term decline (Figure 5.5.2). However, there are several cases of overfishing and failures of stock management in Japan (Kawai et al., 2002).

It should be noted, however, that, although total landings of demersal fish had reached a plateau by the 1970s, those of pelagic fish increased until the late 1980s. Some of these pelagic fish species have naturally fluctuated greatly in stock quantities, even without being fished for several thousand years (Baumgartner et al., 1992). The collapse of the Japanese sardine in the 1990s was almost certainly caused by natural changes in the environment (Watanabe et al., 1995). When stocks are at a low level, the impact of fishing on pelagic fish prevents the stock from recovering (Kawai et al., 2002).

There is a mismatch between demand for and supply of fisheries resources from the food security viewpoint. In the case of Japanese fisheries, the total allowable catch (TAC) exceeded the allowable biological catch (ABC) for some fish, including sardine, and the actual catch exceeded ABC in some years (Figure 5.5.3). In contrast, the actual catch is much smaller than the ABC in some species, including Pacific saury and anchovy. It should be noted that the total ABC of these species is more than 2 million tons (Figure 5.5.4). However, economic demand for these species is low in Japan, whereas economic demand for overfished species,

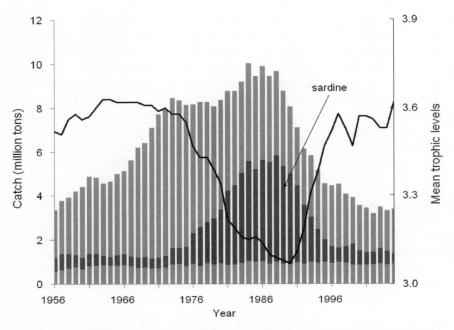

Figure 5.5.2 The mean trophic level of fisheries landings (bold line) and the total landings (bars) of Japanese fisheries.

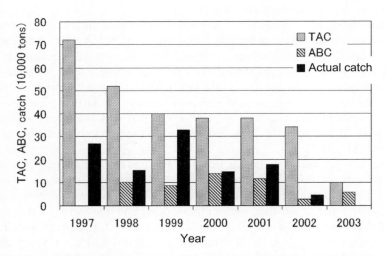

Figure 5.5.3 Allowable biological catch (ABC), total allowable catch (TAC) and actual catch of Japanese sardine, 1997–2003.
Source: Fisheries Research Agency, Japan, unpublished.

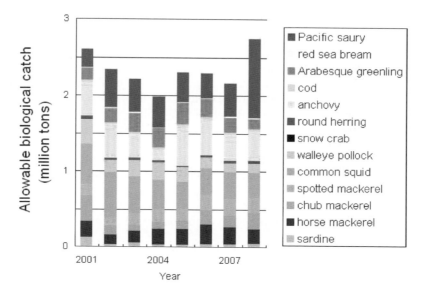

Figure 5.5.4 Total allowable biological catch for several major species (please see page 477 for a colour version of this figure).
Notes: from top to bottom of each bar: Pacific saury, red sea bream, Arabesque greenling, cod, anchovy, round herring, snow crab, walleye pollock, common squid, spotted mackerel, chub mackerel, horse mackerel and sardine.
Source: Fisheries Research Agency, Japan, unpublished.

including tuna and chub mackerel, is still high, partly because of the overcapitalization of fishing vessels and a stiff market. Moreover, the Japanese do not use jellyfish despite the fact that it frequently appears in Japanese coastal regions and the Chinese eat it.

5-5-3 Fisheries co-management in the Shiretoko World Natural Heritage site

Marine management in Japan is characterized by seeking a balance between sustainable use and ecosystem conservation and by involving the co-management of fishers' organizations (Makino et al., 2009). Co-management is defined as the sharing of responsibilities between governmental institutions and groups of resource users (Persoon et al., 2005). In many countries, environmental management has shifted from exclusive state control to various kinds of joint management in which local communities, indigenous peoples and non-governmental organizations share authority and benefits with governmental institutions. Fisheries in Japan face several important challenges: exclusive use by fishers with fishery rights/licences (there are a few exceptions for free fisheries and recreational angling); a lack of full transparency in management procedures; a lack of objective benchmarks or numerical goals in management plans; and strong political pressure from abroad (Matsuda et al., 2009). Matsuda et al. (2009) reported these characteristics to explain why coastal fisheries exist at the Shiretoko World Natural Heritage Site.

Shiretoko was registered as the third World Natural Heritage Site in Japan because of its formation of seasonal sea ice at some of the lowest latitudes in the world, and its high biodiversity and many globally threatened species. The United Nations Educational, Scientific and Cultural Organization (UNESCO) and IUCN required the natural resource management plan of the Shiretoko site to be sustainable, but the national government of Japan guaranteed that no additional regulations would be included in the plan.

The Japanese Ministry of the Environment (MOE) organized a Scientific Council (SC) to draft a proposal nominating Shiretoko as a World Natural Heritage candidate. The proposal was reviewed by the IUCN, which then sent a letter dated 20 August 2004 to the Japanese government recommending an increase in the level of conservation of marine waters and an investigation into the effects of dams on wild populations of salmonids. This document is referred to here as "the first IUCN letter". It was not until a local newspaper revealed the existence of this letter that the government of Japan notified the SC of the receipt of the letter. The Japanese government did not call a meeting of the SC to discuss the letter; however, the SC voluntarily compiled a document advising the Japanese government on how to respond to the first IUCN letter. This step was probably taken because the SC members recognized that this review process was historically important to establish Scientific Council action on future nature conservation or restoration projects in Japan. The SC recommended additional, essential mitigation of river structures, further marine conservation efforts and the formation of marine and river structure working groups. The MOE ignored this advice and replied that further regulation of the walleye pollock fisheries was unnecessary.

IUCN sent another letter, dated 2 February 2005, to the MOE and explicitly requested the expansion of the marine registered area and an expedited marine management plan. This document is referred to here as "the second letter". After this letter was received, the MOE convened the SC, which compiled its recommendations and formed two working groups: the Marine Working Group and the River Structure Working Group. The Marine Working Group included several SC members and

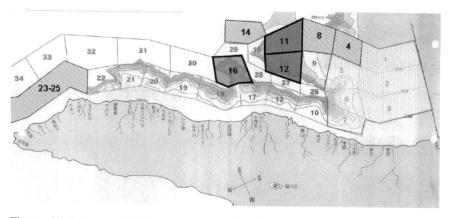

Figure 5.5.5 Seasonal fishing-ban areas of walleye pollock fishery in Shiretoko World Natural Heritage Site.
Notes: Grids 11, 12 and 16 are spawning grounds; grids 4, 8, 11, 14 and 23–25 are seasonal fishing-ban areas since 1995; grids 1–3 and 5–7 are seasonal fishing-ban areas since 2005.
Source: Map provided by Rausu Fisheries Cooperative Association.

other fisheries scientists, and it invited members of regional fisheries cooperative associations (FCAs) as observers.

Officials from the Ministry of the Environment, the government of Japan and the Hokkaido prefectural government rejected the possibility of future fisheries regulation at the Shiretoko site. The SC acknowledged the existing efforts of fishers voluntarily to regulate their fishing efforts, and the FCAs agreed to the expansion of the heritage area without any top-down regulation (Makino et al., 2009).

In accordance with the advice of the SC, the MOE replied to the IUCN and agreed to the major points of the second letter. In addition, the Rausu FCA voluntarily expanded the seasonal fishing-ban area for the 2005 fishing season (Figure 5.5.5). The contribution of fishers to the review process for the Shiretoko World Natural Heritage Site was indispensable, because they were the only group to satisfy the requests by the IUCN and UNESCO to increase the level of conservation of marine ecosystems. UNESCO registered Shiretoko as a World Natural Heritage Site in 2005.

The adaptive management plan involves voluntary activities by local resource users that are suitable for use within a local context, can respond to ecological and social fluctuations and can be efficiently implemented by increased legitimacy and compliance. Such an approach is suitable for developing coastal countries where a large number of artisanal fishers catch a variety of species using various types of gear.

The objective of the Multiple Use Integrated Marine Management Plan formulated by the Ministry of the Environment, the government of Japan and the Hokkaido prefectural government (see Matsuda et al., 2009) for the Shiretoko site is to ensure a balance between the conservation of the marine ecosystem and stable fisheries through the sustainable use of fisheries' resources in the marine component of the heritage area. The target area of this plan is the marine component, which extends up to 3 km from the coastline. The premise of the plan involves legal restrictions relating to the conservation of the marine environment, marine ecosystems and fisheries, as well as voluntary restrictions on marine recreation and community-based marine resource management carried out by fishers.

The management plan defines measures to conserve the marine ecosystem, strategies to maintain major fisheries resources, monitoring methods for those resources and policies for marine recreation. The management plan details the vast food web structure of the Shiretoko site (Figure 5.5.6) and includes fisheries yields for 10 categories of major fisheries resources (Figure 5.5.7). Adaptive management plans usually determine criteria and feedback control measures for indicator species. For example, management plans monitor and enforce conservation actions to achieve numerical goals within a limited amount of time. Management plans usually devise action plans to achieve these numerical goals or to maintain thresholds for indicator species. However, the Marine Management Plan for the Shiretoko site does not include any thresholds or numerical goals for its indicator species, which are currently only monitored. A crucial short-term goal will be to establish such thresholds and/or numerical goals for these indicator species.

The maximum sustainable yield policy for the entire ecosystem does not guarantee the coexistence of all species (Matsuda and Abrams, 2006). Therefore, the goal of the management plan is twofold: sustainable fisheries and biodiversity conservation. Ecosystem-based fisheries management can use some data from fisheries, such as catch amount and its distribution, catch per unit effort, age structure of harvest, and by-catch data of non-target species. Information on unused fish and oceanographic information gathered by individuals other than fishers (usually government authorities) need to be monitored. It is necessary to establish the impact of fisheries on ecosystem processes. For example, fisheries may have a negative impact on Steller sea lions because walleye pollock, which is eaten by sea lions, is a target species of fisheries. Steller sea lions are a threatened species and are important from a conservation viewpoint. These species are controllable by several conservation measures (Matsuda et al., 2009).

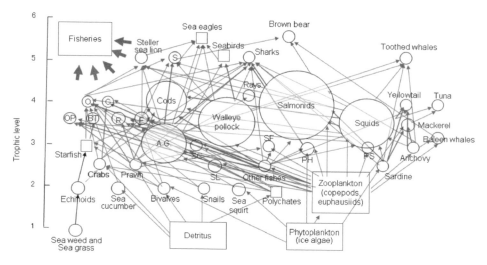

Figure 5.5.6 Flow diagram for the Marine Management Plan of the Shiretoko World Natural Heritage Site.
Notes: AG: Arabesque greenling, G: greenling, O: octopus, OP: ocean perch, BT: bighand thornyhead, R: rockfish, F: flatfish, S: seal, PS: Pacific saury, SL: sand lance, SC: saffron cod, SF: sandfish, PH: Pacific herring. Circles and squares represent utilized and unutilized organisms, respectively.
Source: Adapted from Makino et al. (2009).

If major fisheries resources decrease, these stocks should be conserved and the target resource should be changed. In the Shiretoko area, the target species was changed from walleye pollock to chum salmon in the early 1990s. Target switching in fisheries is effective in multi-species fisheries management (Matsuda and Katsukawa, 2002). In addition, the fishing-ban area for walleye pollock changed with stock quantities in 1995 and 2005 (Figure 5.5.5).

Unlike fisheries in developed countries, there is no centralized top-down management in traditional fisheries. Although Japan was modernized in the second half of the nineteenth century, the country still has a decentralized co-management system involving fishers and the government. There are many artisanal fishers in Japan (Makino and Matsuda, in press). The transaction cost of fisheries management is one of the strongest arguments against top-down management systems. In a co-managed system, the costs of monitoring, enforcement and compliance can be shared between the government and local fishers (Makino and Matsuda, 2005).

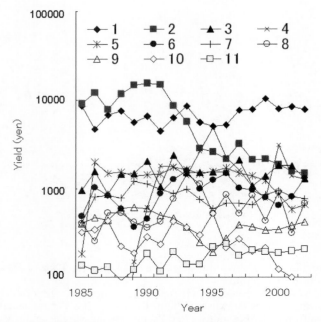

Figure 5.5.7 Fisheries yields for 11 major exploited taxa in the Shiretoko-daiichi, Utoro and Rausu Fisheries Cooperative Associations.
Notes: 1: chum salmon, 2: walleye pollock, 3: kelp, 4: common squid, 5: bighand thornyhead, 6: Pacific cod, 7: Arabesque greenling, 8: pink salmon, 9: sea urchin, 10: scallop and 11: *Octopus dofleini*.
Source: Ministry of the Environment, Government of Japan and Hokkaido Prefectural Government (2007).

5-5-4 Conclusion

Possible future perspectives on sustainable ecosystem services were introduced that offer a possible alternative theory of sustainable fisheries yield. Modern fisheries management and co-management theory were also discussed. The Japanese use many types of fisheries resources. The mean trophic level of Japanese fisheries is much higher than the world average, and it has not decreased. Moreover, the MTL has shortcomings as an indicator of overfishing. The potential availability of fisheries resources in Japan's Exclusive Economic Zone is still large, and Pacific saury and anchovy are under-used. This suggests that warnings about world fisheries do not reflect the degradation of the marine ecosystem, but rather are caused by a mismatch between supply of and demand for fish species. Co-management in Japanese coastal fisheries has worked well, at least in the Shiretoko World Natural Heritage Site. All of these different circumstances suggest new insights into sustainable fisheries.

Acknowledgements

Sincere thanks are owed to Drs D. Beard, J. C. Castilla, Y. Hiyama, M. Ijima, Y. Katsukawa, M. Makino, Y. Sakurai, M. Sano and A. Yatsu for their valuable comments, the members of the Rausu Fisheries Cooperative Association, the Scientific Council for the Shiretoko World Natural Heritage Site, the Hokkaido Prefecture and the Japanese Ministry of the Environment for providing unpublished data. This work is partly supported by a grant from the Global Environment Research Fund (H-092) of the Ministry of the Environment, Japan, and the Global Centers of Excellence program (E-03) of Japan's Ministry of Education, Culture, Sports, Science and Technology (MEXT).

REFERENCES

Baumgartner, Tim R., Andy Soutar and Vicente Ferreira-Bartrina (1992) "Reconstruction of the History of the Pacific Sardine and Northern Anchovy Populations over the Past Two Millennia from Sediments of the Santa Barbara Basin, California", *CalCOFI Reports* 33: 24–40.
Castilla, Juan C. and Omar Defeo (2005) "Paradigm Shifts Needed for World Fisheries", *Science* 309: 1324–1325.
Costanza, R., R. d'Arge, R. de Groot, S. Farber, M. Grasso, B. Hannon, K. Limburg, S. Naeem, R. V. O'Neill, J. Paruelo, R. G. Raskin, P. Sutton and M. van den Belt (1997) "The Value of the World's Ecosystem Services and Natural Capital", *Nature* 387: 253–260.
Delgado, Christopher L., Nikolas Wada, Mark W. Rosegrant, Siet Meijer and Mahfuzuddin Ahmed (2003) *Fish to 2020: Supply and Demand in Changing Global Markets*. Worldfish Center Technical Report 62. Washington DC: International Food Policy Research Institute; Penang, Malaysia: WorldFish Center.
Hampton, John, John R. Sibert, Pierre Kleiber, Mark N. Maunder and Shelton J. Harley (2005) "Fisheries: Decline of Pacific Tuna Populations Exaggerated?", *Nature* 434: E1–E2.
Hannesson, Rögnvaldur (1996) *Fisheries Mismanagement: The Case of the North Atlantic Cod*. Oxford: Blackwell Scientific Publications.
Kawai, Hiroaki, Akihiko Yatsu, Chikako Watanabe, Takumi Mitani, Toshio Katsukawa and Hiroyuki Matsuda (2002) "Recovery Policy for Chub Mackerel Stock Using Recruitment-per-Spawning", *Fisheries Science* 68: 961–969.
Maguire, Jean-Jacques, Michael Sissenwine, Jorge Csirke, Richard Grainger and Serge Garcia (2006) "The State of World Highly Migratory, Straddling and

Other High Seas Fishery Resources and Associated Species", FAO Fisheries Technical Paper 495.

Makino, Mitsutaku and Hiroyuki Matsuda (2005) "Co-management in Japanese Coastal Fishery: Institutional Features and Transaction Cost", *Marine Policy* 29: 441–450.

Makino, Mitsutaku and Hiroyuki Matsuda (in press) "Fisheries Diversity and Ecosystem-based Co-management", in R. Ommer, I. Perry, P. Cury and K. Cochrane (eds), *Coping with Global Changes in Social-Ecological Systems*. Oxford: Wiley-Blackwell.

Makino, Mitsutaku, Hiroyuki Matsuda and Yasunori Sakurai (2009) "Expanding Fisheries Co-management to Ecosystem-based Management: A Case in the Shiretoko World Natural Heritage Area, Japan", *Marine Policy* 33(2): 207–214.

Matsuda, Hiroyuki and Peter A. Abrams (2006) "Maximal Yields from Multi-Species Fisheries Systems: Rules for Systems with Multiple Trophic Levels", *Ecological Applications* 16: 225–237.

Matsuda, Hiroyuki and Peter A. Abrams (2008) "Can We Say Goodbye to the Maximum Sustainable Yield Theory? Reflections on Trophic Level Fishing in Reconciling Fisheries with Conservation", in J. L. Nielsen, J. J. Dodson, K. Friedland, T. R. Hamon, J. Musick and E. Verspoor (eds), *Reconciling Fisheries with Conservation: Proceedings of the Fourth World Fisheries Congress*. American Fisheries Society Symposium Series 49. Bethesda, MD: American Fisheries Society, pp. 737–744.

Matsuda, Hiroyuki and Toshio Katsukawa (2002) "Fisheries Management Based on Ecosystem Dynamics and Feedback Control", *Fisheries Oceanography* 11(6): 366–370.

Matsuda, Hiroyuki, Mitsutaku Makino and Koji Kotani (2008) "Optimal Fishing Policies That Maximize Sustainable Ecosystem Services", in K. Tsukamoto, T. Kawamura, T. Takeuchi, T. D. Beard, Jr and M. J. Kaiser (eds), *Fisheries for Global Welfare and Environment: Memorial Book of the 5th World Fisheries Congress 2008*. Tokyo: TERRAPUB, pp. 359–369.

Matsuda, Hiroyuki, Mitsutaku Makino and Yasunori Sakurai (2009) "Development of Adaptive Marine Ecosystem Management and Co-management Plan in Shiretoko World Natural Heritage Site", *Biological Conservation* 142: 1937–1942.

Matsuda, Hiroyuki, Yasuto Takenaka, Tetsukazu Yahara and Yuji Uozumi (1998) "Extinction Risk Assessment of Declining Wild Populations: In the Case of the Southern Bluefin Tuna", *Researches on Population Ecology* 40: 271–278.

Ministry of the Environment, Government of Japan and Hokkaido Prefectural Government (2007) *The Multiple Use Integrated Marine Management Plan and Explanatory Material for Shiretoko World Natural Heritage Site*, December. Available at: <http://dc.shiretoko-whc.com/data/management/kanri/seawg_kanri_en.pdf> (accessed 23 March 2010).

Mori, Mitsuyo, Toshio Katsukawa and Hiroyuki Matsuda (2001) "Recovery Plan for the Exploited Species: Southern Bluefin Tuna", *Population Ecology* 43: 125–132.

Myers, Ransom A. and Boris Worm (2003) "Rapid Worldwide Depletion of Predatory Fish Communities", *Nature* 423: 280–283.
Pauly, Daniel, Villy Christensen, Sylvie Guénette, Tony J. Pitcher, U. Rashid Sumaila, Carl J. Walters, R. Watson and Dirk Zeller (2002) "Towards Sustainability in World Fisheries", *Nature* 418: 689–695.
Persoon, Gerard A. van Est, M. E. Diny and Percy E. Sajise, eds (2005) *Co-Management of Natural Resources in Asia: A Comparative Perspective*. Copenhagen: Nordic Institute of Asian Studies Press.
Pikitch, E. K., C. Santora, E. A. Babcock, A. Bakun, R. Bonfil, D. O. Conover, P. Dayton, P. Doukakis, D. B. Fluharty, B. Heneman, E. D. Houde, J. Link, P. A. Livingston, M. Mangel, M. K. McAllister, J. Pope, K. J. Sainsbury (2004) "Ecosystem-Based Fishery Management", *Science* 305: 346–347.
Secretariat of the Convention on Biological Diversity (2006) *Global Biodiversity Outlook 2*, Montreal, Canada.
Watanabe, Yoshio, Hiromu Zenitani and Ryo Kimura (1995) "Population Decline of the Japanese Sardine *Sardinops melanostictus* Owing to Recruitment Failures", *Canadian Journal of Fisheries and Aquatic Sciences* 52: 1609–1616.
World Resources Institute, ed. (2005) *Ecosystems and Human Well-being: Synthesis*. St Louis, NO: Island Press.

6
Bridging between sustainability and governance

6-1
The concept of sustainability governance

Nobuo Kurata

Although there is no clear definition of sustainability governance, it is possible to agree that it is decision-making of a local participatory type with the aim of building a sustainable regional society. Sustainability governance is not governance of the top-down type under the leadership of an administration. Nor is it "management". Rather it is a bottom-up style of decision-making that involves various participating citizens and stakeholders.

6-1-1 Sustainability governance

The term "sustainability governance" might bring to mind processes whereby regulations, legislation, ordinances and guidelines are created under the leadership of government offices. However, the conception of "sustainability governance" used here is not governance of the top-down type but a participatory type.

Indeed, it is easy to point out that encouraging stakeholder "participation" is often tricky. Stakeholders have different values and opinions, which might make reaching consensus difficult. In order to encourage decision-making that is participatory, it is necessary to find out and clarify stakeholders' values and to analyse the structure of the problems facing them and their understanding of a situation.

Finding the most appropriate type of participatory decision-making for sustainability governance has been one of the prime objectives of the

---

Designing our future: Local perspectives on bioproduction, ecosystems and humanity,
Osaki, Braimoh and Nakagami (eds),
United Nations University Press, 2011, ISBN 978-92-808-1183-4

research reported here. It necessitates introducing the opinions of, for example, the local populace and of non-profit organizations into actual "regional governance". This requires some kind of model of consensus-building that takes the opinions of consumers, farmers and the local populace into consideration. For this purpose, two Hokkaido cities, Furano and Obihiro, were selected for research on the model of local governance.

6-1-2 Sustainability science and values

In addition to the environment, which is a major factor in the concept of sustainability, public health and economic growth also have to be considered in order to create a sustainable policy.

In discussing environmental problems, a distinction is sometimes made between "strong sustainability" and "weak sustainability". "Weak sustainability" is based on contemporary mainstream economic theories, and its proponents maintain that the existing industrial structure does not have to be changed. Simply put, it is "economic sustainability". "Strong sustainability", on the other hand, places more value on obligations to future generations and on the conservation of natural resources than on economic development. This is the understanding of sustainability here. The value of the natural environment is an important element in this concept of sustainability, which requires the building of a recycling society and the formation of a social system that will not burden future generations.

However, in designing a sustainable society, it is also necessary to meet the basic needs of current generations. Strong sustainability may be valued in this society, but policies can never be "sustainable" if its own basic needs are not met.

Thus, the concept of "sustainability" is a kind of normative concept. The base of sustainability science is a fundamental normative value, that of the natural environment. Previously, the natural sciences have not been concerned with values, and natural scientists have been required to take a neutral stance on values. But scientists engaged in sustainability science, including environmental scientists, cannot possibly conduct research without incorporating some form of social value or political affairs. For example, conservation ecology is inseparable from value judgements.

6-1-3 Building a sustainable society

In order to make rational regional decisions, it is necessary to grasp the real conditions of the region and to know what kinds of stakeholders are involved. For participatory decision-making, it is necessary to articulate and clarify the problems confronting regional populations. It is also necessary to analyse the structure of various sustainability values, including environmental values, for rational decision-making about local sustainability. Therefore, to solve local problems requires a technique for making the relationships between actors and values visible. With this technique, it is possible to understand the structure of the relationships between stakeholders and the values they hold, and to clarify the purposes and values that should be shared among stakeholders.

Scientific knowledge, although undoubtedly necessary for political decision-making, is still woefully insufficient. Indeed, experts' judgements and ecological knowledge are indispensable in decision-making on regional environmental conservation and sustainable regional development. Moreover, the ecological field of regional research is often accompanied by scientific uncertainty, which requires consideration of how to treat uncertain scientific knowledge in society. Scientists (including social scientists) and engineers do not always have the necessary scientific knowledge and information to solve regional problems. Even if they do possess rational solutions for regional environmental problems, one fixed scientific answer may not always be possible owing to the lack of necessary information. Therefore, it is essential to consider how to make social decisions based on "uncertain" scientific knowledge.

Finding solutions to regional environmental problems often requires some kind of local knowledge. Laypeople may have local knowledge that can contribute to local decision-making, so it is necessary to present this knowledge to the experts. In order to share local knowledge, new techniques for local decision-making and consensus-building must be tried.

However, there are some obstacles to participatory local decision-making. The relationships between stakeholders in a region may not be very close. In Furano, for instance, communications between agricultural groups and people engaged in the tourism industries are inadequate. It is thus necessary to hold workshops among stakeholders who lack the communication network necessary to achieve consensus. It is also necessary to employ a coordinator who can facilitate regional communication.

In most cases, disagreements between laypeople are based on some kind of rational thinking. However, when appropriate facilitation is lacking, there is a danger that such disagreements might descend into irrational disputes. Some social rationality is necessary for "regional governance" or for regional decision-making, so developing appropriate methods of consensus-building to maintain social rationality is another problem that needs to be examined. The problem of locating sustainability in a deliberative democracy must be considered.

BIBLIOGRAPHY

Kagaya, Seiichi (2008) "Tokachi-gawa ni okeru Shiminsanka ni yoru Kasenkanyko Seibi Sakutei Sisutemu", in Yoshinobu Kumada and Kayoko Yamamoto (eds), *Kankyoshimin ni yoru Chiikikankyoshigen no Hozen*. Tokyo: Kokinsyoin, pp. 197–214.

Motoda Yuka, Kudo Yasuhiko, Shiroyama Hideaki, Kato Hironori and Tsuji Nobuyuki (2009) "Chiho Jichitai no Jizokukanousei ni kansuru Kankei Acta no Mondai Kouzou Ninshiki: Hokkaido Furanoshi wo Jirei to shite" [Problem Perception of Main Local Actors on Sustainability in Furano City, Hokkaido], *Shakaigijyutsu Kenyku Ronbun Syu* 6: 121–146.

6-2
Problem-structuring for local sustainability governance: The case of Furano

Yuka Motoda, Yasuhiko Kudo, Nobuyuki Tsuji, Hironori Kato and Hideaki Shiroyama

6-2-1 Considering sustainability governance at the local level

Policies made and implemented by local governments directly affect how citizens lead their daily lives, making it necessary to learn how to govern sustainability at the local level. As section 6-1 has argued, however, "sustainability" is a complex, broad concept that requires a number of issues to be considered if it is to be transformed into concrete activities.

First, sustainability requires multi-sectoral policies for its realization. For example, in order to sustain and improve our lives we need policies that address environmental problems, promote industrial activities, prepare for ageing-associated problems, and so on. In addition, the design of the necessary fiscal management systems, decision-making processes and policy implementation systems that shape and maintain them must be argued at the institutional level (Denser, 2002; Ueta 2004).

Secondly, the wide range of sustainability-related policies suggests the existence of many interested actors. Recent arguments concerning governance indicate that governments are no longer the only actors forming and stewarding the rules regulating the public realm: actors from civil society and the private sector also manage economic, political and social processes within society, making it important to analyse how different actors are heard through the governance process (Pierre and Peters, 2000).

Thirdly, in relation to the first and second points, the pursuit of sustainability at the local level brings together different actors with divergent ideas and opinions concerning which issues are important. This leads to confusion over the issues at stake and the subsequent implementation of incoherent policies (Ueta, 2004).

Using Furano City as a case study, this section aims to find ways to govern sustainability at the local level by:
1. using interviews to understand how major local actors structure problems preventing the realization of sustainability,
2. identifying policy problems to be tackled by the city administration;
3. investigating feasible policies to address these problems so as to find out what the actors expect from one another.

6-2-2 Methodology and background

This study adopts a "problem-structuring method" for its enquiry. Detailed explanations of this method have been offered elsewhere,[1] so it suffices to give a brief outline of how the method was applied when designing the interviews for this study.

The first task was to map out the key issues, relevant policy sectors and major actors concerned with the city's sustainability governance process. Annual reports, policy plans, statistical materials and other policy papers published by the city government[2] were used to understand what they had done in order to realize and maintain local-level sustainability. Several policy sectors were selected that had strongly influenced local sustainability. As well as analysing the problems facing each sector, the relationships among the sectors were examined. Interviews with city employees working in these sectors deepened understanding of the city government approach to local sustainability.[3] Table 6.2.1 shows the policy sectors and the selected actors. Agriculture has been the key Furano industry since Hokkaido was first opened up at the end of nineteenth century.[4] Farmers and cooperatives quickly introduced large-scale mechanized farming systems to the region and they responded swiftly to national agriculture policies, switching from rice production to upland planting in the 1960s. Today, Furano City is one of the largest producers of carrots and onions in Japan. Tourism has long been expected to lead the city's economic activities. Resources include a famous ski resort area first developed in the 1960s, a long history as a film location for the popular television programme "Kita no kuni kara", beautiful scenery and an abundance of good food. Agriculture and tourism both affect the natural environment and energy use, suggesting the city needs to consider these sectors when analysing sustainability at the local level. Moreover, commerce was chosen as the third policy sector for investigation because it

Table 6.2.1 Outline of the interviews: Sectors, actors and date of implementation

Policy sector	Reasons for selecting sector	Actor	Date of implementation
Agriculture	Has been Furano's main industry since Hokkaido's reclamation era began	• Private farming household • Member of Agricultural Committee • Staff of Furano Agriculture Cooperative	Late Nov. – early Dec. 2007
Tourism	Is expected to play a leading role in the economy in Furano	• Private dairy farming household • Staff of Furano Tourism Association • Member of local Hotel Association • Manager of a major hotel	Late June 2008 Late Nov. – early Dec. 2007
Commerce	Is expected to make linkages between agricultural and tourism activities	• Staff of Furano Chamber of Commerce and Industry • Member of Committee on Renovation in the City Centre • Restaurant owner	Late June 2008
Waste disposal and recycling	Has successfully established a comprehensive garbage separation system resulting in a high recycling rate (93%) of domestic garbage	• Member of federation of neighbourhood associations • Member of Furano Citizens' Environment Council	Late Nov. – early Dec. 2007
Social welfare	Rapidly ageing population	• Staff of welfare section of Furano government • Care worker in a private welfare institution • Manager of a private hospital (specializing in internal medicine)	Late Nov. – early Dec. 2007 Late June 2008
Mass media	Offer information on the local political situation	Chief editor of local newspaper	Late Nov. – early Dec. 2007

was expected to create linkages among the activities that arose from agriculture and tourism.

However, if Furano inhabitants are to lead long and healthy lives, waste disposal and recycling policies also need analysis. These policies reveal how the city copes with its resource-related problems, an obvious concern for long-term sustainability. Furano City government's long efforts to reduce waste disposal and establish a strict recycling system – requiring citizens to sort their garbage into 14 different categories – merit special attention here. Thus waste disposal and recycling were also selected as the fourth policy sector. At the same time, what effects an ageing population has had on certain Furano activities was examined. Although not directly incorporated into the analysis, the local mass media were included to obtain information on the local political context at work.

Secondly, in preparation for the interviews, figures were created that showed how the interviewers predicted the interviewees would structure the problems to be tackled with regard to realizing local-level sustainability (for further information on how to create these figures, see Figure 6.2.1). In so doing, information from various sources was utilized: interviews with city officials; reports prepared by the interviewees; homepages; arguments developed in similar case studies; and so on. Each interviewee was informed in advance of the purpose and outline of her/his interview as well as the underlying hypotheses displayed in her/his own problem-structuring figure.

Following the problem-structuring method during the interviews, the problem-structuring figures were shown and interviewees were asked to identify any issues that would prevent them from living long healthy lives in the future (i.e. achieving sustainability in their lives) and the necessary actions for addressing them. The hypotheses demonstrated in the problem-structuring figures were corrected in front of the relevant interviewees when necessary, in line with their suggestions. (Figures 6.2.2 and 6.2.3 offer an example of how the figures created beforehand were modified as the interviews proceeded.)

Three essential questions were asked in each interview: what the interviewees' goals or aims were; what would prevent them from pursuing these goals and aims; what they would expect from whom in order to achieve these goals and aims. In addition, the interviewees were asked to identify any problems or solutions they felt were missing from the prepared problem-structuring figures. In this manner, opportunities were guaranteed for each interviewee to present her/his viewpoint free from the context set by the hypotheses. At the same time, any questions and uncertainties were clarified whenever necessary during the interviews. It

1. Definitions for the nodes and lines of the problem-structuring figure

◇ : the aims and goals of the relevant actor
○ : state of affairs, conditions and environment around the relevant actor
□ : actions taken by the relevant actor
↑ : action/influence inducing positive change
⇡ : action/influence inducing negative change

2. How to make problem-structuring figures?

Imagine a case in which the relevant actor relates its aims and goals to its surrounding environment when taking actions as follows:

The actor's aim is to achieve 'sound farm management' in order to have a stable, healthy life over the long term. The 'cold climate' in the actor's farming area easily causes 'cold weather damage'. To adapt to the surrounding natural environment, the actor promotes the 'development of mechanized farming systems'. However, this in turn results in 'increased production costs', which hampers the 'increased income from farming'. Since such 'increased income from farming' contributes to the realization of 'sound farm management', it can be said that 'development of mechanized farming systems' hampering 'increased income from farming' has a negative influence on the realization of 'sound farm management'.

This actor's problem-structuring figure can be presented as follows:

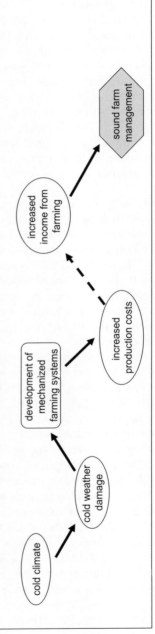

Figure 6.2.1 Creating problem-structuring figures.

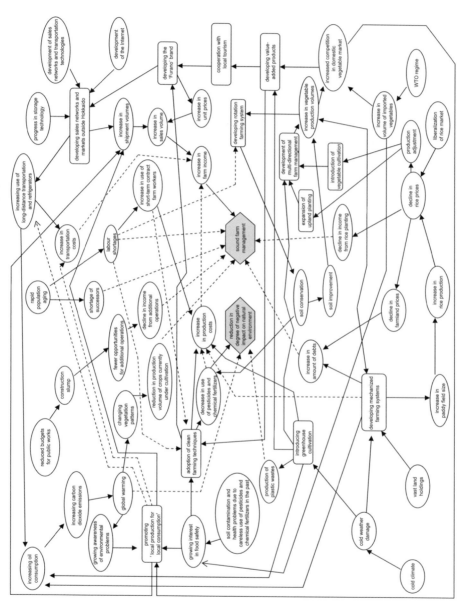

Figure 6.2.2 Problem-structuring figure of member of Agricultural Committee: Before the interview.

293

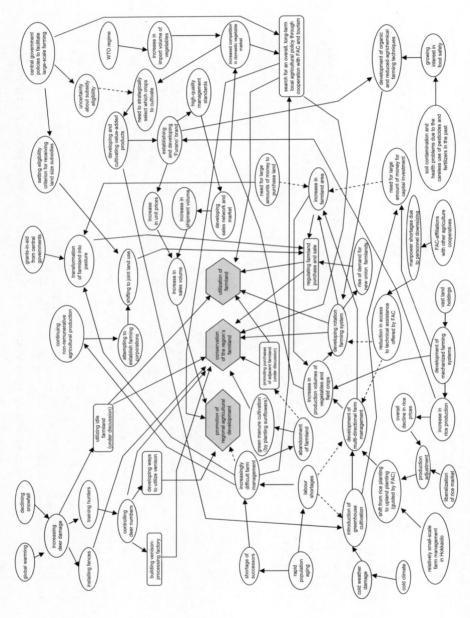

Figure 6.2.3 Problem-structuring figure of member of Agricultural Committee: After the interview.

usually took about an hour and a half to two hours to conduct each interview, the average length designated by the problem-structuring method.

The problem-structuring method has the advantage of helping interviewees relate their problems and the actions necessary to address them more concretely by referring to their arguments on the hypotheses displayed before them. This merited special attention in the interviews that dealt with the concept of sustainability: sustainability is such a broad, complex concept that interviewees would often get confused if asked to state their viewpoints on it without any clues.

The interviews took place over two stages, as shown in Table 6.2.1. The first group of interviews was held between late November and early December 2007. A total of 10 actors from 4 policy sectors (agriculture, tourism, waste disposal and recycling, and social welfare) and the local mass media were selected and interviewed at this stage. Realizing that more information would be necessary to deepen the analysis on local sustainability governance in Furano city, a second group of interviews was conducted in late June 2008. At this stage commerce was chosen as the fifth policy sector for investigation, and three actors were selected from this sector. Moreover, to grasp the ongoing changes triggered by an ageing society, two more actors were added from the welfare sector to the list of interviewees. To follow up the multidirectional development of agriculture, the aim was to obtain information on cooperation between dairy farming and agriculture from a dairy farmer.

6-2-3 Actors' perceptions: How they structure problems on sustainability

Agriculture

Growing numbers of private farming households in Furano have recently adopted either organic or low-agrichemical farming techniques. The farm with which an interview was conducted is a good example of this.

This farming household strives to make farm management a worthwhile activity, producing high-quality crops at low cost with minimum energy consumption. This is achieved by keeping pesticide use to a minimum; improving the soil through less frequent tractor use; and cooperating with nearby dairy producers in land use and soil improvement.

This farming household markets and sells most of its high-quality crops directly to customers, largely bypassing the channels offered by the Furano Agricultural Cooperative (FAC). Word of mouth is the primary means of advertising, with the women of the household using mail, telephone and the Internet for sales promotion. However, some products are shipped through the FAC distribution chain. Because evaluation of the FAC-branded crops in the market has a direct influence on the farmhouse's products, the farm asks the FAC to maintain a higher standard of quality control.

The role of the FAC is to promote agriculture within its region – of which Furano City is a part – to the benefit of its members and the regional economy. As such, it is one of the most important actors in Furano. In contrast to the private farming household mentioned above, the FAC has established a large-scale network to ship the crops produced by its members. It also provides farmhouse management and technical assistance.

In addition, the FAC cooperates with the local tourism industry, by developing farmland in ways that contribute to the formation of beautiful pastoral landscapes as well as to soil improvement. The FAC also goes to great lengths to promote low-agrichemical farming methods. Concerned that private farmers who employ organic farming techniques without *any* use of pesticides might potentially cause harm to neighbouring farmland, the FAC strongly argues that it is necessary to adopt consistent methods of farming.

The Agricultural Committee, a part of the Furano City government, is primarily concerned with the conservation of the region's farmland. However, owing to the ageing population, many Furano farming households suffer from labour shortages. This results in ever-increasing areas of abandoned farmland, in effect thwarting the Agricultural Committee from achieving its objective. That purchasing such abandoned farmland in Hokkaido is prohibitively expensive simply exacerbates the problem. Thus, even well-off farming households would find it difficult to amalgamate nearby abandoned land. To address these constraints, the Agricultural Committee seeks to develop frameworks – including establishing farming corporations – within which joint land use could be facilitated.

The impact of the central agricultural policy introduced in 2007 by the Liberal Democratic Party – Komei Party coalition government to support good farm management was another concern during the interview period. Even though the policy was thought to facilitate large-scale farming by offering subsidies to those farms cultivating designated products on farmland areas larger than the criterion set by the central government, the details of the whole system had not yet been clearly specified. With the change of central government in 2009, the Agricultural Committee will have to keep a close eye on any new agricultural policy in order to develop its own farmland management policy in Furano.

According to the Agricultural Committee, personnel downsizing in conjunction with FAC affiliations with other agricultural cooperatives has resulted in staff shortages and a commensurate reduction in the technical assistance it offers to its members.

For a dairy farming household, it is necessary to offset the costs accumulated as a result of the high unit land values and to make its business more economically viable. It cultivates dent corn and cooperates with field husbandry farmers through the switch between cultivated land for dent corn and that for onions, which is of benefit to both sides. Land integration is considered essential if efficient farm management is to be pursued further. International factors have greatly influenced the household's business. For instance, if the yen is revalued, dairy products find it difficult to compete with imported counterparts in the domestic market. In order to buffer any negative impacts of the international economy, the farmhouse plans to reduce the amount of feed crop it produces by decreasing fertilizer inputs, so as to boost the price of milk while reducing its yield.

Tourism

The Furano Tourism Association (FTA) seeks to develop tourism in Furano, which is presently dominated by two separate external factors: the construction of a "skiing town" under the leadership of a major resort developer in the 1960s; and the boom in the popularity of the television programme "Kita no kuni kara". Recent declines both in the size of the domestic skiing population and in the popularity of "Kita no kuni kara" have forced the FTA to establish its own tourism policy for the future. It has listed the following as its immediate tasks: developing services for foreign tourists; organizing an extensive tourism network covering the nearby Asahiyama Zoo; diversifying services to correspond to the seasonality of tourist movements. To maximize the economic effects, the FTA aims to increase the number of tourists lodging in Furano and to attract tourists to the city year-round. To achieve this, the FTA is aware that it has to cooperate with the agricultural sector – through green tourism or agricultural study tours – and the commercial sector to develop a comprehensive package of measures to promote tourism in Furano. However, just how the FTA will establish close links with both sectors remains to be seen, since previous attempts have been sporadic at best.

Broadly speaking, there are two types of lodging accommodation in Furano: large hotels run by major organizations; and small privately owned hotels and inns. The latter belong to the Hotel Association. According to the interviewee in this study, the shabby facilities of the smaller hotels do not meet customer demands for Western-style rooms equipped with beds, bathrooms and so on, and are thus under-utilized. That many of their managers find themselves getting too old to refurbish these outdated facilities exacerbates the problem. At the same time, lodging facilities in Furano are forced to reduce running costs in order to compete with other sightseeing areas – Okinawa is a strong potential rival – in the Japanese domestic tourism market. Since the costs of building and maintaining lodging facilities are fairly uniform across Japan, the managers in Furano have to cut their profit margins in order to set their prices low enough to be competitive. As a means to fend off such pressure to lower prices, the Hotel Association is trying to develop value-added services targeted at elderly people.

Another interviewee was the manager of a major hotel. Part of a nationwide resort development group, this hotel is the key actor among Furano's lodging facilities. To offer satisfactory services to its guests, the hotel regularly renovates its facilities and has constructed a new annex and drilled a hot spring. Such efforts have allowed the hotel to attract more guests but, for its adjacent skiing facilities to be commercially viable in the winter season, it needs more customers than it can accommodate. It needs the small hotels and inns nearby to attract more skiing guests.

The hotel has established close ties with local businesses: it is a member of the FTA and the Furano Chamber of Commerce and Industry (FCCI), purchases farm products from the FAC, places orders with local companies wherever possible, and employs local residents as its workforce.

Commerce

The role of the FCCI is to offer its members information and support for their commercial activities. Moreover, it should lead the local economy into the future. With its head active in the administrative and fiscal reform activities of Furano City government, the FCCI regards itself as a good partner of the city government. However, it expresses dissatisfaction with the city government's slow responses to its appeals. It thus plans to develop its own projects, such as establishing a shopping mall in the city centre, while strengthening its conventional activities, including lobbying. At the same time, the FCCI has to strategically utilize the projects subsidized by the central government because both the FCCI and the city government suffer from a lack of resources.

Commerce is expected to provide linkages between the two major industries in the Furano City economy – agriculture and tourism. The FCCI and FAC have started jointly processing "local" agricultural products, and many FCCI members belong to the FTA. A further link between commerce and tourism will centre around developing products targeted at tourists, mostly in the form of locally processed goods. To make the most of the "Furano" image, it will be necessary for the FCCI to establish a large-scale production system and sales network through which locally sourced materials are processed and sold.

As with agriculture and tourism, rapid population ageing has become a serious problem in commerce, with about 50 per cent of the shops and stores in Furano reported to have no successors. These businesses have no other option but to close down when their current managers retire. Local shopping areas inevitably decline as they fail to attract customers and new investments. Cooperative relationships with large-scale suburban stores headquartered outside the region must be established to cope with this vicious circle.

Those engaged in commercial activities deem it necessary to renovate the city centre if Furano City is to continue to promote itself as a sightseeing area: it has to attract visiting tourists into the city centre. The Committee on Renovation in the City Centre (CRCC), which was established to address these urgent needs, has been working with the FCCI to develop a private-sector-driven land-use plan for a vacant lot in the city centre. This time, priority has been given to private initiatives because of the poor reviews given to previous renovation projects around Furano station led by the city government. Local consultancies have been recruited at early stages of planning. A highlight of the plan is the construction of a shopping mall offering an array of high-quality foods produced in the Furano area. This mall is expected to entice visitors away from Route 38 to explore the city centre around Furano station. It is hoped this will make the city centre more accessible to both tourists and local residents alike. For this plan to succeed, the communication services offered by the station require substantial improvement. For the mall to operate sustainably, the CRCC plans to create a close relationship with the FAC, develop its own sales networks and strengthen cooperation with local restaurants and major hotels.

The Furano restaurant industry attracts two types of customer: local residents and tourists from outside the area. The majority of customers at a restaurant whose owner was interviewed here consist of tourists in summer and local people in winter. The restaurant offers a wide variety of dishes at various prices in order to accommodate different demands from the two customer types. Articles in women's magazines, followed by word of mouth and the Internet, have proven to be the main modes of advertising for the restaurant.

The owner opened his restaurant in Furano because he could attract customers in both summer and winter. He purchases fresh produce directly from local farms and fishers. Although the FTA sometimes introduces tourists to the restaurant, the owner thinks that the relationship between the FTA and newcomers like himself is underdeveloped. He argues that Furano needs to nurture its image by creating its own culture rooted in rural life. For this purpose, he insists that the city government should play a role in promoting regional development.

Waste disposal and recycling

The daily activities of Furano's neighbourhood associations are the cornerstone of its successful waste disposal and recycling services. Examples of the strategies devised by neighbourhood associations to persuade their members to sort their garbage into 14 categories include: participating in garbage separation seminars offered by the city government; asking members to write their names on their garbage bags when they deposit them at the collection points; having some of their members stand at the collection sites to check how others put out the garbage.

However, interviews revealed that the amount of waste has not declined despite the best efforts of the neighbourhood associations. Moreover, some people find it difficult to sort their garbage because many items do not clearly indicate the raw materials from which they are produced. Plastic garbage illustrates this point well: the confusion about which plastics belong in which garbage category is such that the disposal and recycling facility cannot process garbage from Furano smoothly. Another challenge raised in the policy sector is how to address illegal dumping. Tourists as well as local people are known to dump their waste illegally in the mountains or the woods.

The Furano Citizens' Environment Council (FCEC) was set up in January 2003 to protect and look after the urban environment by mobilizing and coordinating the initiatives of a wide range of actors, including Furano citizens and local enterprises. The FCEC insists that Furano citizens make continuous efforts to protect their environment, by offering them opportunities to understand the current environmental problems they face and to analyse possible measures to address them.

Social welfare

The Furano City government admits that a wide gap exists between what citizens demand as social welfare services and what the government actually offers them. The growing elderly population has led to increased demands for welfare services, but the local government lacks the revenue to adequately fund policies to meet the demands. Some from the agricultural sector argue that creating an immigration policy to encourage middle-aged people to move to Furano would promote regional development. Such a policy would, however, be an enormous financial burden on the government, since it would inevitably result in more citizens needing social welfare services.

According to a staff member from the welfare section of the city government interviewed for the study, local social workers play an important role in understanding how the elderly are faring. At the same time, local networks, underpinned by the neighbourhood associations, are active in taking care of their elderly neighbours on a day-to-day basis by, for example, clearing snow from in front of their houses or taking their garbage to the collection sites. The city government hopes that non-profit organizations (NPOs) and volunteers can also contribute to improving welfare services for the elderly. But in reality, as economic stagnation continues, fewer and fewer people have the time to engage in welfare-related activities. The dual structure of the city – a vast farming area surrounding an urban centre – precludes the efficient supply of public services. To overcome this, the government encourages elderly persons living in rural areas to move to the urban centre. Whether such actions can turn Furano into a "compact city" remains to be seen, since few seem willing to respond to the government's requests.

According to a care worker in a private welfare institution, only a horizontal network consisting of trained community leaders who can provide coverage going well beyond that presently offered by the public services will be able to address the difficulties faced by elderly people living alone. Drawing on the successful experience of an organized walk along the riverside footpath, which gave participants the opportunity to get to know one another and to forge close ties, the care worker argued that it is possible to design new types of services aimed at promoting both the welfare of local residents and the development of tourism.

The current problems facing people in Furano – including limited employment opportunities for the young and deteriorating employment conditions – are undermining the ties that have traditionally united local communities. If Furano City is to continue its pursuit of regional development in the years to come, it will be essential to mobilize the generations who are still in work to support community-building.

The private hospital selected for the interview helps local residents lead healthy lives by offering them medical care. With 15 per cent of the city's total population aged 75 years and older, the hospital recognizes that Furano is becoming an ageing society but it does not think that it suffers from the problems of ageing more than any other area. Most of its patients either are suffering from lifestyle-related diseases or are elderly (one-third of patients are 65 years old and above). How the elderly people living in a rural area can secure access to medical services has recently become an important policy issue.

The manager of the hospital, who is also a medical doctor, recognizes local concerns about the quality of medical facilities in Furano, which in his view is adequate. Local hospitals are classified on the basis of the severity of patients' symptoms. When patients fall very seriously ill or develop symptoms other than those of internal disease, the hospital refers them to higher medical institutions in Asahikawa. Similarly, in the area of emergency medical care, the local hospitals offer primary care to patients on a rotating schedule, while the central municipal hospital handles any cases requiring secondary medical care. The manager argues that, in order to further improve the quality of medical services in Furano, the city government should develop a more comprehensive emergency care system and reduce the burden currently shared by individual local hospitals.

6-2-4 Analyses: What are the issues? Who expects what from whom?

Issue identification

Considering the information thus obtained, several issues were then identified that need to be overcome if the people of Furano are to lead sound, healthy lives in the years to come. First the potential risks faced by the people living in Furano were looked at, then the issues were identified by examining possible means to mitigate these risks.

The potential risks defined in this analysis are: uncertainty about the agricultural production environment owing to increasing internationalization of the domestic market, frequent shifts in central agricultural policies and changes in consumer preferences; a reduction in the size of the domestic skiing population; a decline in the popularity of "Kita no kuni kara"; the rapid decline of the central city shopping area; and further transformation into an ageing society, leading to a shortage of successors in all sectors. As a possible strategy to lessen these risks, the utilization of existing resources should be pursued first and foremost. Further development of the "Furano" brand and diversification of business activities in each industry will serve as the backbone of the strategy.

Table 6.2.2 lists the six issues identified through this analysis. Reflecting the above-mentioned resource utilization strategy, cooperation among the different sectors assumes great importance. Accordingly, the basis for good cooperation will be established by enhancing the quality of individual business activities in each sector. Adequately financing public services and private initiatives will require raising revenues. Development of the local culture is expected to improve the quality of life of Furano people on the one hand, and promote the regional economy on the other. Finally, or all these issues to be tackled, better coordination among institutions as well as capacity for those aiming to facilitate cooperation with others must be developed.

Table 6.2.2 What should be done to achieve sustainability governance in Furano? Groups of issues identified from analysing information obtained from the interviews

Group 1: Enhancing the quality of business activities in each sector

Strengthening management bases for agriculture
 Example: Promoting the amalgamation of farmland and pastureland for efficient farm management and area-wide land conservation
Building basic infrastructure for tourism
 Example: Renovating shabby lodging accommodation to provide services to satisfy customer needs at a fair price and to increase utilization of the entire city's lodging capacity
 Example: Creating barrier-free environments for better public amenities to be enjoyed by all people, including elderly persons

Group 2: Promoting cooperation between two different industries/sectors

Promoting cooperation between the agricultural and tourism sectors
 Example: Facilitating initiatives by rural women for business diversification
 Example: Encouraging retired persons to engage in small-scale agriculture as new providers of agricultural study tours
Promoting cooperation between the tourism and waste disposal and recycling sectors
 Example: Organizing visits to highly acclaimed recycling facilities as a new tour package and promotion of a "recycling culture"
Promoting cooperation between the tourism and welfare sectors
 Example: Organizing "health tours" targeted at elderly persons both living in and visiting Furano
Promoting cooperation between the education sector and other individual industries/sectors
 Example: Facilitating "dietary education" for knowledge building about local agriculture and its products, thereby contributing to menu planning targeted at tourists by referring to local dishes

Group 3: Promoting multi-sectoral cooperation among agriculture, tourism and commerce

Promoting "local production for local consumption"
 Example: Developing new products and services targeted at tourists by utilizing locally produced agricultural commodities, accompanied by a building up of supply capacity and sales networks
Building capacity to cooperate in concrete terms
 Example: Accumulating consultation experience and developing institutional arrangements for a renovation plan of the city centre

Group 4: Enhancing the fiscal base

Creating new sources of revenue
 Example: Organizing new tour packages including visits to recycling facilities and health tours
Generating revenues to finance activities to nurture a sound environment
 Example: Building new facilities including toll car parking in the city centre and tourist spots in order to generate revenue

Table 6.2.2 (cont.)

Group 5: Nurturing and utilizing local culture

Nurturing "Furano hospitality" and extending it to people both living in and visiting to Furano
 Example: Improving public amenities by establishing barrier-free environments, reforming medical care systems, and maintaining strict garbage recycling systems
Developing new products and services attractive to tourists staying in Furano
 Example: Organizing festivals and events by mobilizing local initiatives in ceramic art, music and drama as well as the food service industry (through "dietary education")

Group 6: Promoting institution-building and human resource development

Promoting institutional frameworks for cooperation
 Example: Setting up rules, procedures and relevant facilities, including offices for implementing the city centre renovation plan
Developing the capacity of actors involved
 Example: Training coordinators and leaders to facilitate networking and cooperation

Mutual expectations

It is necessary to take into account existing relationships among actors in order to develop viable strategies for the issues. The agricultural and tourism sectors work together in very few instances. This is at odds with the frequent argument that agriculture and tourism work in tandem to lead the Furano economy. Agricultural actors insist on handling the necessary tasks to promote cooperation with their tourism counterparts. For example, the FAC considers that it bore most of the burden when it organized agricultural study tours. For those in agriculture to call for more future cooperation initiatives from the tourism sector should come as no surprise. In particular, agricultural actors expect more local agricultural products to be utilized when the tourism sector develops new services. Likewise, tourism actors expect more support from agriculture. For instance, they would like local farm households to host a greater number of tourists participating in agricultural study programmes. They also expect more agricultural produce to be made available to the Furano tourism industry, thereby requiring changes to the current FAC distribution policies.

At the same time, actors from both sectors recognize the interrelatedness of their activities: landscape improvements would offer beautiful scenery to tourists and high-quality farm produce to farmers. They also agree to work in tandem to provide tours in which participants gain experience in farming, develop and exploit local products, and utilize the "Furano" brand. In order to further facilitate such initiatives, both sectors keenly feel the need to develop a mediator to provide both sides with information and to coordinate their often-conflicting roles. Whereas some expect that the Furano government will assume this function, others argue that NPOs should do so. Much has to be done before the actors of either sector can build any mediator capacity and develop institutions within which the mediators can fulfil their roles.

Before the urban city centre can be renovated, the commerce sector needs to establish cooperative relationships with the agriculture and tourism sectors. For example, in its plan to construct a shopping mall, the FCCI expects to continuously attract both tourists and local residents to the mall if it sells farm products supplied by the local agricultural sector. However, judging from the sporadic nature of the current links with the agricultural and tourism sectors, an institutional framework underpinning cooperation among the three sectors for the mall is still in its infancy. The restaurant in this study also appreciates its close ties with the two other sectors, through which it can obtain agricultural products from farmers and utilize information provided by the FTA. At the same time, it differs from the FCCI in its definition of local culture: the FCCI argues that both rural and urban lifestyles have to contribute to the formation of Furano culture; the restaurant owner insists that the people in Furano should focus more on the rural aspects of their lives to develop their own culture.

The labour shortage triggered by the rapid ageing of the population has constrained business development of the three sectors, each of which has failed to make it known what it expects from the city government to address the problem of an ageing population. On the other hand, the government officer argues that the people in Furano can lead sound, healthy lives thanks to the welfare services offered by the city government. At the same time, there is a possibility that mutually beneficial linkages have been established between the social welfare and tourism sectors, as was indicated earlier by the care worker in a private welfare institution. It should be noted, however, that a personnel shortage might erode the very foundation on which the countermeasures to the problem of ageing were developed: the question of how to mobilize a wider range of citizens when their number is in decline must be addressed.

It can be said that Furano's clean living environment is maintained by the activities of the actors involved in the waste disposal and recycling services. Those involved in tourism, for example, may regard such an environment as a source of hospitality to be offered to their visitors. Alternatively, the value of local farm produce may be boosted if increasing numbers of consumers regard the clean environment as the key factor in determining its quality. The actors involved in the disposal and recycling services, on the other hand, need the cooperation of their tourism counterparts in order to stop illegal dumping by tourists.

Whereas the FAC expects the city government to end its dependence on the FAC for planning and implementing agricultural policies, the government appreciates the FAC's financial contributions in the form of tax payments. Some actors in the tourism sector insist that, if their industry's financial contribution to the city coffers is properly valued, it should enjoy a stronger voice in the government's relevant fiscal policies, including the development of barrier-free public services, which appear attractive in the eyes of elderly tourists. However, the city government expects a greater financial contribution from the tourism industry in order to reduce its budget deficit. In response to expectations that the Furano government should act as a mediator between agriculture and tourism, the local government argues that it will not lead but will facilitate discussions between the two sides by offering to them a place to interact.

In managing both social welfare activities and the waste disposal and recycling services, the city government is attempting to develop and utilize citizen networks. The actors involved in these activities, however, find it necessary to develop networks whose functions and coverage far exceed those set up by the city government. At the same time, they admit that the government should continue to play a role in formulating the burden-sharing rules for waste disposal and in developing local medical systems.

The results of the analyses shown above were presented to a symposium held in Furano on 12 September 2008 to obtain feedback from the "ordinary citizens" of Furano. The basic line of argument developed thus far was accepted as relevant by the participants.

6-2-5 Concluding remarks: Implications from the analyses

The analyses suggest that, to realize local-level sustainability governance, it is necessary for the actors involved in the process to foster better cooperation with one another. However, the analysis of mutual expectations revealed that, even though the interviewees agree that cooperation among actors from different policy sectors is important, their actual commitment to promoting such cooperation is often weak. For example, actors in agriculture and tourism both assume that someone other than themselves should facilitate cooperation between the two sectors. In addition, the actors in both sectors expect greater contributions from their counterparts to promote cooperation while simultaneously regarding their own contributions as adequate. By way of conclusion, the study now investigates two possible factors preventing actors from working together.

First, the actors must decide with whom they will work in tandem. For example, they have to learn about the problems in the sector(s) to which their potential partners belong, analyse how these potential partners frame their interests into specific demands, and develop good knowledge about how relationships between potential partners and other actors might affect the activities of the former. Whether or not the actors are willing to bear such high costs and launch the process of building cooperative relationships with the partners they finally select depends on the degree of trust they have in these partners. Only when the actors are sure that their commitment to the partners will be well rewarded will they make it. Here the actors and their counterparts fall into "social dilemma" situations in which individual interests hamper cooperative actions.

To overcome such situations, the actors must understand that promoting cooperation with their partners also serves their own interests, and thereby become motivated to do so. At the same time, the actors must develop trust that their counterpart will do them no harm (Yamagishi, 1990). The literature on social psychology offers a wealth of examples in which mutual confidence among actors with competing interests is successfully built. Drawing on the lessons from these examples, it can be said that a viable strategy towards confidence-building is to develop a small network through which a limited number of actors interact with each other in an informal manner. As mutual confidence among actors is enhanced, the network will expand. In due course the network becomes large enough to facilitate cooperation between different sectors (Doba and Shinoki, 2008; Seiyama and Umino, 1991).

A second factor hindering the process to promote cooperation with partners is that the actors may become discouraged from participating if they are not well informed about which type of activities are at issue, on the one hand, and about who bears what kind of burden and responsibility, on the other. However, it is quite difficult to offer actors information on these points beforehand and to urge them to participate in the process, because the ways in which cooperation actually takes place become clear only *after* the actors begin to participate. With the aim of overcoming this difficulty as much as possible, the actors could refer to similar attempts to promote cooperation among different sectors in other areas or frequently hold discussions with their partners, thereby developing a shared image of future cooperation as a guide for their involvement hereafter.

Acknowledgements

The original, Japanese version of this section first appeared as: Yuka Motoda, Yasuhiko Kudo, Nobuyuki Tsuji, Hironori Kato and Hideaki Shiroyama (2009) "Chihoujichitai no jizokukanousei ni kansuru kankei akutah no mondaikozoninshiki: Hokkaido Furano-shi wo jirei toshite [Problem

Perception of Main Local Actors on Sustainability in Furano City, Hokkaido]", *Shakaigijutsukenkyu Ronbunshu* 6: 124–146.

Notes

1. See, for example, Kato et al. (2007), or the chapter by Kato in Komiyama, Takeuchi, Shiroyama and Mino (eds) (forthcoming), *Sustainability Science: A Multidisciplinary Approach*. Tokyo: United Nations University Press.
2. Many of these materials can be obtained from the government's website: <http://www.city.furano.hokkaido.jp/> (accessed 23 March 2010).
3. These interviews were conducted in June 2007 at Furano City Hall. The authors would like to thank these city employees for their generous cooperation on that and other occasions throughout the study.
4. For further information on the overview of the city, see the appendix in the chapter by Tsuji, Kudo and Tanaka in Komiyama, Takeuchi, Shiroyama and Mino (eds) (forthcoming), *Sustainability Science: A Multidisciplinary Approach*. Tokyo: United Nations University Press.

REFERENCES

Denser, Simon (2002) *The Principles of Sustainability*. London: Earthscan.

Doba, Gaku and Mikiko Shinoki, eds (2008) *Kojin to shakai no soukoku: shakaiteki jirenma apuroch no kanousei* [Beyond Conflicts between State and Individuals: How Can Social Dilemmas Approach Be Applied?]. Tokyo: Mineruba Shobo.

Kato, Hironori (forthcoming) "Problem-Structuring Methods Based on Cognitive Mapping", in Hiroshi Komiyama, Kazuhiko Takeuchi, Hideaki Shiroyama and Takashi Mino (eds), *Sustainability Science: A Multidisciplinary Approach*. Tokyo: UNU Press.

Kato, Hironori, Hideaki Shiroyama, Yoshinori Nakagawa and F. Fukuyama (2007) "Problem Structuring in Transport Planning: Cognitive Mapping Approach", *Proceedings of the 11th World Conference on Transport Research*. Berkeley, CA.

Pierre, Jon and B. Guy Peters (2000) *Governance, Politics and the State*. Basingstoke: Palgrave Macmillan.

Seiyama, Kazuo and Michiro Umino, eds (1991) *Chitujomondai to shakaiteki jirenma* [Problems of Order and Social Dilemmas]. Tokyo: Habesuto-sha.

Tsuji, Nobuyuki, Yasuhiko Kudo and Noriyuki Tanaka (forthcoming) "Field Study in Sustainability Education: A Case from Furano City, Hokkaido, Japan", in Hiroshi Komiyama, Kazuhiko Takeuchi, Hideaki Shiroyama and Takashi Mino (eds), *Sustainability Science: A Multidisciplinary Approach*. Tokyo: UNU Press.

Ueta, Kazuhiro (2004) "Jizokukanou na chiikishakai" [Sustainable Local Society], in Kazuhiro Ueta, Naohiko Jinno, Akira Morita, Mari Ohsawa and Takehiko Kariya (eds), *Jizokukanou na chiikishakai no dezain: seizon to amenithi no koukyoukukan* [Designing Sustainable Local Society: Public Space for Living and Amenity]. Tokyo: Yuhikaku.

Yamagishi, Toshio (1990) *Shakaiteki jirenma no shikumi: "hitori gurai no shinri" no maneku mono* [How Social Dilemmas Work: What Happens When Individual Interests Prevail]. Tokyo: Saiensu-sha.

6-3
Decision-making in sustainability governance

Seiichi Kagaya

6-3-1 Introduction

Sustainability is the balance between human society and the environment for future generations. To achieve this equilibrium, several complex problems need to be solved. Meanwhile, governance involves many organizations and stakeholders in addition to the government that takes decisions affecting others. Sustainability governance can be defined as a framework within which the global environment for future generations is discussed and then determined. Specifically, it is beneficial to make collective and comprehensive decisions in collaboration with the public sector, the private sector and civilian society. Collaboration among these sectors is necessary to tackle the broad and complex challenges of sustainability, and decision-making within such a framework is usually a group decision-making process. The group should comprise all stakeholders who have an interest in a particular decision, either as individuals or as representatives of a group (Hemmati, 2002).

Participation requires all stakeholders to have a voice in influencing decision-making and also ensures legitimacy in the governance system (Renn et al., 1995). After holding several stakeholder workshops, detailed policies are constructed that play important roles in the sustainability governance decision-making process. This process has been a frequent topic of discussion and has led to the introduction of strategic environmental assessment (SEA) (Annandale et al., 2001). SEA is a systematic and comprehensive process for evaluating the environmental effects of a

Designing our future: Local perspectives on bioproduction, ecosystems and humanity, Osaki, Braimoh and Nakagami (eds),
United Nations University Press, 2011, ISBN 978-92-808-1183-4

policy, plan or programme (PPP) and its alternatives, at the earliest appropriate stage of the publicly accountable decision-making process, thereby ensuring full integration of relevant biophysical, economic, social and political considerations (Partidario, 1999). The potential of SEA to improve governance has also been discussed (Kidd and Fischer, 2007), based on its ability to increase transparency, participation and inclusiveness by advocating a participatory and structured assessment process. In SEA, communication, participation and reporting have important roles to play by introducing the perspectives and inputs of different stakeholders into the PPP-making process.

The objectives of this study are to: introduce the notion of sustainability governance into decision-making; establish a method for assessing the alternative plans of the many stakeholders; develop a procedure for determining the most appropriate plan for sustainability governance.

6-3-2 Sustainability governance for planning

The basic concept of sustainability governance

As mentioned above, sustainability governance involves utilizing all of a society's capabilities and mechanisms, including central and local governments, companies, universities and even individual citizens, to solve sustainability-related problems. The 1987 Brundtland Report, *Our Common Future*, defined sustainable development as development that meets the needs of the present without compromising the ability of future generations to meet their own needs. According to this report, sustainability draws on the continuity of economic, social, institutional and environmental aspects to provide the optimum outcome for human and natural environments both now and in the indefinite future.

Sustainability affects every level of organization, from the local neighbourhood to the entire planet. So sustainability governance can be conceptualized as comprising every level of organization and stakeholder. To evaluate sustainability, the decision-making process should involve the PPP procedure and other appropriate indicators. In PPP, the policy is a broad statement of intent, comprising an objective and a broad course of action to meet it; the plan is a specific strategy outline for policy implementation; and the programme lists highly specific proposals or instruments for policy implementation. PPP creates the hierarchy of the decision-making process and also results in a tiered relationship. Thus, the SEA approach provides decision-makers with better information on the impacts of alternatives in a proactive and systematic manner.

Comprehensive ideas of governance should contribute to such analyses, and synthetic indicators are important for appraising sustainability. Sustainability indicators differ from general economic indicators in that they reflect broader concepts, including interactions between the environment, the economy and the particular society. Synthetic sustainability indicators were introduced for three fields: natural environments, living environments and the interfaces between them. Natural environments consist of such items as green fields and clean water. Living environments include safety, comfort and amenities. Finally, interface items comprise resource use and environmental loads (Kagaya et al., 1995).

Decision-making using analytical strategic environmental assessment

In Japan, either national or local governments undertake infrastructure improvements, for example in transportation systems and river environment systems, based on the SEAs. The more advanced the plan, the more complex the decision-making framework, and the greater the dependence on public involvement. As a result, conflict often arises between the administration and the community owing to the multiplicity of opinions. Therefore, it is essential to ensure that governance includes a communication system to obtain common information from each group. Workshops may be organized within the governance framework as a communication support group. As shown in Figure 6.3.1, stakeholder groups can be divided into three subgroups: administration, community and the support group(s). Efficient collaboration between these groups can enhance sustainability governance planning.

This planning process was developed via SEA analysis. It includes the preparation and use of any evaluation findings in publicly accountable decision-making (Therivel and Partidario, 1996). In this study, three procedural phases were devised from Stage 1 to Stage 3:

Stage 1: First, problem finding and checking are discussed in terms of brainstorming and morphological methods. Next, structural models are constructed via fuzzy structural modelling (FSM) (Kagaya, 1991; Kagaya et al., 1995). These models are used in policy-making for river environment projects (Kagaya et al., 2004).

Stage 2: During this stage, project planning is prioritized. Several alternative plans are selected by discussing how best to achieve the project aims. Next, the most appropriate plan is determined in terms of fuzzy multi-criteria analysis. The benefits are then estimated by the contingent valuation method (CVM) to confirm the economic validity of the analysis (Kagaya et al., 2005a; Kagaya et al., 2007).

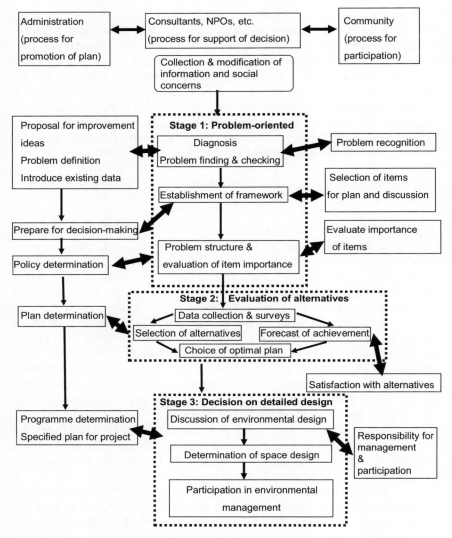

Figure 6.3.1 Planning process based on SEA.

Stage 3: The outline of the design is discussed with regard to basic factors related to river environments. Conjoint analysis is then used to evaluate several designs in terms of specific attributes. Next, a concrete design is settled upon via the governance framework, which is composed of certain stakeholders (Kagaya and Uchida, 2006). A workshop is then

organized within the governance framework for stakeholder discussions, usually involving administrative organizations, local residents, technical experts, consultants and non-profit organizations (NPOs). The role of the workshop is important in preparing the following:
(1) *Determination of overall framework*: Based on the government's aims, the overall project framework is discussed and the relevant factors determined. This is vital for understanding the policy and beginning the project.
(2) *Determination of evaluation of alternatives*: Based on the entire plan, several factors are selected and a number of alternatives arising from them identified.
(3) *Determination and confirmation of the optimal alternative*: Based on public opinions and the workshop discussions, the optimal plan for river environment improvement is confirmed. Moreover, the workshop discusses the detailed design of the river basin environment. By carefully considering all opinions, a realistic design can finally be determined.

6-3-3 Empirical case study

Environmental improvement planning in a river basin

A spatial plan for the river basin is very important in considering sustainable development in the region, because the river basin is a fundamental space for both the living and the natural environments. The establishment of a sustainability governance framework is necessary for future strategic planning. Specific stakeholder workshops made up of residents, planners, NPOs and administrators were set up to create good governance. Next, factors that effectively contribute towards understanding inhabitants' concerns are introduced. Finally, the decision-making system settled on for a sustainable river environment is established. The practical procedure for using the planning system shown in Figure 6.3.1 was constructed as shown in Figure 6.3.2. Table 6.3.1 shows the supporting technologies introduced for the practical application. In Stage 1, brainstorming (BS), the K-J method (one of the seven management and planning tools used in Total Quality Control) and FSM were adopted to build the problem structure and evaluate items of importance. The fuzzy integral method was used to establish a synthetic concept for improvement of the river environment (Yager, 1999; Zahariev, 1991). Next, conjoint analysis was applied in order to decide the most appropriate alternative river environment plan for Stage 2 (Hanemann, 1984; Louviere, 1994). Moreover, in

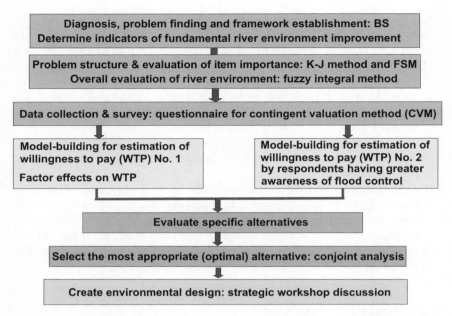

Figure 6.3.2 Practical application of the river environment improvement process.

Table 6.3.1 Several supporting technologies in systems analysis

Supporting technology of systems analysis	Outline and characteristics
Brainstorming (BS)	Heuristic and creative discussions of problems and their formulae (used at Stage 1)
K-J method: morphological analysis	As a problem-structuring and problem-solving technique, the arrangement of problems using an affinity diagram (at Stage 1)
Fuzzy structural modelling (FSM)	Problem-oriented structuring method in terms of fuzzy logic (at Stage 1)
Fuzzy integral method: multi-criteria analysis	Synthetic alternative evaluation combining the performances and weights of several criteria by fuzzy integral analysis (at Stage 1)
Contingent valuation method (CVM)	Survey-based economic technique for the valuation of non-market resources such as environmental preservation or the impact of contamination (at Stage 1)
Conjoint analysis	Statistical technique used in market research to determine how people value different features that make up an individual (at Stages 2 and 3)

Stage 3, the workshop discussed and created the strategic river environment design (Kagaya et al., 2007).

Since the new river law was established in Japan in 1997, it has become necessary to introduce synthetic planning, including river environment, flood control and water use, into the river improvement project. The planning system, which takes into account opinions from the river basin, should be added to the list of river environment improvement and conservation tasks, i.e. improvement plans should be strategic and promoted in response to the needs of the region's inhabitants.

Many discussions to develop the methodology have been conducted over the years. The most important task is to introduce public participation into the planning process and to complete the decision process during planning. Previous research demonstrated the need for such a planning process. However, a more advanced strategic planning process is still required. It is essential to construct governance in which the various stakeholders discuss and modify their ideas. Here, governance is defined as a government-administered framework for dealing with the processes and systems by which a stakeholder group or society operates.

The existing condition of the study area

In order to establish these approaches, the Aioi-Nakajima district was selected, a mid-stream location on the Tokachi River in Hokkaido, Japan, as shown in Figure 6.3.3. The capacity of the river discharge flow in the district is insufficient in its present state. During times of heavy rainfall,

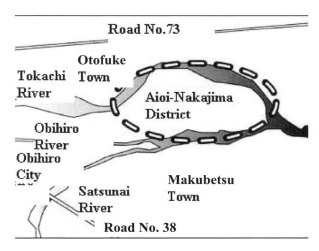

Figure 6.3.3 Map of the Aioi-Nakajima district.

Table 6.3.2 Outline of the questionnaire survey conducted to ascertain the basic attitudes of the inhabitants and workshop members

Age group	20–29	30–39	40–49	50–59	>60
% of inhabitants	7	20	22	28	24
% of workshop members	8	13	46	23	8
Distance between dwelling and district (% of workshop members)	<4 km (8%)	4–5.9 km (17%)	6–7.9 km (28%)	8–9.9 km (23%)	>10 km (25%)
Consequences for flood control (% of workshop members)	Very important (41%)	Important (47%)	Fairly important (7%)	Not important (5%)	
Consequences for conservation of natural environment (% of workshop members)	Very important (52%)	Important (38%)	Fairly important (7%)	Not important (3%)	

the river flow is obstructed, severely affecting upstream areas. An outline of the district's characteristics is given below:

(1) The Aioi-Nakajima district of the Tokachi River has a planned flood flow of 7,100 $m^3 s^{-1}$. Presently, however, the existing secured flow against flooding is only 3,200 $m^3 s^{-1}$, well below the planned (safe) flow.
(2) The Aioi-Nakajima district is composed of a large sandbank, making smooth flow impossible and causing upstream water levels to increase rapidly when flooding occurs. Obihiro City is at considerable risk from flood damage owing to levee overflow.
(3) In the Aioi-Nakajima district, life and property are not directly threatened because no inhabitants live there when flooding occurs. However, river improvements to prevent overflows and to reduce flood damage are necessary because the many trees flourishing along the river are potential obstacles to the flood flow. Moreover, a large river meander would not hinder the smooth flow downstream.
(4) However, the preservation of important habitats and species, such as the nesting sites of rare swallow species, is also important.

At the beginning of the planning process a questionnaire survey was conducted in the urban and suburban areas of Obihiro City, Otofuke Town and Makubetsu Town, to ascertain the basic attitudes of both the inhabitants and the workshop members (see Table 6.3.2). In this survey, the resident respondents were selected equally from almost every generation and occupation. In contrast, the workshop participants were from various occupations, but most (46 per cent) were aged between 40 and 50 years (Kagaya et al., 2005b). Most of the participants were equally affected by the possible consequences for flood control and for conservation of the natural environment.

Preference structure models of river basin problems

During the first stage, brainstorming (BS) was introduced to generate unlimited ideas concerning sustainable river improvement. After the BS discussion, the ideas that emerged were arranged by the K-J method, a morphological approach. The ideas were classified into the following six categories: (1) flood damage prevention (flood control), (2) river site access, (3) familiarity with the natural environment of the river, (4) promotion of land use within the district, (5) maintenance of the natural environment and (6) protection of flora and fauna.

Next, the importance of each item was hierarchically evaluated via the FSM technique and compared with the opinions of residents and workshop members. Based on the discussion and adjustment of ideas via the K-J method, the hierarchical structure of the improvement items was analysed.

Table 6.3.3 shows the results obtained from the survey of residents and workshop members. In both cases, flood damage prevention was most important, with maintenance of the natural environment and protection of flora and fauna ranked second. Both respondent groups considered planning for fundamental improvement to be significant. Familiarity with the river, riverside access and river land-use improvement were ranked bottom. There were no large differences between residents and workshop members.

Evaluation of alternatives in workshops

Following the workshop discussions, the alternative proposals were: excavating a new waterway, excavating an intermediate water channel and

Table 6.3.3 Results of the FSM analysis

Respondent	Preference structure by degree of importance
Residents in the river basin	A > E = F > B = C = D
Workshop members	A > E > F > B = C = D

Notes: A: prevention of flood damage (flood control), B: access to the riverside, C: familiarity with the natural environment of the river, D: promotion of land use in the district, E: maintenance of the natural environment, F: protection of flora and fauna; >: superior, = indifference.

Table 6.3.4 Alternatives discussed in the workshop

Alternative plan	Content	Advantages and disadvantages
1. Excavation of new waterway with straight line	Excavation with straight line Modification of waterway width	Impossible to access sandbank Risk of blockage Risk of destruction of bird nests
2. Excavation of mid-scale riverbed within existing waterway	Intermediate excavation with straight line Construction of some ponds within river	Use intermediate bed only at times of flooding Use ponds for both flood control and water use Possible access to sandbank
3. Widening of existing waterway	Expansion of existing waterway Construction of floodwater storage area	Deforestation Risk of erosion of riverside Use ponds for both flood control and water use
4. Expansion of existing waterway and upstream improvement	Expansion of existing waterway Construction of floodwater storage area Change in waterway upstream	Large-scale deforestation Risk of erosion of riverside Use ponds for both flood control and water use
5. Excavation of new low-head waterway	Excavation of new shallow waterway	Risk of blockage owing to sedimentation High maintenance costs Water runs along the waterway all the time

widening the existing waterway. All the alternatives are shown in Table 6.3.4.

The opinions of residents and workshop participants concerning the river improvement plan were ascertained via workshop discussions and the planning support system. The alternative projects, which consisted of five main plans, were evaluated comparatively. The selected evaluation items (factors) included: (1) flood damage prevention, (2) riverbed access, (3) familiarity with the natural environment of the river, (4) promotion of land use within the district, and (5) maintenance of the natural environment. Table 6.3.5 shows these items were evaluated on a continuous scale from 0 to 1. This means that, the larger the numerical value, the higher is the expectation of the item. The degree of importance in relation to the item is also shown in Table 6.3.5.

Table 6.3.5 Evaluation of alternative projects based on the Choquet integral

(1) Degree of importance for computing fuzzy measure (corresponding to the five factors)

Factor 1	Factor 2	Factor 3	Factor 4	Factor 5
$\alpha_1 = 0.25$	$\alpha_1 = 0.20$	$\alpha_1 = 0.16$	$\alpha_1 = 0.13$	$\alpha_1 = 0.35$

(2) Achievement score of alternative project on each factor

Alternative project	Factor 1	Factor 2	Factor 3	Factor 4	Factor 5
1. New channel excavation	0.8	0.1	0.5	0.1	0.5
2. Mid-scale riverbed excavation	0.9	0.8	0.8	0.5	0.6
3. Channel widening	0.3	0.5	0.8	0.8	0.1
4. Widening of upstream control channel	0.3	0.5	0.8	0.7	0.1
5. New low-head channel excavation	0.7	0.3	0.6	0.5	0.3

(3) Results of comprehensive evaluation

Alternative project	1	2	3	4	5
Evaluation value (order)	18.03 (3)	29.52 (1)	17.01 (4)	16.51 (5)	18.93 (2)

A comprehensive evaluation was performed using the Choquet integral, with results also presented in Table 6.3.5. Based on this analysis, Alternative 2 was recommended as the optimum plan, indicating that a mid-scale riverbed excavation should be undertaken. Another analysis estimated that the cost of this improvement plan would be JPY 1.5 billion.

Analysis using the contingent valuation method (CVM)

As mentioned in the previous section, workshops were used to discuss plans for basic river improvement, and ultimately the optimal plan was found in just such a workshop. The next stage involved ascertaining whether the plan would be acceptable to inhabitants. The following assumptions were presented to inhabitants: (1) the river basin improvement would involve a new waterway through the sandbank, 400 meters wide and 2 meters deep; and (2) a proportion of the budget would be used to incrementally widen the river basin area over 20 years.

On this basis, inhabitants were asked how much they would pay to support this project. In the survey, the payment card method was adopted, by

Table 6.3.6 Parameters in the model based on the contingent valuation method

Contents	Unit	Parameter	T-value	Significance	Mean
Proposed sum of WTP, β	yen	−0.0007	−17.000	***	–
Annual income, γ	1 to 5	0.1727	2.253	**	2.302
Constant, α	–	2.3077	6.536	***	–
Concern about flood control	1 to 7	−0.8469	−9.305	***	1.711
Age group	1 to 5	0.1857	3.169	***	3.459
Likelihood ratio	0.481				
Hit rate (%)	84.64				

** 5% significance; *** 1% significance

which it is easy for respondents to choose the appropriate value. It is essential for the project to determine the willingness to pay (WTP). A logit model, based on random utility theory, was used as the estimate model. In this way, it was possible to evaluate the inhabitants' outcomes for river improvement in relation to their attitudes and opinions, and to exclude opposition to expenditure funded by taxes.

The CVM model discussed previously was estimated by using the maximum likelihood method, and the result is presented in Table 6.3.6. In this model, the proposed WTP, annual income and concern about flood control all made strong contributions. In particular, concern about flood

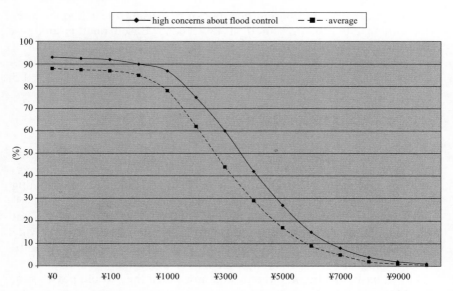

Figure 6.3.4 Differences in willingness to pay, depending on concerns about flood control.

Table 6.3.7 Total willingness to pay (WTP)

	Amount (JPY)
Annual WTP per household	2,682
Accumulated WTP per household over 20 years	37,900
Annual total WTP of residents in the district	279,000,000
Accumulated WTP of residents in the district over 20 years	3,944,000,000

control had a significant influence on the estimated WTP. In this case, the smaller the discrete number representing the category of flood, the higher the concern about flood control. Figure 6.3.4 demonstrates WTP values in the highly concerned group and for all respondents. The median difference between the two groups is JPY 1,235. Inhabitants with high concerns about flood control also value high WTP. At the same time, such inhabitants have experience of voluntary activity and flood drill activity.

Finally, WTP was estimated for the total value in the whole river basin area (see Table 6.3.7). Specifically, WTP per household was JPY 2,682 per year, and the present WTP value is JPY 37,900 for 20 planning years. The inhabitants therefore accepted the project because total WTP exceeded the total cost of the long-term plan and the provisional cost/benefit ratio was approximately 2.63 over 20 years.

Results of evaluation of alternatives by conjoint analysis

As mentioned above, it is necessary for inhabitants to support river environment improvement. Here, the specific alternatives were assessed in terms of several attributes. Table 6.3.8 shows four attributes and their category levels. Using these attributes and their levels, eight alternatives were proposed and presented to the respondents. This allowed the inhabitants' preferred alternatives to be determined. In this survey, visualized alternatives were adapted so that they could be more easily understood.

Figure 6.3.5 shows the partial utility of each attribute. For all attributes, the inhabitants evaluated the degree of nature-friendliness. Considering the importance of each attribute, trees in a waterway (B) and roads along the riverside (D) were evaluated highly. At the same time, the utility of

Table 6.3.8 Attributes and their levels of conjoint analysis

Attribute (factor)	Level 1	Level 2
A. Forests along riverside	Nature friendly	Conventional
B. Trees in waterway	Nature friendly	Conventional
C. Ponds in waterway	Nature friendly	Conventional
D. Roads along riverside	Nature friendly	Conventional

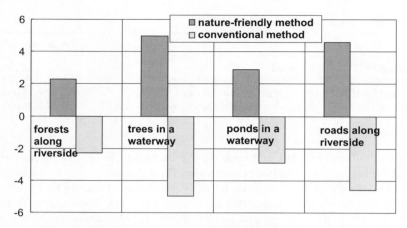

Figure 6.3.5 The partial utility of each attribute.

the nature-friendly method was rated higher on every attribute than that of the conventional method.

Figure 6.3.6 represents the importance of each attribute for inhabitants with prior experience of flood drill activity. The concerns for the maintenance of trees in the waterway and of roads along the riverside are high compared with the other attributes. This is because inhabitants are responsible for maintaining the forests and roads in the river basin.

Workshop discussions on specific design

In Stages 1 and 2, the basic planning system was discussed. As a result, the following basic policy and plan were considered: (1) the river

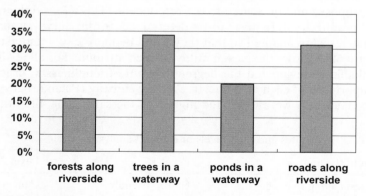

Figure 6.3.6 The importance of each attribute for inhabitants with prior experience of flood drill activity.

Table 6.3.9 Results of the subgroup discussion

Content of design	Subgroup				
	A	B	C	D	E
1. Existing elm trees	Y	Y	U	Y	Y
2. Riverbed girdle	Y	N	Y	Y	U
3. Riverside wood	Y	Y	Y	Y	Y
4. Right-slope gradient	U	Y	Y	Y	Y
5. Existing ponds	Y	U	Y	U	Y
6. Sandbank wood	Y	U	U	U	Y
7. Left-slope gradient	U	U	U	Y	U
8. Observatory	U	Y	Y	Y	U
9. Access roads	U	U	U	N	Y

Notes: Y: positive opinion, N: negative opinion, U: undecided (for example, most groups have positive opinions about existing elm trees).

environment should be improved to allow the safe flow of flood water; (2) the woods located on riversides or floodplains should be left undisturbed as much as possible; (3) the floodplain should be improved and assigned as a space for inhabitants to interact with nature; and (4) access roads should be improved without destroying the natural environment.

In Stage 3, the specific design was examined in the workshop discussions. Then the provisional design was drawn up. Table 6.3.9 shows the design content, as discussed in the workshops.

To evaluate the content, each workshop was separated into five subgroups (A–E) and each group continued debating the introduction of detailed designs in the project area. The administrative sector provided budget and technical information. The design was repeatedly refined following workshop discussions. It was therefore vital for each party to have a system for dialogue among the stakeholders at every stage. The group was governed in its decision-making by majority rule.

Workshop members from the residents' group required that:
(1) The riverside woods and elm tree forests should be preserved in their existing condition as much as possible.
(2) A riverbed girdle in and out of the shallow cut-off channel should be established and the gradient of the new channel slope should be between 1:2 and 1:10, depending on its role.
(3) The number of ponds should not decrease.
(4) An observatory should be built, but roads onto the riverbed should not be constructed.

The consensus reached by the whole workshop is given in Table 6.3.10.

Table 6.3.10 Consensus of the workshop

1. The channel should be planned so as to preserve the elms completely
2. The location of the riverbed girdle should be according to the plan
3. Riverside woodland should be preserved as much as possible
4. The slope gradient of the right riverbank should be 1:10
5. The scale and location of ponds should be maintained in their existing condition
6. Woodlands on the sandbank should be planned to make use of the remaining linden trees
7. The slope gradients of the left riverbank should be 1:0.5 where there are woods and 1:2 without woods
8. The observatory should be established at the top of the bank
9. An access road should not be built on the riverbed.

6-3-4 Conclusion

The following points are pertinent in terms of analysing and modifying the improvement plan for the particular river environment:

(1) After the workshop discussions, it was agreed that excavating a mid-scale riverbed in the existing waterway was the most appropriate improvement system, providing the best balance of flood control and environmental protection in terms of the fuzzy integral method.
(2) Inhabitants recognized this idea through a CVM-based questionnaire. There were strong policy concerns and the total WTP value was larger than the cost of the proposed improvement project.
(3) Based on the primary comprehensive plan, several alternative projects were proposed to inhabitants.
(4) A method for evaluating alternative river improvement projects was developed. Following conjoint analysis, the nature-friendly method was found to produce the optimal project.
(5) The optimal plan for the river environment facilitated design decisions because discussions among stakeholders allowed for differing opinions and enabled a consensus to be reached.

In this way, a system of decision-making was created by connecting administrative organizations and experts with the community, which is a stage of sustainability governance. Specifically, the workshop can make an excellent contribution to the process of governance. It is very important to take account of the differing opinions of stakeholders via such a platform.

Some effective methods for supporting the system of decision-making were also introduced, including the fuzzy integral method, the contingent valuation method and conjoint analysis. These are appropriate for analysing the ideas and opinions of inhabitants and to guide scientific information in the joint stage of decision-making.

In a future study, a pilot system should be advanced and refined, adding additional discussion stages and other useful methods. In addition, the communication of risk in the field of river improvement should be established via a comprehensive system simulation.

Acknowledgements

Much of this section is based on S. Kagaya, E. Ikoma and K. Uchida, "Evaluation Method of Alternatives for River Improvement Project with Public Participation", *Proceedings of 19th PRSCO Regular Meeting*, Tokyo, Japan, 2005; available on CD-ROM.

REFERENCES

Annandale, D. D., John Bailey, et al. (2001) "The Potential Role of Strategic Environmental Assessment in the Activities of Multi-lateral Development Banks", *Environmental Impact Assessment Review* 21: 407–429.

Brundtland, G. H. (1987) "Chairman's Foreword", in World Commission on Environment and Development, *Our Common Future*. Oxford: Oxford University Press.

Hanemann, W. H. (1984) "Welfare Evaluation in Contingent Valuation Experiments with Discrete Responses", *American Journal of Agricultural Economics* 66(3): 332–341.

Hemmati, M. (2002) *Multi-Stakeholder Processes for Governance and Sustainability*. London: Earthscan Publications Ltd.

Kagaya, S. (1991) "Use of Inhabitant Opinions on Regional Infrastructure Planning under Fuzzy Environment", *Proceedings of Fuzzy Symposium* 7: 95–98 (in Japanese).

Kagaya, S. and K. Uchida (2006) "Evaluation Method on Perception of Quality of Life in Terms of Urban Transportation System Improvement", *Studies in Regional Science* 36(2): 471–485.

Kagaya, S., S. Kikuchi and K. Sato (1995) "An Application to the Evaluation of Infrastructure Planning in Terms of Fuzzy Decision Making", in *Proceedings of the FUZZ-IEEE/IFES'95*, pp. 607–614.

Kagaya, S., E. Ikoma and K. Uchida (2005a) "Evaluation Method of Alternatives for River Improvement Project with Public Participation", *Proceedings of 19th PRSCO Regular Meeting*, Tokyo, Japan, available on CD-ROM.

Kagaya, S., K. Uchida and T. Hagiwara (2005b) "Dialogue System of Decision-making for Sustainable Planning in River Basin Improvement", *Sustainable Development and Planning II* 2: 1107–1118.

Kagaya, S., K. Uchida, T. Adachi and Y. Nakayama (2004) "A Dialogue Method of Strategic Environmental Assessment for River Improvement Planning System", *Studies in Regional Science* 34(1): 153–172 (in Japanese).

Kagaya, S., et al. (2007) "Sustainability Governance for Planning River Environment", *Proceedings of the International Conference of Social Management Systems*, Yinchang, China, available on CD-ROM.

Kidd, S. and T. B. Fischer (2007) "Integrated Appraisal in North West England", *Environment and Planning C* 25(2): 233–249.

Louviere, J. J. (1994) "Conjoint Analysis", in R. P. Bagozzi (ed.), *Advanced Method of Marketing Research*. Cambridge, MA: Blackwell, pp. 223–259.

Partidario, M. (1999) "Strategic Environmental Assessment: Principles and Potential", in J. Petts (ed.), *Handbook of Environmental Impact Assessment*, Volume 1. Oxford: Blackwell Science.

Renn, Ortwin, et al. (1995) *Fairness and Competence in Citizen Participation: Evaluating Models for Environmental Discourse*. Dordrecht: Kluwer Academic Press.

Therivel, R. and M. R. Partidario, eds (1996) *The Practice of Strategic Environmental Assessment*. London: Earthscan Publications Ltd.

Yager, R. R. (1999) "Criteria Aggregations Function Using Fuzzy Measure and the Choquet Integral", *International Journal of Fuzzy Systems* 1(2).

Zahariev, S. (1991) "Group Decision Making with Fuzzy and Non-Fuzzy Evaluation", in J. Kacprzyk and M. Fedrizzi (eds), *Multi-person Decision Making Using Fuzzy Sets and Possibility Theory*. Dordrecht: Kluwer Academic Publishers, pp. 188–197.

7
How to sustain social, cultural and human well-being

7-1
Ecology, sustainability science and "knowing" systems

Osamu Saito and Richard Bawden

7-1-1 Introduction

Sustainability science can be characterized by its interdisciplinary approach to addressing complex, dynamic problems. It also serves the quest for advancing both useful knowledge and informed action by creating a dynamic bridge between basic and applied research (Clark, 2007). However, because of its short history as a research discipline and the complexity and diversity of sustainability issues, many researchers from conventional disciplines, including ecologists, are finding it difficult to share their knowledge and to contribute to the development of a common framework to promote sustainability studies.

Most of the sustainability problems that challenge human society and the environment are fundamentally ecological in nature. Therefore, policy decisions for a sustainable environment need to be informed by the knowledge and understanding of ecological systems, which means that there is wide scope for ecologists to contribute to the field of sustainability science. This study reviews the relationships between ecology and sustainability science, examines the similarities and differences between them as scholarly areas of scientific endeavour, and explores a paradigm for sustainability from a systems perspective, all as key concepts for a closer collaboration between the two disciplines.

Designing our future: Local perspectives on bioproduction, ecosystems and humanity,
Osaki, Braimoh and Nakagami (eds),
United Nations University Press, 2011, ISBN 978-92-808-1183-4

7-1-2 Ecology and sustainability science: Objectives and core questions

Ecology

In 1869, Haeckel defined "ecology" as "the entire science of the relations of the organism to the surrounding exterior world, to which relations we can count in the broader sense all the conditions of existence" (cited in Friederichs, 1958: 154). Moore (1920: 3) described ecology as "the science dealing with the environment", and Eggleton (1939: 56) defined it as "the science which treats of the interrelationships of organisms and their complete environment". The environment in this case embraces both the physical (abiotic) environment (e.g. temperature, water availability, wind speed and soil acidity) and the biotic environment, which includes competition, predation, parasitism and cooperation. After reviewing the historical development of the concept of ecology, Friederichs (1958: 157) proposed that "ecology is the science of the superindividual complexes of nature (life units or ecosystems) or the science of animals and plants as members of the whole nature".

In general, there are four identifiable subdivisions of scale that ecologists investigate: (1) considering the response of individuals to their environments; (2) examining the response of populations of a single species to the environment and considering the processes; (3) investigating the composition and structure of communities; and (4) identifying the processes occurring within ecosystems (Mackenzie et al., 1998).

A major omission in such studies for the majority of ecologists is consideration of the nature and role of human beings within such ecosystems. As Berkes (1999) has stated with such eloquence, within ecology human beings either are frequently regarded as somehow "un-natural" components of "natural eco-systems" or are placed into such mythological categories as the "Ecologically Noble Savage", the "Intruding Wastrel" or the "Fallen Angel". To this deficiency must be added the contestable issue of the ontological status and organization of "nature itself", and whether or not the notion of ecosystems as cybernetically regulated, stability-seeking entities that can evolve in all of their wholeness can be empirically validated beyond mathematical representation. And what a sloppy concept "nature" turns out to be in such circumstances, "society" too for that matter, and yet the new field of sustainability science claims that it seeks to "understand the fundamental character of interactions between nature and society" (Kates et al., 2001: 641).

Today, although demands for "ecosystem services" such as food, clean water and fuel are growing, "human actions are at the same time diminishing the capability of many ecosystems to meet these demands" (Millennium Ecosystem Assessment, 2003: 1). Global environmental challenges represented by climate change and biodiversity loss have driven many ecologists to engage in international research projects such as the Millennium Ecosystem Assessment and the Intergovernmental Panel on Climate Change. In this sense, ecology is expected to contribute not only to our understanding of natural ecosystems but also to the search for options that can better achieve core human development and sustainability goals in what are increasingly conceptualized as "coupled social–ecological systems" (Berkes and Folke, 1998).

Sustainability science

A range of definitions and core questions regarding sustainability science are given in Table 7.1.1. Based on these descriptions, it can be concluded that sustainability science emphasizes (1) understanding dynamic interactions between nature and society; (2) problem-driven and problem-solving efforts; (3) multi-criteria assessment and systems approaches (and thus the study of nature and society as natural and human systems respectively); (4) multi-, inter- or transdisciplinary characteristics; (5) adaptive management and social learning; and (6) collaboration between scientists and practitioners. (1) is a major target of this discipline, (2)–(5) are related to the approach and method (Figure 7.1.1) and (4)–(6) are concerned with interaction and collaboration between not only disciplines but also science and society. The mobilization of scientific knowledge to link research· with action for sustainable development is particularly acute for many sustainability problems (Kristjanson et al., 2009). These attributes are more or less required by other fields today, but their importance is much higher for sustainability science.

Regarding interactions between nature and society (1), sustainability science considers both "how social change shapes the environment and how environmental change shapes society" (Clark and Dickson, 2003: 8059). In this sense, it often attempts to assess not only the magnitude of pressure that causes environmental impacts, but also the resilience and vulnerability of the systems affected by that pressure.

Since "[s]ustainability science is a field defined by the problems it addresses rather than by disciplines it employs" (Clark, 2007: 1737) and because it is a young and emerging field, a clear definition of its boundaries is elusive at present. Indeed, one of the conspicuous features of sustainability science may be the fact that its boundaries and the scope of its core questions, assessment criteria and membership are in substantial flux (Clark and Dickson, 2003).

Table 7.1.1 Definitions and core questions of sustainability science

References	Description of sustainability science
Kates et al. (2001: 641–642)	"A new field of sustainability science is emerging that seeks to understand the fundamental character of interactions between nature and society. Such an understanding must encompass the interaction of global processes with the ecological and social characteristics of particular places and sectors." "CORE QUESTIONS OF SUSTAINABILITY SCIENCE (1) How can the dynamic interactions between nature and society – including lags and inertia – be better incorporated into emerging models and conceptualizations that integrate the Earth system, human development and sustainability? (2) How are long-term trends in environment and development, including consumption and population, reshaping nature–society interactions in ways relevant to sustainability? (3) What determines the vulnerability or resilience of the nature–society system in particular kinds of places and for particular types of ecosystems and human livelihoods? (4) Can scientifically meaningful 'limits' or 'boundaries' be defined that would provide effective warning of conditions beyond which the nature–society systems incur a significantly increased risk of serious degradation? (5) What systems of incentive structures – including markets, rules, norms, and scientific information – can most effectively improve social capacity to guide interactions between nature and society toward more sustainable trajectories? (6) How can today's operational systems for monitoring and reporting on environmental and social conditions be integrated or extended to provide more useful guidance for efforts to navigate a transition toward sustainability? (7) How can today's relatively independent activities of research planning, monitoring, assessment and decision support be better integrated into systems for adaptive management and societal learning?"
Clark and Dickson (2003: 8059–8060)	"In seeking to help meet this sustainability challenge, the multiple movements to harness science and technology for sustainability focus on the dynamic interactions between nature and society, with equal attention to how social change shapes the environment and how environmental change shapes society." "Sustainability science is not yet an autonomous field or discipline, but rather a vibrant arena that is bringing together scholarship and practice, global and local

Table 7.1.1 (cont.)

References	Description of sustainability science
	perspectives from north and south and disciplines across the natural and social sciences, engineering and medicine. Its scope of core questions, criteria for quality control and membership are consequently in substantial flux and may be expected to remain so for some time."
Komiyama and Takeuchi (2006: 2–3)	"We approach the problem of sustainability at three levels of 'system' – global, social and human . . . All three systems are crucial to the coexistence of human beings and the environment, and it is our view that the current crisis of sustainability can be analyzed in terms of the breakdown of these systems and the linkages among them."
	"The ultimate purpose of sustainability science is to contribute to the preservation and improvement of the sustainability of these three systems. Although sustainability science has its origin in the concept of sustainable development, we propose that it is, in reality, a much more multifaceted concept."
Clark (2007: 1737)	"Sustainability science is a field defined by the problems it addresses rather than by disciplines it employs."
	"From its core focus on advancing understanding of coupled human–environment systems, sustainability science has reached out with focused problem-solving efforts targeted to urgent human needs."
	"Sustainability science is thus most usefully thought of as neither 'basic' nor 'applied' research. Rather, it is an enterprise centred on the 'use-inspired basic research'."
Jäger (2008)	"Transdisciplinary sustainability science implies that the problems to be solved are not predetermined by the scientific community and need to be defined cooperatively by science and society. This kind of science connects problem definition, searching for solutions and implementation of solutions in a recursive (iterative) societal negotiation and learning process."
	"Sustainability science involves participatory processes of negotiation in which interests and thus power constellations play a major role as well. Consequently, apart from expert and professional competence, transdisciplinarity requires a high degree of social competence and a willingness among scientists to be aware of the values inherent to scientific knowledge and to make values explicit."

ECOLOGY, SUSTAINABILITY SCIENCE AND "KNOWING" SYSTEMS

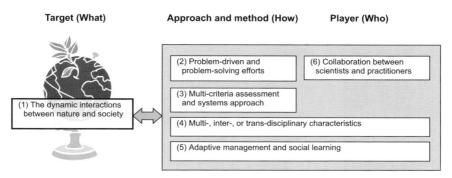

Figure 7.1.1 Major target and characteristics of sustainability science.

Common issues and research priorities

The Ecological Society of America (ESA) founded the Sustainable Biosphere Initiative (SBI) in 1991 to promote the continued development of ecological science and its integration into decision-making and education, linking the ecological research and management communities. The SBI calls for three basic directions – research, education and environmental decision-making – and proposes three research priorities (Table 7.1.2).

Review results of the selected manuscripts in connection with research priorities addressed by SBI (Lubchenco et al., 1991; Table 7.1.2) and the core questions of sustainability science (Kates et al., 2001; Table 7.1.1) are summarized in Table 7.1.3. In this review, over 100 manuscripts, which

Table 7.1.2 Three research priorities of the Sustainable Biosphere Initiative

1. *Global change*, including the ecological causes and consequences of changes in climate; in atmospheric, soil and water chemistry (including pollutants); and in land- and water-use patterns
2. *Biological diversity*, including natural and anthropogenic changes in patterns of genetic, species and habitat diversity; ecological determinants and consequences of diversity; conservation of rare and declining species; and effects of global and regional change on biological diversity
3. *Sustainable ecological systems*, including the definition and detection of stress in natural and managed ecological systems; restoration of damaged systems; management of sustainable ecological systems; the role of pests, pathogens and disease; and the interface between ecological processes and human social systems

Table 7.1.3 Reviewed results of the selected manuscripts

	Manu-scripts issued	Manu-scripts reviewed	Research priorities (Lubchenco et al., 1991)			Core questions (Kates et al., 2001) (see Table 7.1.1)						
			Global change	Biological diversity	Sustainable ecological systems	(1)	(2)	(3)	(4)	(5)	(6)	(7)
Sustainability-related journals												
Sustainability Science	50	12 [24%] (100%)	8 (67%)	0 (0%)	6 (50%)	8 (67%)	5 (42%)	7 (58%)	3 (25%)	1 (8%)	1 (8%)	1 (8%)
Sustainability: Science, Practice, & Policy	37	9 [24%] (100%)	1 (11%)	2 (22%)	5 (56%)	2 (22%)	1 (11%)	3 (33%)	1 (11%)	3 (33%)	7 (78%)	6 (67%)
PNAS-Sustainability Science	115	58 [50%] (100%)	16 (28%)	24 (41%)	21 (36%)	40 (69%)	16 (28%)	29 (50%)	2 (3%)	15 (26%)	8 (14%)	11 (19%)
Total	202	79 [39%] (100%)	25 (32%)	26 (33%)	32 (41%)	50 (63%)	22 (28%)	39 (49%)	6 (8%)	19 (24%)	16 (20%)	18 (23%)
Applied ecology-related journals												
Ecological Applications	—[a]	25[b] (100%)	4 (16%)	5 (20%)	12 (48%)	21 (84%)	5 (20%)	7 (28%)	4 (16%)	10 (40%)	3 (12%)	5 (20%)
Journal of Applied Ecology	—[a]	2 (100%)	1 (50%)	0 (0%)	1 (50%)	1 (50%)	0 (0%)	1 (50%)	0 (0%)	0 (0%)	2 (100%)	1 (50%)
Ecological Research	—[a]	6 (100%)	2 (33%)	1 (17%)	4 (67%)	5 (83%)	0 (0%)	1 (17%)	0 (0%)	2 (33%)	0 (0%)	3 (50%)
Total	—	33 (100%)	7 (21%)	6 (18%)	17 (52%)	27 (82%)	5 (15%)	9 (27%)	4 (12%)	12 (36%)	5 (15%)	9 (27%)

Notes:
[a] Because of the long history of each journal, the exact number of manuscripts is not available.
[b] Manuscripts published since 2005.

could be classified under both sustainability science and ecology, were selected from three major journals on sustainability science (*Sustainability Science*; *Sustainability: Science, Practice, & Policy* [*SSPP*]; and the *Sustainability Science* section of *PNAS* [*Proceedings of the National Academy of Sciences of the United States of America*]) and three ecology-related scientific journals (*Ecological Applications*; *Journal of Applied Ecology*; and *Ecological Research*). The manuscripts were selected by looking up the words "ecology", "ecological science" and "ecosystem" in sustainability-related journals and by looking up "sustainability" and "sustainability science" in applied ecology-related journals.

The review results displayed in Table 7.1.3 illustrate that 50 per cent of the *PNAS-Sustainability Science* manuscripts are relevant to ecological science and management, and the share of ecological research in both *Sustainability Science* and *SSPP* is 24 per cent. This reflects the editorial policy and scope of each journal on one hand and implies the strong affiliation between ecology and sustainability sciences in *PNAS* on the other. In total, the three sustainability-related journals cover the research priorities evenly, but the coverage of core questions is uneven. Merely 8 per cent of the reviewed manuscripts cover the core question (4) concerning "limits" or "boundaries". Defining scientifically meaningful "limits" or "boundaries" for more sustainable, coupled socio-ecological systems is a challenging task for sustainability scientists. This is partly because, in the present state of the dynamics of coupled socio-ecological systems, "there is fundamental uncertainty about both the structure of the system and the measures of its performance" (Perrings, 2007: 15180), and partly because such tasks usually entail time-consuming decision-making processes, cooperation with policymakers and stakeholders, and consequently larger social responsibility.

Among the applied ecology-related journals, *Ecological Applications* (issued by ESA) contains many papers relevant to sustainability science (Table 7.1.3), which suggests that many ecologists have shown interest in, and have contributed to, this field. In contrast, the contributions of the *Journal of Applied Ecology* and *Ecological Research* to sustainability science are smaller. The research focus of the reviewed manuscripts in these journals is mainly on sustainable ecological systems (Table 7.1.3), including the detection of stress in natural and managed ecological systems, restoration of damaged systems, management of sustainable ecological systems and the interface between ecological processes and human social systems (Lubchenco et al., 1991). In addition, core question (4) again indicates the lowest share among the seven core questions. This underlines the difficulty in defining thresholds and demarcation or "boundary work" (Gieryn, 1995; Swart and van Andel, 2008).

7-1-3 The systems idea and the nature of knowing systems

Systemic development as a paradigm for sustainability

The application of theories and practices from the so-called systems movement (Checkland, 1981) provides a powerful framework for the potential integration of ecology with sustainability science, because both adopt the concept of the system as an organizing principle (for nature by ecologists, and for both nature and society by sustainability scientists). Both disciplines concern themselves, to a greater or lesser extent, with the relationships within specific systems and between systems and the environments (as higher-order environmental systems or supra-systems) in which they are embedded. As illustrated in Figure 7.1.2, managed ecosystems (e.g. agricultural systems) can be conceptualized as existing at the interface of two different but interconnected environmental supra-systems: the biophysical environment and the sociocultural environment. The boundary of ecology, as already mentioned, is usually limited within the biophysical environment, with the essential unit of study being natural ecosystems themselves as they are influenced by the environments in which they are embedded. Sustainability science, on the other hand, is concerned with managed ecosystems themselves as well as with both biophysical and sociocultural environments, which it also regards as systems (or supra-systems) in which the system of focus is embedded. Sustainability science is thus expressly concerned with the interface between focal social systems and natural systems, and its boundary as a science is accordingly much wider than that of ecology. It is also con-

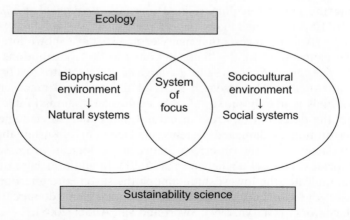

Figure 7.1.2 Ecology and sustainability science from a systems perspective.

cerned with the management of ecosystems rather than their mere understanding.

Within both ecological and sustainability sciences, the idea of "a system" is generally taken to mean some form of formal entity. From this perspective, systems are whole entities that have properties that cannot be understood by any study of their parts in isolation from each other or from the system as an integrated whole: systems have integrity. It is frequently stated that systems are greater than, or at least different from, the sum of their parts. This means that systems are conceptualized as embedded whole entities with emergent properties.

What is frequently overlooked in both ecological and sustainability sciences is the characterization of the "knowing" or "cognitive" systems that bring other "systems of focus" into being. Yet it is the characteristics of these "knowing systems" that are central to the whole endeavour of understanding the form and dynamics of such systems and the system/ environment interactions that are in turn central to both systems ecology and sustainability science (Bawden, 2007). A focus on "knowing systems" would thus seem to be an essential aspect of any framework for integrating ecology with sustainability science.

The nature of knowing systems and worldviews

The process of learning is not simply the accumulation of knowledge but the use of knowledge, transformed from experience, to inform adaptive action. Learning thus involves processes in both the concrete and the abstract, while also involving both "finding out" and "taking action". As human beings, we experience in our totality our emotions, attitudes and values, which all come to bear on how we then try to make sense of a subject before designing, planning and finally acting (Kolb, 1984) (Figure 7.1.3).

However, we do all of this from particular positions or "worldview perspectives", which reflect three sets of highly idiosyncratic (but rarely appreciated) beliefs and/or assumptions concerning (a) the nature of nature (ontology), (b) the nature of knowledge and the process of knowing (epistemology) and (c) the nature of human nature, particularly with respect to moral values (axiology). Using simple distinctions between two fundamentally different ontological beliefs as well as two epistemological beliefs, a matrix drawn from two axes can be generated that represents four quite different worldview positions as the four respective "cells" of the matrix (Figure 7.1.4). A technocentric worldview reflects assumptions about the objectivity of knowledge and reductionist ontology. This assumes that any whole entity can be understood by studying its parts, and it is the perspective that is assumed by scientists from most disciplines.

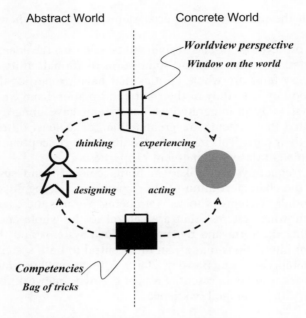

Figure 7.1.3 Image of the learning process.

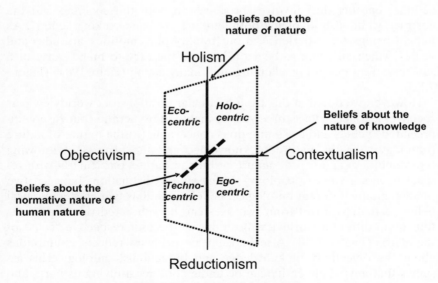

Figure 7.1.4 A worldview matrix.

ECOLOGY, SUSTAINABILITY SCIENCE AND "KNOWING" SYSTEMS 339

Holocentricity, in absolute contrast, reflects an acceptance of both the contextuality of knowledge and the holistic nature of nature.

Most ecologists assume an ecocentric worldview that accepts holistic ontology but embraces an objectivist position on knowledge. And finally, within any community there are invariably individuals who, as egocentrics, focus their perspectives on very personal considerations, often reflecting a mystical dimension that plays a major part in many religions. Although all these four worldview perspectives are equally valid, depending on the context of the situation being explored, holocentricity is the most appropriate to the focus of sustainability science as it is being articulated.

It is worth briefly digressing here to explore a distinction often made in the systems literature between the so-called "hard" systems approaches on the one hand and "soft" systems approaches on the other, for this is of great significance to further the development of sustainability science in particular. The basic distinction between the two approaches lies in the difference between an ecocentric perspective (the "hard school") and the holocentric perspective (the "soft school") – and in practice these two views are expressed in two different methodological approaches (Figure 7.1.5).

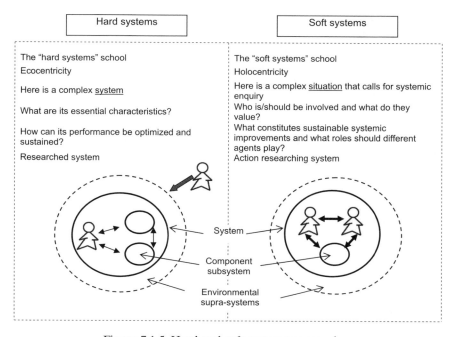

Figure 7.1.5 Hard and soft systems approaches.

Whereas systems ecologists clearly subscribe to the "hard ecocentric school", the situation with respect to the sustainability scientists is far less clear. Although there would be very clear advantages for the latter to adopt both approaches (and in particular to embrace the soft holocentric perspective), a review of the literature would suggest that most remain firmly within the hard ecocentric school. This has particular significance with respect to the vital importance played by human values in the quest for sustainability. As Thompson (2004) has so poignantly argued, sustainability is not simply a question of what can be designed to persist into the future, but also, and most emphatically, what should be allowed to persist. This latter question has foundations in ethical, aesthetic and even spiritual beliefs and values. These are normative positions that influence our judgements and they are likely to vary very considerably from individual to individual, community to community, culture to culture, etc., and they are vital components of both our personal worldviews as individuals and our collective perspectives as social collectives.

A point to be emphasized here is that it is difficult enough for any individual to shift his or her worldview perspective from one position to another, and even more so when the aim is to seek a consensus on collective perspectives. This is of serious consequence when the demand is for scientists and citizens to assume the much more holistic positions that both ecology (ecocentricity) and sustainability science (preferably holocentric) demand. What is essential is that we learn about the characteristics of these worldviews and appreciate both their differences and the challenges of shifts (essentially cognitive transformations) from one to another. We need to know from which position we are arguing during conversations about what constitutes improvements in any "system in focus".

Accordingly, it is important that each of us learns to learn "in three dimensions" – representing three levels of cognitive processing (Kitchener, 1983) – in which we address (a) learning about the matter to hand, (b) learning about how we are going about that learning and (c) learning about the limits to our learning imposed by the particular assumptions that we are holding that comprise our worldview perspective. In this manner, the process of learning can itself be conceptualized as a three-level "knowing system" (Figure 7.1.6). Such a "learning" or "knowing" system is (or at least should be) an essential subsystem of any "system of focus". In other words, scientists (as learners/researchers) are always component subsystems of any system that they are intent on studying.

What we do in the world is a reflection of how we see it. If we want to change what we do in this world, we have to change the way we see. If we are going to commit to sustainability – the belief in doing things in a more sustainable, ecologically responsible and ethically defensible way –

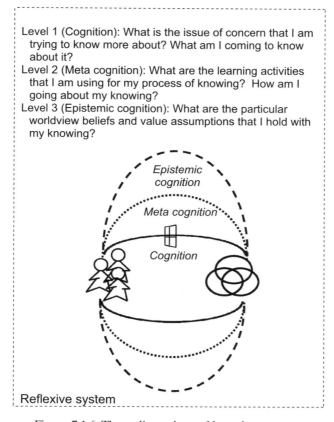

Figure 7.1.6 Three dimensions of knowing systems.

we need to appreciate interconnections and contextual knowledge, and we need to be much more communalistic both about ourselves as a species (human beings) and about our relationships with the rest of nature. Under the conventional technocentric worldview, or paradigm, technical

Table 7.1.4 Paradigm shift from technocentric to holocentric

Technocentric		Holocentric		
Technical feasibility	A shift in paradigm ➡	Technical feasibility	+	Ecological responsibility
Economic viability		Economic viability	+	Ethical defensibility
Social desirability		Social desirability	+	Spiritual sensitivity
Political acceptability		Political acceptability	+	Cultural amenability
Productivism		Productivism		Sustainabilism

feasibility, economic viability, social desirability and political acceptability have been emphasized, with progress and development in these parameters being of central concern. All of those are still important in the current (and future) quest for sustainable ways of existing, but metrics will also be needed for ecological responsibility, ethical defensibility, spiritual sensitivity and cultural amenability (Table 7.1.4).

7-1-4 Conclusion

This section has reviewed the definitions and core questions of sustainability science, with particular regard to the relationship with ecology. Although it is difficult to clearly define the boundaries of the field, owing to its embedded problem-driven nature, six characteristics or requirements of sustainability science were identified (Figure 7.1.1). Recent manuscripts selected from sustainability- and ecology-related journals were then analysed to examine the relationship between sustainability research and ecological research.

Acknowledging that our future depends on how humans learn to manage natural processes and resources, ecologists must take a step forward in sustainability research and "play a greatly expanded role in communicating their research and influencing policy and decisions that affect the environment" (Palmer et al., 2005: 4). Historically, the establishment of ecology was an attempt to integrate all the fields of biology (Moore, 1920) and yet it has consistently struggled with "the biology" of humans and the role that they play in the dynamics and organization of natural systems. Similarly, the establishment of sustainability science may be considered as an attempt to integrate all fields of ecological and social sciences. Such integration and alliance will enable cross-fertilization of sustainability science and ecological science. Practically, introducing the perspectives and approaches of sustainability may contribute to the reframing of a curiosity-driven ecological study and shed light on a new problem in the context of global, regional or local sustainability. In conjunction, ecological studies can add significant knowledge to the understanding of complex and diverse ecosystems and their interactions.

The results of a review of selected manuscripts and the above discussion suggest four critical bridging concepts between the two fields:
(1) *Contextualization*: Regionalized or localized context, an applicative context and extended forms of accountability are more important than the classic approach of pure, curiosity-driven, autonomous science (Swart and van Andel, 2008). When comparing sustainability science with conventional ecological research, social, human and sometimes even political contexts are taken into account.

(2) *Co-benefits and trade-offs*: Assessing coupled human–ecological systems by multiple impact criteria including climate change, biodiversity, ecosystem functioning, human health and local economies, and balancing these different benefits and risks (Graham and Wiener, 1995; Steffan-Dewenter et al., 2007), rather than optimizing based on a single criterion.

(3) *"Threshold work" and "boundary work"*: Threshold work implies defining sustainably meaningful limits or thresholds to provide effective warning of conditions and feedbacks (Kates et al., 2001). Boundary work explores the interface between science and society and seeks ways to span the boundaries and stimulate collaboration of stakeholders, including scientists, toward sustainability goals (Kristjanson et al., 2009; Swart and van Andel, 2008).

(4) *Adaptive management and social learning*: Independent activities of research planning, monitoring, assessment and decision support are integrated into systems for adaptive management and societal learning (Kates et al., 2001). Scientists encourage policymakers, resource managers and citizens to acquire knowledge about socio-ecological relationships, facilitating information flow among policy actors and creating a shared understanding among them (McLain and Lee, 1996).

Furthermore, a fifth aspect demands to be added to this list that relates to the nature of learning and knowing as the basis for responsible action:

(5) *Collective responsibility and knowing systems*: As individuals, and as collectives, everything we do reflects the worldview perspectives that we hold. These are frequently in conflict. So if we, as members of the human race, want to improve our relationships with each other and with the world about us, we need to devise a method that appreciates different ways of knowing and different sets of values, assumptions, moralities and so on. We also need to possess the capability of articulating the characteristics of our own worldviews, and then facilitate the development of and engagement with multi-stakeholder knowing systems in response to particular situations.

As the above has indicated, a fourth vital category ought to be added to Berkes' (1999) typology of the role of human beings in ecology: the "Ecologically Noble Savage", the "Intruding Wastrel" and the "Fallen Angel" ought to be accompanied by the intrinsically three-dimensional "Knowing/Valuing Being". Through their cognitive competencies, humans are capable of coming to know about matters to hand that concern them, knowing how they come to know this and knowing the epistemic contexts in which these two "lower-order" processes operate as interacting subsystems within learning systems. Because these learning systems are themselves subsystems of any system of interest under review for sustainability or sustainable development, the nature and development dynamics of "learning systems" are aspects of "human biology" that need to be embraced and understood by ecologists and sustainability scientists alike.

REFERENCES

Bawden, R. J. (2007) "Knowing Systems and the Environment", Chapter 15 in J. Pretty, A. S. Ball, T. Benton, J. S. Guivant, D. R. Lee, D. Orr, M. Pfeffer and H. Ward (eds), *The SAGE Handbook of Environment and Society*. Los Angeles: Sage, pp. 224–234.

Berkes, F. (1999) *Sacred Ecology: Traditional Ecological Knowledge and Resource Management*. Philadelphia: Taylor & Francis.

Berkes, F. and C. Folke, eds (1998) *Linking Social and Ecological Systems: Management Practices and Social Mechanisms for Building Resilience*. Cambridge: Cambridge University Press.

Checkland, P. B. (1981) *Systems Thinking Systems Practices*. Chichester: John Wiley and Sons.

Clark, W. C. (2007) "Sustainability Science: A Room of Its Own", *PNAS* 104(6): 1737–1738.

Clark, W. C. and N. M. Dickson (2003) "Sustainability Science: The Emerging Research Program", *PNAS* 100(14): 8059–8061.

Eggleton, F. E. (1939) "Fresh-Water Communities", *American Midland Naturalist* 21: 56–74.

Friederichs, K. (1958) "A Definition of Ecology and Some Thoughts about Basic Concepts", *Ecology* 39(1): 154–159.

Gieryn, T. F. (1995) "Boundaries of Science", in S. Jasanoff, G. E. Markle, J. C. Petersen and T. Pinch (eds), *Handbook of Science and Technology Studies*. Thousands Oaks, CA: Sage, pp. 393–443.

Graham, J. D. and J. B. Wiener (1995) *Risk vs. Risk – Tradeoffs in Protecting Health and the Environment*. Cambridge, MA: Harvard University Press.

Jäger, J. (2008) "Sustainability Science and the Challenges of Transdisciplinary Research", paper presented at the 2008 Berlin Conference on the Human Dimensions of Global Environmental Change/International Conference of the Socio-Ecological Research Programme, 22–23 February, unpublished.

Kates, R. W., William C. Clark, Robert Corell, J. Michael Hall, Carlo C. Jaeger, Ian Lowe, James J. McCarthy, Hans J. Schellnhuber, Bert Bolin, Nancy M. Dickson, Sylvie Faucheux, Gilberto C. Gallopin, Arnulf Grübler, Brian Huntley, Jill Jäger, Narpat S. Jodha, Roger E. Kasperson, Akin Mabogunje, Pamela Matson, Harold Mooney, Berrien Moore III, Timothy O'Riordan, Uno Svedin (2001) "Sustainability Science", *Science* 292(5517): 641–642.

Kitchener, K. S. (1983) "Cognition, Meta-Cognition, and Epistemic Cognition: A Three-Level Model of Cognitive Processing", *Human Development* 26: 222–232.

Kolb, D. A. (1984) *Experiential Learning: Experience as the Source of Learning and Development.* Englewood Cliffs, NJ: Prentice Hall.

Komiyama, H. and K. Takeuchi (2006) "Sustainability Science: Building a New Discipline", *Sustainability Science* 1: 1–6.

Kristjanson, P., Robin S. Reid, Nancy Dickson, William C. Clark, Dannie Romney, Ranjitha Puskur, Susan MacMillan, and Delia Grace (2009) "Linking International Agricultural Research Knowledge with Action for Sustainable Development", *PNAS* 106(13): 5047–5052.

Lubchenco, J., Annette M. Olson, Linda B. Brubaker, Stephen R. Carpenter, Marjorie M. Holland, Stephen P. Hubbell, Simon A. Levin, James A. MacMahon, Pamela A. Matson, Jerry M. Melillo, Harold A. Mooney, Charles H. Peterson, H. Ronald Pulliam, Leslie A. Real, Philip J. Regal, Paul G. Risser (1991) "The Sustainable Biosphere Initiative: An Ecological Research Agenda: A Report from the Ecological Society of America", *Ecology* 72(2): 371–412.

Mackenzie, A., A. S. Ball and S. R. Virdee (1998) *Instant Notes in Ecology.* Oxford: Bios Scientific Publishers.

McLain, R. J. and R. G. Lee (1996) "Adaptive Management: Promises and Pitfalls", *Environmental Management* 20(4): 437–448.

Millennium Ecosystem Assessment (2003) *Ecosystem and Human Well-being: A Framework for Assessment.* Washington DC: Island Press.

Moore, B. (1920) "The Scope of Ecology", *Ecology* 1(1): 3–5.

Palmer, M. A., Emily S. Bernhardt, Elizabeth A. Chornesky, Scott L. Collins, Andrew P. Dobson, Clifford S. Duke, Barry D. Gold, Robert B. Jacobson, Sharon E. Kingsland, Rhonda H. Kranz, Michael J. Mappin, M. Luisa Martinez, Fiorenza Micheli, Jennifer L. Morse, Michael L. Pace, Mercedes Pascual, Stephen S. Palumbi, O. J. Reichman, Alan R. Townsend, Monica G. Turner (2005) "Ecological Science and Sustainability for 21st Century", *Frontiers in Ecology and the Environment* 3(1): 4–11.

Perrings, C. (2007) "Future Challenges", *PNAS* 104(39): 15179–15180.

Steffan-Dewenter, I., Michael Kessler, Jan Barkmann, Merijn M. Bos, Damayanti Buchori, Stefan Erasmi, Heiko Faust, Gerhard Gerold, Klaus Glenk, S. Robbert Gradstein, Edi Guhardja, Marieke Harteveld, Dietrich Hertel, Patrick Höhn, Martin Kappas, Stefan Köhler, Christoph Leuschner, Miet Maertens, Rainer Marggraf, Sonja Migge-Kleian, Johanis Mogea, Ramadhaniel Pitopang, Matthias Schaefer, Stefan Schwarze, Simone G. Sporn, Andrea Steingrebe, Sri S. Tjitrosoedirdjo, Soekisman Tjitrosoemito, André Twele, Robert Weber, Lars Woltmann, Manfred Zeller, and Teja Tscharntke (2007) "Tradeoffs between Income, Biodiversity, and Ecosystem Functioning during Tropical Rainforest Conversion and Agroforestry Intensification", *PNAS* 104(12): 4973–4978.

Swart, J. A. A. and J. van Andel (2008) "Rethinking the Interface between Ecology and Society: The Case of the Cockle Controversy in the Dutch Wadden Sea", *Journal of Applied Ecology* 45: 82–90.

Thompson, P. B. (2004) "Sustainable Agriculture: Philosophical Framework", in R. M. Goodman (ed.), *Encyclopedia of Plant and Crop Science.* New York: CRC.

7-2
The preservation and creation of a regional culture for glocal sustainability

Tatsuji Sawa

7-2-1 Introduction

After World War II, Japan experienced unprecedented economic development, growing extremely rich and assuming a major role among the developed nations. However, since the collapse of its "bubble" economy, and despite the information technology bubble and the longest post-war period of economic prosperity in Japan's history, the country has faced massive problems such as the impoverishment of regional areas and growing disparity in incomes. Japan's situation is somewhat ambivalent: on the one hand, it enjoys economic prosperity, while, on the other hand, it faces accelerating impoverishment. At the end of Shigeru Mizuki's *History of the Showa Period*, Mizuki depicts the disgruntled figure of the "salaried worker [who] is enriched neither in mind or body. It seems that only the companies have been enriched", and that "most salaried workers do not possess houses and are hounded relentlessly by stress. Can we truly say that Japan is a wealthy nation?" (Mizuki, 1994: 264). What is it that is lost? Mizuki's words indicate a loss of home, of a place to return to.

Perhaps one cannot reclaim what has already been lost, but this section discusses what we can do in the time we have left.

7-2-2 The turning point in the 1960s

Regional impoverishment and income disparity are not problems that began after the war. In his novel *The Blackbird in the Rising Sun*, Paul

---

Designing our future: Local perspectives on bioproduction, ecosystems and humanity,
Osaki, Braimoh and Nakagami (eds),
United Nations University Press, 2011, ISBN 978-92-808-1183-4

Claudel, who lived in Japan as the French ambassador, wrote: "the traditional characteristics of the Japanese spirit are a feeling of reverence, self-effacement before objects of respect, and modest thoughtfulness offered to all surrounding life, to all things" (cited in Haga, 2002: 201). However, in the 1990s, when Jean Baudrillard visited Japan, he remarked, "Japan is so rich, because its people are so poor" (cited in Sawa, 1999: 86), expressing an opinion seemingly opposite to Claudel's. Why then have the Japanese, once extremely noble despite their relative poverty, now become poor in body and spirit? And when was it that they turned down this path? I believe that the point of this divergence was the 1960s, the period that separates Claudel on the one hand and Mizuki and Baudrillard on the other. This focus should probably be slightly widened to encompass the period from 1955 to around 1970, which includes the end of the post-war recovery period, the onset of rapid economic growth and a concurrent shift in focus from politics to economics.

Raising objections to modernity

There is a tendency to see the 1960s as a major turning point (Tominaga, 2009; Unno, 2008; Yamamuro, 2009). In particular, just before and after 1968 there were major upheavals in all parts of Japan and across all fields – politics, economics, society, culture and art. These can also be considered as reactions to the modernization that had taken place during the past few hundred years. Shinichi Yamamuro suggests the following:

> By examining the problems of the various social strains and individual burdens of instability brought about by rapid economic growth, and by focusing on the problems neglected in the process of rapid economic growth such as the exclusion of environmental pollution and isolated regions, we can reconsider from a starting point the nature of humanity, and what it means for individuals to exist as literary or intellectual figures. These were compelling issues raised through the 1960s. (Yamamuro, 2009: 328)

Two contradictory orientations are evident here: the pre-modern, which seeks to return to a time prior to this loss of community, and the post-modern, which seeks to transcend modernity. Yamamuro was commenting on global trends, but the 1960s represented just such a turning point for Japan.

The year 1968 was precisely 100 years after the Meiji Restoration, falling between the 1964 Tokyo Olympics and the 1970 Osaka Expo. It was a period when modernization came to be questioned once more. The year 1967 saw Ryōkichi Minobe become the governor of Tokyo. Reformist governors appeared all over Japan. Protests were held to prevent Prime

Minister Eisaku Sato's trip to the United States (the Haneda conflict). The year 1968 saw May revolutions in Paris, the Prague Spring, the student occupation of Yasuda Hall at The University of Tokyo, a conflict over USS *Enterprise*'s Sasebo port call, Yasunari Kawabata's Nobel Prize, the signing of the Nuclear Non-Proliferation Treaty (Japan signed it in 1970), the JPY 300 million robbery, the Kwon Hyi-ro incident and the Kanemi Yushō disease. The year 1969 saw the Tokyo University Yasuda Hall battle, the inauguration of the Associations to Indict Minamata Disease and Apollo 11's first manned landing on the moon. The year 1970 saw the Osaka Expo, Yodo-go hijacking incidents and Yukio Mishima's ritual suicide at the Ichigaya Self-Defense Force base. From the late 1960s to the start of the 1970s, various forces emerged in both positive and negative directions, as well as directions unrelated to either. From just some of the events of this period, it is clear that it represented a major turning point; the maturing of modernity in Japan was accompanied by a crisis of orientation between the post-modern and the pre-modern, as well as serious reflection on the meaning of modernism itself.

Countless novels and critical literary works have dealt with this period: Hidemi Suga's *Revolutionary, Too Revolutionary* (Suga, 2003) and *1968* (Suga, 2006), Inuhiko Yomota's *High School 1968*, Ryu Murakami's *69 Sixty Nine*, Haruki Murakami's *Norwegian Wood* and many others. The number of books dealing with this period is remarkable, which indicates how fascinating the period and its mysterious atmosphere were. Reading Inuhiko Yomota's *High School 1968*, despite the author's glorification of this period, one also senses its impending decline – a certain atmosphere of decadence is conveyed underlying the glory. We see junior high and high school students reading Marx and Engels, believing firmly in revolution and participating in political groups with guidance from university student activists. However, behind all this, they were also being baptised into the culture of free jazz at the jazz cafes of Shibuya and Shinjuku and indulging in decadent art and culture. Ryu Murakami examines identical themes of activism in his book *69 Sixty Nine*. Although set contemporaneously in 1969, Haruki Murakami's *Norwegian Wood* depicts a very different atmosphere. It is extremely interesting that the two Murakamis, both established authors and both representative of contemporary Japan, managed to produce such vastly different works, both set in the same spring of 1969. This serves, perhaps, to further indicate the heightened sense of ambivalence peculiar to the period.

On the other hand, Shunya Yoshimi makes a good case for the "levelling out" of daily life through the 1960s. Yoshimi describes how, as Japan's rapid economic growth narrowed the various economic disparities between farming villages and cities, between managers and employees, and between occupations, "the equal opportunities afforded by public education were guaranteed to some extent, which led to higher levels of education, the expansion of the white-collar class, large-scale migration from rural to urban areas, and the sudden advent of the nuclear family, as well as the deconstruction of traditional notions of order, which in turn drove a levelling out of daily life on a national level" (Yoshimi, 2009: 50). In this way, the middle class expanded, leading to the widely held notion that all Japanese people belong to the middle class.

7-2-3 The collapse of regional structures and standardized regional development

Throughout the 1960s, various movements emerged to raise objections to modernization. There was a certain ambivalence in that over-concentration and the disparities between regional and rural came to be viewed as more problematic, while economic disparities were gradually disappearing. What was the reason behind this? The author Taichi Sakaiya (under his real name Kotarō Ikeguchi) wrote *Regional Framework in Japan* during his time as a bureaucrat in the Ministry of International Trade and Industry. Ikeguchi suggested that Japan's regional composition has since ancient times been twofold in nature, having a "centralized and concentric form, while at the same time being an elliptical construction possessing two centres (one eastern, one western) with regard to economic trends and actual accumulation" (Ikeguchi, 1967: 277). With the Meiji Restoration, this was perfected into a centralized system, and after World War II, as everything became even more concentrated in Tokyo, this elliptical structure underwent a crisis of collapse. Identical themes are evident in Sakyō Komatsu's novel *Japan Time Travel: Regional Society Transfigured*, also published in 1969. For Komatsu, after the Meiji Restoration and the subsequent hundred years of modernization, Tokyo assumed a completely central position as "Edo", which served to emphasize the symbolic urban centrality of Kyoto and the economic centrality of Osaka (Komatsu, 1969).

As Sakaiya and Komatsu suggested, the completion of the centralized system of authority (in other words, the collapse of Japan's regional structure and of regional "home towns", both of which had existed since ancient times) took place in the 1960s. This meant the loss of places to which people felt they could return. In addition to the relative narrowing of disparities accompanying the "levelling out" of daily life in the 1960s, there was the problem of the national government's aim to eliminate regional disparities. The state's approach to regional development demonstrated a uniform and mono-dimensional view of regional areas, despite the diversity of these regions and their independent cultures. The comprehensive national development plans were a major factor in today's evident regional impoverishment.

Uniform, mono-dimensional perspectives

In the midst of this ambivalent situation – on the one hand, various disparities were acknowledged as problems, but on the other hand these disparities were also disappearing – a comprehensive national development plan was first announced in 1962. Behind this plan lay the problems of income disparities and over-urbanization that accompanied the shift towards rapid economic growth. The ultimate purpose of the comprehensive plan was to "construct a welfare state so that all citizens, in all regions, could achieve prosperous, stable lives and enjoy modern conveniences" (Economic Planning Agency, 1962). Reading between the lines, one might say that its purpose was to achieve uniform development and growth for regional areas.

At the time the comprehensive plan was formulated, various ministries and agencies each submitted their own comprehensive plans, but, as Ikeguchi writes, "development across the entire country was simply divided into several stages, each aiming for the same kind of construction development. It was never tailored to the individual characteristics and qualities of each regional area" (Ikeguchi, 1967: 91). This was not self-contained development suited to each region. Accordingly, "one particular type of regional formation was simply mass-produced within the national organic body" (Ikeguchi, 1967: 91). Ikeguchi's statement indicates the most significant reason the various comprehensive plans (announced every 10 years or so), even those specifically aimed at reviving regional areas, did not lead to any actual regional development and revival. Given that these were 10-year government plans, it is not surprising that they were aimed at constructing a vigorous economy and society. However, in the 35 years that elapsed between the first comprehensive plan, which aimed to be "a comprehensive solution from a national economic perspective for various productivity problems caused by urban over-expansion and productivity disparities in regional areas" (Economic Planning Agency, 1962), and the "Grand Design for the 21st Century" (the Fifth Comprehensive National Development Plan) in which the emphasis shifted from a mono-polar to a multi-polar model for national infrastructure (Ministry of Land, Infrastructure and Transport, 1998), and again in the 50 years that elapsed between the first comprehensive plan and the National Spatial Plan announced in 2009, the impoverishment of regional areas and the over-centralization of Tokyo were not resolved. In fact, they continue to worsen each year.

The futurologist John Naisbitt predicted a global paradox: as globalization proceeds, governance will integrate in an upward direction and disperse in a downward direction. In Japan's case, however, only upward integration has been accelerated. Perhaps the reason for this is the uniform, mono-dimensional perspectives on regional formation that has pervaded each national comprehensive plan, as well as the one-step-at-a-time public works spending that presumes the absolutely centrality of the "nation to region to citizen" system, the seemingly natural premises of which continue to be unchallenged. It is easy to criticize this relationship of absolute subordination between the regional and the national but, as Ikeguchi indicates, "all Japanese people really need to properly reconsider and understand the reality that lies before our eyes: the massive obesity that afflicts the capital and the anaemia that afflicts regional Japanese society" (Ikeguchi, 1967: 285). One must not forget that the responsibility here lies in the lack of interest of each and every Japanese person.

Tourism and associated problems

Under the comprehensive plan, tourism promotion and regional development were seen as linked. Post-industrial society has greatly hailed tourism, culture and art, and in recent years, along with the creative boom, these have been positioned as central to regional promotion. However, this was already being argued 50 years ago. In the comprehensive plan, a central measure to promote tourism was "an emphasis on the wide-ranging formation of tourist areas and development of tourist routes by establishing tourism infrastructure such as highways, railways, airports, ports etc. in order to drive tourism development", as well as "establishing accommodation and leisure facilities such as youth hostels, government-operated hotels and vacation villages, natural parks, highway parks". In other words, looking at the stated objectives of this plan, a great deal of emphasis on defunct and unnecessary construction and government-operated facilities is evident.

The author Junichiro Tanizaki recorded some passing thoughts in the *Mainichi* newspaper in December 1962, under the title "Thoughts on Kyoto". Tanizaki, while recollecting time spent at the Senkantei in Shimogamo in Kyoto, was also lamenting Kyoto's gradual and ongoing modernization.

> [J]ust a few years ago, I enjoyed taking my leisure at Saga, but on discovering municipal housing built near Rakushisha [a famous tea hut established by Matsuo Bashō's disciple Kyorai Mukai, where Bashō wrote his Saga chronicles] and the spoiling of the Ogurayama scenery, I was greatly surprised. Likewise, with the beautiful bamboo forests around there, which seemed to have been cut down overnight. The avenue of pines near Ninna-ji temple had also been felled.

Would the walls of Ryōan-ji be demolished next, perhaps, to make way for a tourist highway leading through the Kinugasayama foothills to Kinkaku-ji? (Tanizaki, 1962)

Tanizaki also criticized the fact that "driveways, ropeways, summit observation decks and cheap hotels are suddenly now the established course. Despite this, we do not seem to be able to afford to construct a single satisfactory toilet." He further opined, "Kyoto, with its countless fires and earthquakes, was at least spared the horrors of war. We need to realize that if humans can only refrain from making these pointless, unnecessary additions, we would not risk losing the beauty of this town, which has been ongoing since the Heian period." With these words, Tanizaki was making an appeal for the preservation (without unnecessary development) of culture, cultural assets and the city as a whole. In a song that remains relevant today, we hear his lament about "the elevated tracks that cut across Nihombashi and crowd out the big Tokyo sky". This is classic Tanizaki.

Tanizaki was not writing specifically with the comprehensive plan in mind, but in his words we can read for ourselves his intense disapproval of the sort of tourism promotion schemes that were part of the comprehensive plan. More than anything, Tanizaki's thoughts have the power to move us even when we read his words today, or perhaps all the more when we read his words today. When tourism development places great emphasis on establishing infrastructure such as roads and accommodation facilities but fails to install satisfactory toilets and, furthermore, ignores the landscape and all its subtle features that give it meaning, then to some extent it must bear the responsibility for driving this loss of culture.

The etymology of the word "tourism" (*kanko* in Japanese) originally comes from a passage in the *I Ching Kanke*, meaning "to view the light of a land" (*Kuni no hikari wo miru*, in Japanese). When one views the prosperity of a land (which is to say, when one views the best of that land's culture), the act of viewing enables thinking, which in turn enables the devising of policies for its future cultural development. In regard to the term "viewing", Tatsusaburō Hayashiya sought to grasp the essential principles of tourism as follows:

While viewing means looking, at the same time it also bears the meaning of expression. The words are located within a reciprocal master–guest relationship. It follows that within tourism, one land's culture must also be pro-actively expressed. If the expressing party is not pro-active here – if viewing is a one-way affair of the viewing party alone – then tourism is not adequately performing its role. (Hayashiya, 1985: 150)

The pervasive problems of the comprehensive plan explained above and the problems indicated here with tourism promotion schemes share a common root. This is the inevitable problem of short-sighted, top-down policies. Human resources and objects tend to integrate in an upward direction; they become concentrated in Tokyo, in this case, and only the tiniest amount of resources is returned for investment in public works. This "centre vs. periphery" model has proved persistent and difficult for Japan to shake off.

7-2-4 Awareness of the importance of art and culture

In recent years, as the idea of "creativity" has attracted increasing attention through concepts such as the creative city, the creative economy, the creative class, and so on, the government has also been establishing measures to link the promotion of art and culture with economic development. In Japan, following the Fundamental Law for the Promotion of Culture and the Arts enacted in December 2001, the Basic Policy on the Promotion of Culture and the Arts was established through a cabinet decision on 10 December 2002 (the First Basic Policy). It asserts that "the state of the arts can exert a major influence on economic activities, and can contribute to the development of many industries", thus indicating a relationship of positive feedback between culture and the economy.

In the Basic Policy on the Promotion of Culture and the Arts established by a cabinet decision on 19 February 2007 (the Second Basic Policy), culture and the arts were defined as:

> 1. Providing the necessary base for humans to live in a human fashion, 2. Producing feelings of solidarity enabling human relationships, 3. Contributing to realization of high quality economic activities, 4. Contributing to the true development of mankind through developing science, art and information, 5. Acting as a keystone of world peace through sustaining cultural diversity.

The Basic Policy also emphasizes the importance of the social role of culture and the arts, stating:

> possession of arts and culture means the power to attract people and influence societies; thus, it is globally recognized that cultural power is national power. Also within the economic sphere, culture and the arts can produce new demand and increase added value. It is now well known that the economy and the arts share an intimate connection.

It goes on to state:

> culture and the arts are born from the actions of diverse people of all ages and countries, and we continually discover new forms of their value as they are inherited and change. It is thus difficult to measure their value from a purely short-term perspective. The nature of culture and the arts means that short-term economic efficacy is not to be sought for cultural and artistic activities; rather, development of policies must take the long-term and sustainable perspective.

This explains the necessity for policy to be formulated from a long-term perspective, as the relationship of positive feedback between the arts and the economy attracts increased attention.

Creative urban spaces for art and culture: The state of culture and the arts

All culture contains within it the history it has cultivated. Therefore, the destruction of culture may be accomplished in a single moment, and it is almost impossible to reclaim it once it has gone. For this reason, one needs urban policies that emphasize culture and the arts and a cautious attitude to urban creation. Norio Iguchi asserts,

> [W]hen undertaking national planning or urban planning, any planning for physical layout that does not incorporate changes in community and cultural factors and does not consider global shifts rapidly becomes pointless. At present, rather than establishing the "hardware" [infrastructure] or quantitative aspects of urban planning, the emphasis is increasingly being placed on "soft" content, with the process itself also gradually changing from a top-down to a bottom-up emphasis. Due to the opinions and participation of local residents, land values increase in areas where elevations and land applications are subject to restrictions. This form of urban creation does not rely on a "master plan," and consequently it is currently achieving civic empowerment on a major scale. A great deal of tax money is spent on surveys, investigations and advertising activities. If we consider the relocation of the capital, which as governmental business proceeded in a top-down fashion, or the circumstances within which various studies and industrial urban configuration are currently located, this is easy enough to understand. (Iguchi, 2007: 15–16)

Iguchi here indicates the need for a thorough understanding of the people residing in such urban spaces and the meaning of culture in those spaces, and, to this end, the necessity and significance of the bottom-up approach to policy creation.

The extreme difficulty of practising this bottom-up approach to policy is not hard to imagine. Moreover, that which is actually produced by the bottom-up approach is not always necessarily genuine or good. The ancient Roman poet Decimus Junius Juvenalis used the expression "bread and circuses" to mock the society of his day. Vast wealth extracted from countless other lands made Roman citizens so wealthy (in terms of food and entertainment) that they did not have to work. In order to ensure that the citizens took no interest in political matters, the authorities constantly distributed food and provided entertainment. This decadent situation produced social decay and was one reason for the fall of the Roman Empire. This is what Juvenal meant by "bread and circuses".

Patrick Brantlinger, in his work *Bread and Circuses*, suggests that, because of the positive and negative aspects – principally negative – of the mass media, "the modern world has entered a stage much like the decline and fall of the Roman Empire" (Brantlinger, 1983) and that the term "bread and circuses" aptly describes modern society. If, as Brantlinger suggests, contemporary society is in fact "bread and circuses", how should the voices of citizens be incorporated into bottom-up policy creation, as Iguchi indicates? Moreover, it is not impossible that the bottom-up approach could veer entirely off course. Undeniably, questions still exist regarding the necessity of the bottom-up approach.

Cultural losses caused by policy

If policy is viewed and executed only from a one-sided perspective, then culture may be lost even if policies incorporate the opinions of local residents. The designation of a special industrial zone in Nishijin, Kyoto, and the subsequent decline of Nishijin, is a typical example.

The Nishijin area of Kyoto, lined with traditional townhouses, underwent major changes at the start of the 1970s. First, in 1972, the first highrise apartment buildings near Sembon Imadegawa were constructed. Then, in 1974, Nishijin was designated "a special industrial zone for protection and cultivation of traditional industries". Accordingly, building floor-area ratios were relaxed in order to facilitate large-scale expansion. This relaxation of floor-area ratios meant that comparatively spacious land, such as empty factory sites, could be acquired alongside highly convenient transport near the urban centre, which in turn spurred the construction of apartment buildings. There were 273 real estate offices in the region in 1980, but this had grown to 362 by 1985 – an increase of 30 per cent despite the area's small size (Taniguchi, 1993). Japan then entered the period of its "bubble economy" (from December 1986 to February 1991). Needless to say, this spurred massive real estate investment and construction of apartment buildings.

In *Lifestyle and Urban Aspects of a Changing Nishijin*, a textile business manager stated that "managing real estate was always one of the three aspects of the Nishijin textile business: retail sales, real estate, and stocks. In a many-sided business, managing apartments has always been the fastest way to make money" (Taniguchi, 1993: 98). For weavers, whose business is known for rarely continuing past three generations, this trichotomous business model is simple common sense. However, it is a fact that, with the collapse of the bubble economy, two of these aspects – real estate and the stock market – suffered a devastating blow. A great many weaving businesses disappeared because of the collapse of the bubble economy. However, it was not only weavers who were affected. Countless Nishijin commercial businesses, as well as the Kamishichiken pleasure district supporting many Nishijin businesspeople, also suffered a major blow. "The decline of Nishijin, which together with Gion and Kiyomizu boasted a distinctive Kyoto personality built up by the weavers and other businesses that epitomize Kyoto, has meant the loss of a uniquely Kyoto charm. This is the collapse of Kyoto, and the collapse of the lifestyle system that was built upon Kyoto's historical time and urban space" (Taniguchi, 1993: 98). One cannot afford to forget that, as seen here, the decline of a single thing can lead to the destruction of the entire culture that surrounds it.

Of course, the major cause of the decline of Nishijin was the vanishing custom of wearing the traditional Japanese kimono. Cultural works of the post-war period such as Yasunari Kawabata's *The Old Capital* or Tsutomu Mizukami's *Gobancho Yugiriro* already describe the decline both of the kimono culture and of Nishijin, and they paint a grim outlook for the weaving industry. This was self-evident. Because the custom of wearing the kimono would never return, Nishijin was designated as a special industrial zone for regeneration. Ironically, however, it was this designation that struck the death blow for Nishijin. The role of administration is not simply to respond to popular demand. Administration also requires the capacity to deter adverse development.

7-2-5 What to protect and what to create

In an address to the US Congress on 21 February 1990, the last Czechoslovak and the first Czech President Vaclav Havel argued: "Without a global revolution in the sphere of human consciousness, nothing will change for the better in the sphere of our being as humans, and environmental, social, or cultural catastrophe will be unavoidable." However, as Juvenal's mocking term "bread and circuses" expresses, when confronted with easily obtained riches, humanity fails to reflect upon its past failures, seeing only the profit before it. The recent global economic crisis has made this clear. In the face of this shallow lack of human reflection, Havel's remarks are perhaps utopian constructions.

Along with this pessimistic forecast, Natsuki Ikezawa's *Wonderful New World* gives some hope regarding the creation of a truly wealthy society and ways to support it. Tamotsu Aoki, in the Afterword to the pocket edition (Ikezawa, 2003: 716), contrasts this book with Aldous Huxley's *Brave New World*, which depicts the future world as a dystopia in which humans are strictly categorized, bred and controlled. He suggests that Ikezawa's work "clearly moves along opposite lines to Huxley's, and contains truly harsh criticism, even likening contemporary Japanese society to Huxley's world". The protagonist, who attempts to introduce uninterrupted wind power generation in the remote Himalayas, is confident that technology can enrich life but also harbours anxieties about whether its benefits can reach as far as the human heart. One might also say that the work depicts a lack of understanding of these places supported by Japan. This mirrors the current situation in Japan – advances in technology represent its sole starting and ending points, and Japan tends not to be concerned about the aftercare of technological support or any changes occurring after its introduction. The book presents a powerful argument for the importance of considering and backing both physical and spiritual aspects.

If policy remains simply an unconvincing vision that is foisted upon residents, the flow of urban development is halted and cities become devoid of the life they once possessed. To sustain urban spaces, culture must be protected and must be developed. To this end, one must first reclaim that spirit of "warm depth and delicate sensitivity to nature and humanity" found in Kawabata's work *Japan, the Beautiful, and Myself*, which centres on the beauty that Japan has possessed since ancient times (Kawabata, 1969). In Kawabata's own words:

> I dine at the little restaurant next to Kamigamo Shrine and return home amidst the lights of evening. The moonlight is bright on the narrow path along the banks of the Kamo river, and the cries of insects are faint among the grass. I hear the river water flowing over countless gates and dams, and see the moon reflected in the light of the lanterns hanging between the trees lining each bank. (Cited in Haga, 2002: 314)

If we can preserve this kind of landscape, if we can feel it in our hearts, then the preservation and creation of culture is surely under way.

The new erodes the old, without a doubt, and destroys it over time. This is why the old must be preserved and protected. The world that Huxley depicted is a dystopia filled only with the new, whereas Ikezawa's world aims to be a utopian society in which the old exists in harmony with the new. In constructing such a utopia, one must adopt a process of first investigating the history and significance of old things, then drawing up a vision of a future society in which these can harmonize with the new, followed by clarifying exactly what is required for this and then creating those new things.

7-2-6 Glocal sustainability, starting with the tenet "think locally, act locally"

Glocalization is often connected with the principle "think globally, act locally", but this is not enough. It offers Japan no hope of escape from the ongoing vicious cycle of the last 50 years or so. First, Japan needs to start thinking locally and acting locally. Navigating the waves of globalization is difficult but, given that the global paradox is not occurring in Japan, "glocal" is a term of convenience used only superficially. In the midst of all this, it is necessary to consider once more the significance of Japan's regional areas. By doing this, it will be possible to start moving away from subordinate relations (national—regional—non-profit organization—residents) to construct an alternative decision-making system (individuals—family—community—town—prefecture—city). Schumacher (1973) commented on the disappearance of the middle class as well as on the pointless technological shifts and development policies that ignore communities and towns. These serve as a warning of the collapse of society and its cultural base, including those technologies that really should exist. Both administrators and citizens need to take another close, hard look at regional areas.

REFERENCES

Brantlinger, Patrick (1983) *Bread and Circuses: Theories of Mass Culture as Social Decay*. Ithaca, NY: Cornell University Press. Japanese translation by Koike Kazuko, 1986, *Pan to sakasu*, Keisō Shobō.
Economic Planning Agency (1962) *The First Comprehensive National Development Plan*, October.
Haga, Toru (2002) *Shika no Morie: Nihon-shi e no Izanai* [To the Forest of the Poetry]. Tokyo: Chuoukoron-Shinsha Inc.
Hayashiya, Tatsusaburō (1985) *Kyoto Bunka no Zahyo* [Coordinate with Culture in Kyoto]. Kyoto: Jimbunshoin Ltd.
Iguchi, Norio, ed. (2007) *Seijuku Toshi no Kurieitobu na Machizukuri* [Creative Town Planning]. Tokyo: Senden Kaigi.

Ikeguchi, Kotarō (1967) *Nihon no Chiiki Kozo* [Regional Framework in Japan]. Tokyo: Toyo Keizai Inc.
Ikezawa, Natsuki (2003) *Subarashii Shin Sekai* [Wonderful New World] (pocket edition). Tokyo: Chuoukoron-Shinsha, Inc.
Kawabata, Yasunari (1969) *Japan, the Beautiful, and Myself: The 1968 Nobel Prize Acceptance Speech*, translated by Edward G. Seidensticker. New York: Kodansha International.
Komatsu, Sakyō (1969) *Nihon Taimu Toraberu* [Japan Time Travel: Regional Society Transfigured]. Tokyo: Yomiuri Shimbun.
Ministry of Land, Infrastructure and Transport (1998) *The 5th Comprehensive National Development Plan: "Grand Design for the 21st Century"*, October, National Planning Division.
Mizuki, Shigeru (1994) *Showa Shi* [History of the Showa Period], Vol. 8. Tokyo: Kodansha Ltd.
Sawa, Takamitsu (1999) *Keizai Gaku no Meigen 100* [Hundred Golden Sayings in Economics]. Tokyo: Daiamond Inc.
Schumacher, E. F. (1973) *Small Is Beautiful: A Study of Economics as if People Mattered.* London: Blond & Briggs. Japanese translation by Kojima Keizō and Sakai Tsutomu, 1986, *Sumōru izu byūtifuru: ningenchūshin no keizaigaku*, Kodansha Ltd.
Suga, Hidemi (2003) *Kakumeitekina, Amarini Kakumeitekina: 1968 Nen No Kakumei Shiron* [Revolutionary, Too Revolutionary]. Tokyo: Sakuhinsha Ltd.
Suga, Hidemi (2006) *1968*. Tokyo: Chikumashobo Ltd.
Taniguchi, Kōji, ed. (1993) *Lifestyle and Urban Aspect of Changing Nishijin*. Bukkyo University, Nishijin Regional Research Association, Houritsu Bunka Sha.
Tanizaki, Junichiro (1962) "Thoughts on Kyoto", *Mainichi*, December.
Tominaga, Shigeki, ed. (2009) *Seek for Point of Evolution: Study of 1960s*. Tokyo: Sekaishisosha Co.
Unno, Hiroshi (2008) *Nijisseiki* [The Twentieth Century]. Tokyo: Bungeishuju Ltd.
Yamamuro, Shinichi (2009) "Kindai no Honryu to Gyakuryu" [Torrent and Backflow of Modern Japan], in Shigeki Tominaga (ed.), *Seek for Point of Evolution: Study of 1960s*. Tokyo: Sekaishisosha Co.
Yoshimi, Shunya (2009) *Posuto Sengo Shakai* [The Post-World War Era]. Tokyo: Iwanami Shoten, Publishers.

7-3
Sustainability and indigenous people: A case study of the Ainu people

Koji Yamasaki

7-3-1 Introduction

When the topic of sustainability and indigenous people is discussed, one of the first issues to be mentioned is the relationship between indigenous people and nature. Presently, all humans are facing great danger as we try to cope with several ongoing global environmental problems. As we look for various means to tackle this situation, it seems the phrase "indigenous people in harmony with nature" invariably crops up.

This section will first outline the depiction of the Ainu people in relation to this statement and then discuss some problems arising from it. Next, it will explore the current relationship between the Ainu people and the sika deer (*Cervus nippon yesoensis*) on the island of Hokkaido, and sketch the historical changes in that relationship. Finally, it will suggest some issues for future discussion.

7-3-2 Perception of the Ainu people

The groups referred to as hunter–gatherers are those most closely linked with discourses of "people in harmony with nature" and "low environmental impact".[1] It is perceived that peoples descended from a traditional hunting and gathering culture, including the Ainu people, "did not take more than was necessary, used the things they gathered to their

Designing our future: Local perspectives on bioproduction, ecosystems and humanity,
Osaki, Braimoh and Nakagami (eds),
United Nations University Press, 2011, ISBN 978-92-808-1183-4

maximum, had a way of life that was in harmony with nature and did not consume excessively" (Stewart, 1996: 3).

The image of the Ainu people has become more prominent in recent years as a research field. Mentioning all the points in greater detail is beyond the scope of this section. However, research has been carried out into the way non-Japanese people have represented the Ainu, as seen in the changes in the depiction of Ainu people in Western Europe between the sixteenth and twentieth centuries, and the history of research into the Ainu (Kreiner, 1993). There have also been studies analysing the image of the Ainu people in newspaper articles in North America at the start of the twentieth century (Takarabe, 2001).

The most comprehensive research into the depiction of the Ainu people in Japan highlighted the historical changes in the representation of the Emishi and Ezo people in the Middle Ages and the modern depiction of the Ainu (Kojima, 2003). Takashi Kinase (1997) also studied symbolism and politics, focusing on the modern era, and noted that one of the sources of the stereotypical notion that "Ainu = hunting people = people in harmony with nature" is modern research into Ainu culture that promotes as "scientifically" accurate the long-held view of the majority population that the Ainu are "savage and undeveloped". Other extensive research includes analysis of the paintings of the Ainu (Sasaki, 2004), investigations of newspaper articles relating to the Ainu people (Higashimura, 2006; Momose and Yanaka, 2005), and examinations of tourist pamphlets (Saito, 2000). These studies covered a wide range of topics, but one common thread is that, until recently, the image of the Ainu was by and large not one of the actual Ainu people at that time but a construct of the viewpoint of the majority population and linked closely with the social politics of the time.

On the other hand, there have been investigations into the way in which Ainu individuals strategically manipulate self-depictions and the images that Ainu people traditionally hold of themselves. There is also research related to the conflict between the image of the Ainu people projected by those around them and that created by modern life, as well as studies focusing on narratives of individual Ainu who lightly and strategically adopt, reject or manipulate these conflicting images (Kinase, 1998; Okada, 2008; Sekiguchi, 2007). Research focusing on the narratives of individual Ainu has been gaining momentum in recent years, and it is expected to intensify in the future.

Some anthropological studies directly address the theory behind not only the image of the Ainu people that research data have accumulated and but also the assertion that the Ainu people "live in harmony with nature" (Kojima, 2003; Honda, 2007). It has been noted that in schools, too, the subject of Ainu culture is generally dominated by consideration of earlier Ainu society, and that an overemphasis on ecological spirituality is added to depictions of the Ainu worldview and their way of life. This approach hinders our understanding of the modern Ainu people and their culture (Yoneda, 1996).

Based on earlier research into the image of the Ainu people, one can recognize just how vague is the phrase "living in harmony with nature" that is so readily associated with the Ainu when the topic of sustainability arises, and one can understand the need to be cautious about using such words. Today, there are many stances on the Ainu people, giving rise to many assertions, but, whatever the circumstances, it is necessary to reaffirm that depictions born from a reverse image of the observer that idealistically portrays the Ainu as conquering all the observer's shortcomings are no more than an illusion.

7-3-3 The situation of sika deer in Hokkaido

The sika deer (*Cervus nippon yesoensis*) is the animal most closely linked with the Ainu people. This subsection first considers the current situation of the sika deer in Hokkaido.

Around the start of the Meiji Era (1868–1912), the sika deer almost became extinct because of heavy snowfall and hunting. But thanks to protection measures and improvement of the habitat, their numbers began to recover and their territory expanded. From the beginning of the Showa Era (1926–1989) and into the current Heisei Era, the costs of the damage the deer have caused to agriculture have increased dramatically, particularly in Eastern Hokkaido, exceeding JPY 5 billion in 1996 and making it a serious social problem (see Figure 7.3.1). In 1997, the government of Hokkaido established the Council on Sika Deer Management Policy, implementing a number of measures to deal with the sika deer, including conservation and management, to prevent damage to the forestry industry, and for effective utilization of deer meat. As part of these measures, from 1998 onwards, the Conservation and Management Plan for Sika Deer in Eastern Hokkaido and the Conservation and Management Plan for Sika Deer were implemented, and there are currently a variety of measures in place across the whole Hokkaido region based on these plans. These conservation and management strategies use "feedback management", whereby they monitor changes in the current situation and take steps accordingly. In addition, putting up wire fencing to prevent damage has been effective to some degree. Despite these steps, the cost of damage in 2006 amounted to almost JPY 3 billion across the whole of Hokkaido, and there were a number of problems that are still to

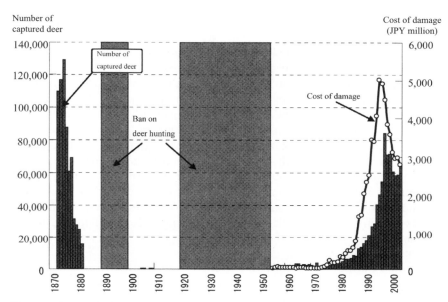

Figure 7.3.1 Variations in the population of captured deer and the cost of damage to agriculture and forests: Hokkaido, 1973–2004.
Source: Adapted from original graph (Kaji, 2006).
Notes: Bar: number of captured deer; line: cost of damage.

be resolved, such as the official number of traffic accidents involving vehicles and sika deer, which rose to 1,206 in 2006 (Hokkaido, 2008).

There has not yet been a scientific study into the reasons for the increase in sika deer numbers. However, in recent years there has been less snow owing to continuing mild winters – hard winters are one of the causes of death for the deer – and it is presumed that the other main reasons are the high breeding capacity of these animals, protective measures for the deer, and the easy availability of highly nutritious pasture grass and agricultural crops (Uno, 2006; Shiro Tatsuzawa personal communication). The increase in the deer population may also be connected with the disappearance of predators such as wild dogs and the Ezo wolf (*Canis lupus hattai*), which became extinct in the Meiji period.

It is also suggested that the rise in the sika deer population has had an adverse effect on the management of other wild animals. Research in the Urahoro region of Eastern Hokkaido on the droppings of the brown bear (*Ursus arctos*) revealed that, as the numbers of sika deer have increased, this has changed the feeding habits of the brown bear. In the period when the number of sika deer rose (between 1988 and 2000), it was clear that the brown bears were feeding on a higher proportion of deer meat throughout the year (Sato et al., 2004). Although it is known that brown bears feed on the sika deer, there is no research regarding what proportion are caught and killed as opposed to those that are found dead and then eaten, but it is possible that the brown bear is eating the remains of carcasses from culls or from the hunting of the deer. It can be surmised that plenty of remains from culls can be found around farmland, and those from hunting kills can be found in forests near roads or rivers. According to Yoshikazu Sato, it is possible that culls carried out to prevent damage to agriculture by sika deer are one of the reasons the brown bear now ventures close to farmland, resulting in more frequent encounters with humans, and thereby the brown bear, rather than being protected from the loss of habitat and its depleting numbers, becomes the target of control (Sato, 2005, 2006). Furthermore, lead pellets left in the remains of the deer after they have been shot have been linked to lead poisoning in eagles.[2]

Against the background of this friction between the deer and human activity, a combination of hunting and trapping was carried out to control the deer population. However, at present one problem being faced is a reduction in the numbers of hunters. The number of hunters in Hokkaido peaked in 1978 at 20,000, but had dropped to 9,000 in 2005 as they aged. In addition, the Yeso sika is an attractive natural resource in Hokkaido; organizations affiliated to the Hokkaido government are seeking ways to market venison and non-meat parts of the deer.

7-3-4 The Ainu and the sika deer

This subsection examines the relationship between the Ainu people and the sika deer. Sika deer historically formed an essential food resource for the Ainu, as noted in the travel writings by Tosaku Hezutsu, who reported that in 1784, when the number of deer decreased because of heavy snowfall, 300–400 Ainu people died of starvation (Hezutsu, 1972).

Prior to the Meiji Restoration, the sika deer migrated over a wide area of Hokkaido, and it is believed that the population varied considerably over the long term because of heavy snowfall and other factors (Inukai, 1952). The problem of the effect that fluctuations in the sika deer population had on the Ainu prior to the Meiji period is an important topic that it is not possible to discuss here, but the importance of the deer as an item of trade also had a major influence on the attitudes (beliefs and rites) of the Ainu people towards it (Akino, 2006; Takahashi, 2004, 2008; Walker, 2001). There are no accurate data on the numbers of sika deer just prior to the Meiji Restoration, but Takeshiro Matsuura ([1863] 1984) wrote that, near Niikappu, "In the distance I could see something red spreading out over around three cho [approximately 3 hectares]. When I asked the Ainu 'What is that?' they took their arrows and ran off in that direction. The plains that we thought were red due to dry grass suddenly moved at the approach of the Ainu. It was all a herd of deer. The deer which ran off in all directions must have numbered in their tens of thousands." This passage suggests it was thought at that time that there was a considerable population of sika deer.

Research by Tetsuo Inukai (1952) indicates the diversity of the Ainu people's deer-hunting techniques. Many different weapons were used by the Ainu for deer hunting: hand-held arrows, spring bows, spears, wooden staffs (clubs) and so forth. In all regions, dogs and beaters were used in the last stages of the hunt. In addition to the well-known example of chasing prey into the snow, rivers or marshes, other documented examples are on the Wakoto peninsula of Lake Kussharo, where the Ainu hunters chased the deer into the lake.[3] In the region of Abashiri, they pursued the deer onto the frozen sea, in Oshima they chased them in the direction of the forest, and in places such as Hidaka, Iburi, Tokachi and Kushiro they also used the sea. In Sikaoi (Tokachi), it is said that hunters would sometimes construct an enclosure made of logs near the village and chase the deer into it. In one region, although it is unclear which one, during the autumn and winter, when the deer have antlers, the people were known to catch deer by pursuing them into thick woods where the deer would become entangled. When hunting large numbers of animals, people used spears and clubs as well as bows.

In the Kitami Bihiro region, during the spring and summer when the deer have lost their antlers or before they are fully developed, people would set a trap on the deer's trail made of rope, with a noose of about 30 cm in diameter, and tie a log on the other end. Once people learned to ride horses, they would sometimes hunt by throwing a lasso from the back of the horse (Inukai, 1952). Another method of hunting involved placing a sharpened stick at an angle facing upwards on the mountain, and when the deer were chased down from the mountain they would be impaled on the stick (Fujiwara, 1985; Kayano, 1978). In 1951, a suspended deer trap made of rope was documented, although this method does not have a long history of use (Watanabe, 1952).

During the period when the exploitation of Hokkaido was just commencing, it is unclear what proportion of all hunting activities by the Ainu people involved poisoned arrows, but, with hindsight, judging by the reaction of the Ainu to the Hokkaido Deer Hunting Regulations, it was without doubt quite a large proportion. The poison for the arrows

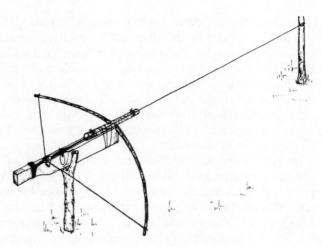

Plate 7.3.1 A spring bow (*Source*: Kayano, 1978).

was made from aconite root. The poison was applied to the tip of the arrow and would enter the bloodstream of the animal that had been hit, causing it to die quickly.

Poison arrows were used with hand-held bows and spring bows. When using hand-held bows, the hunters might creep up on the herd from the downwind side. Or they might use a deer call, or startle the deer from above, using dogs and beaters to drive the deer into an ambush in the valley below where hunters would shoot them (Inukai, 1952; Watanabe, 1952). Spring bows were also used, not only for sika deer but also for brown bears and foxes (Plate 7.3.1). As of 1951, according to verbal information from the Saru River region, hunters normally took about 10–15 spring bows with them on a hunt. Using a hunting hut as a base, they would determine an area – either a mountain ridge or a valley – and set up the bows, selecting positions from which the arrow was most likely to hit its target on the animal trail. Three or four bows could be set in a day. One person might set a total of 20 to 30 bows. Once the positioning of the spring bows was completed, after fixing markers on the trees as a signal to indicate the relevant area around the valley or the mountain ridge, the hunters would go around and examine them every day or two. The area where the spring bows were set was decided on each year, and it was normal to check with other people before setting them up (Watanabe, 1952).

As the Meiji Era dawned, the exploitation of Hokkaido began in earnest, and the depletion of the sika deer increased. In the six years from

1873 to 1878, the total number of deer hunted was 574,462, or between 60,000 and 130,000 per year (see Figure 7.3.1). The skins were mainly exported to France and the antlers to China (Inukai, 1952; Kaji, 2006). At the same time, there were major changes in land use in Hokkaido. In particular, the flatlands of broadleaf foliage were mostly converted to agricultural fields (Nishikawa, 1995).

The Hokkaido Development Commission was deeply concerned about the drop in the population of sika deer owing to overhunting, and so the Hokkaido Deer Hunting Regulations were put into effect on 11 November 1876. These regulations assumed that the sika deer were an important natural resource and they were implemented for the purpose of managing the deer population. A taxable licence system was established, limiting the number of people permitted to hunt deer in Hokkaido to 600 (although with the condition that the Ainu people were exempt from the tax "for the time being"). The hunting season was four months, lasting from 1 November until the last day of February, and hunting deer with poison arrows was prohibited. These regulations were a prototype for the wildlife protection and population control that came later. They were created under the guidance of a foreign adviser (*o-yatoi gaikokujin*), Horace Capron, and paved the way for the management of resources in Japan (Tawara, 1979). However, there was much criticism of the details and the application of the regulations (Momose, 2003; Tawara, 1979; Yamada, 2001, 2006).

Shinichi Yamada (2001) analysed the events that took place until the regulations were repealed. According to him, the process of devising these rules confirmed the view held by the foreign consultant, Capron, and the Development Commissioner, Kiyotaka Kuroda, that hunting with poison was savage, and did not stop at the issue of hunting with poison arrows. By its very nature, the prohibition on poison arrows denied the ethnographic characteristics of Ainu culture and incorporated "civilizing" measures. Another factor, regarding accidents involving spring bows, was that the implicit rules shared by the Ainu people were ignored in order to lend support to the regulations.

It is also important to note that there was major unease and opposition from the local Ainu people immediately prior to the declaration of the regulations, which were issued on 24 September 1876 by the Sapporo office of the Commission. The Ainu of the three counties of Saru, Chitose and Yuufutsu submitted a petition requesting a postponement of the implementation of the ban on the use of poison arrows, explaining that they were not accustomed to using hunting guns, which had been suggested as a replacement for poison arrows, and that there would be people who would have difficulty hunting deer with anything other than a spring bow.

In Mitsuishi, Urakawa, Samani and Tokachi, the objections were acknowledged; however, "moral persuasion" of the Ainu failed to gain their agreement.

The Commission went ahead and issued the Hokkaido Deer Hunting Regulations without any changes to the basic policy, and the prohibition on poison arrows was enacted.[4] A change from poison arrows to hunting guns and a switch to "new trades" were encouraged. People who could not comply with the new policy were treated as destitute and given aid. Hibiki Momose (2003) notes that the ruling "did not encourage independence but was an act which tried to solve a problem from a different approach called 'protection'".

After the declaration of the Hokkaido Deer Hunting Regulations, because the regulations allowed people who had obtained a licence to act without restraint, the influx of hunters from other regions of Hokkaido, which was already an issue, accelerated and led to the destruction of the autonomously regulated system of the hunting range and the traditional methods in each region. Many people who obtained a licence under the Hokkaido Deer Hunting Regulations were Ainu but, because the numbers were limited on a first-come first-served basis, there were many instances of people who applied but did not get a licence. In addition, despite the prohibition on poison arrows, there were many Ainu who could not or would not use hunting guns. However, guns had begun to gain acceptance – either they were on loan or sold by the Sapporo office of the Commission, or they were obtained independently (Yamada, 2006).

The Development Commissioner, who considered the use of deer meat to be one way of promoting the economy in Hokkaido, established a venison-canning factory in Bibi, Chitose, in 1878. In 1879, the Commission also set up a manufacturing plant for saltpetre, an ingredient for preserving raw deer meat. As part of these actions to promote the economy, the Hokkaido Deer Hunting Regulations were revised on 29 June 1878. The main points of revision were: (1) a newly established category of hunting for pleasure, where previously the regulations had applied only to commercial hunting; (2) an increase in the number of licences for all Hokkaido, from 600 to 780; (3) recognition of regional differences in the hunting season, and extension of the season from the previous four months to six months; (4) the criteria for licence issuance, the regulation of hunting areas and the imposition of penalties under the Wildlife Hunting Regulations (Yamada, 2006). It is said that the increase in licences by 180 and the lengthening of the season by two months were intended to make it easier for the canning factory to obtain its raw materials, so at that point the wildlife protection measures came one step closer to industrial development measures (Tawara, 1979). However, not long after the establishment of the canning factory at Bibi, in 1880 and 1881 production was stopped, and in 1884 the factory was closed. The main reason for the closure was that, between January and February 1879, there was record snowfall and the sika deer were much depleted in numbers (see Figure 7.3.1).

There was heavy snowfall again in 1881, and the sika deer population decreased further. Owing to the diminishing numbers of sika deer, the Ezo wolf began to prey on domesticated animals, and this was linked in turn to the destruction and extinction of the Ezo wolf. As the sika deer neared extinction, there was a ban on hunting them from 1890 to 1900. Between 1905 and 1919, the number hunted totalled 2,000 animals – between 16 and 498 per year. In 1920, it was again prohibited to hunt them, and the prohibition lasted until 1956 (Kaji, 2006).

From the middle of the twentieth century, the Ainu people's stories of deer hunts decreased rapidly. In recent years, there has been research based on archaeological and documentary resources (Akino, 2004, 2006; Takahashi, 2004, 2008) on the deer spirit-sending ceremonies, but detailed descriptions of the deer hunt are surprisingly few.

7-3-5 Seeing the present from a historical perspective

The previous subsections looked at the historical relationship between the Ainu and the sika deer. This discussion covered topics such as the diversity of former deer-hunting methods, the ban on hunting with poison arrows by the Hokkaido Development Commission, the objections of the Ainu people to this prohibition, the destruction of traditional hunting grounds, and two hunting bans (1890–1900, 1920–1956) caused by the sudden reduction in the deer population owing to heavy snowfall in 1879. It is thought that these events had a major effect on Ainu culture in relation to the sika deer.

One of the characteristics of the culture of indigenous peoples is a strong connection with the land, and this applies also to the Ainu culture. Against a history of rapid reduction in the population of sika deer, the long hunting bans and the absence of deer as an object of worship, one can unfortunately only give a pessimistic response to the question of whether or not the practice of hunting deer has been sufficiently handed down by Ainu culture. That the unilateral prohibition on poison arrows ushered in sudden changes to the lifestyle of the Ainu people cannot be overemphasized. These days, hunting technology using bows and spring bows can be seen only in reference books and museums.

But has Ainu culture lost everything connected to its relationship with the deer? As stated in the previous subsections, when hunting with poison arrows was prohibited under the Hokkaido Deer Hunting Regulations, this situation was, of course, forced on the Ainu by the Commission, but there were many Ainu people who accepted hunting guns (Yamada, 2006). It is true that the traditional hunting technology of poison arrows has been lost; however, it would be overly simplistic to conclude that deer hunting and the uses thereof, including traditional knowledge regarding the deer's migratory habits in each region, have also been completely lost because of these methods. Following the prohibition on poison arrows, it cannot be denied that deer hunting with guns possibly became integral to Ainu culture.[5]

Unfortunately, at present there is very little written in detail about the Ainu's deer hunting with guns.[6] This dearth of material is closely linked to the circumstances that led to the near-extinction of the Yeso sika, as noted earlier, and the images of the Ainu people held by anthropologists. In other words, the hunting of deer with guns, something seen as no longer being part of Ainu culture, may have escaped the attention of researchers along with all of the traditional knowledge that goes along with such matters. Certainly, it can be maintained that the brown bear and Blakiston's Fish Owl have become the focus of much attention as symbols of Ainu ritualism, somewhat overshadowing the deer (Akino, 2004). This trend overlaps the steep drop in numbers of sika deer from 1879 onwards and the two hunting bans (1890–1900, 1920–1956); since the second half of the twentieth century there has been a dramatic reduction in research on the Ainu people in relation to deer hunting.

On 13 September 2007, the United Nations Declaration on the Rights of Indigenous Peoples was adopted by the United Nations General Assembly, and it is clear that this document will have a major effect on the Ainu people in the future. Articles 25 to 32 of this Declaration lay out the rights of indigenous peoples with regard to land, territory and natural resources. For the Ainu, it is thought that movements to demand these rights will increase, and it is only a matter of time before the issue of the sika deer is brought up. This is because the situation of the sika deer in Hokkaido, described above, is continuing, and the government is looking for ways to increase the number of hunters and the uses for the deer that have been caught.

This is where the image of the Ainu people "living in harmony with nature" finds grounds to rear its head. However, if one looks at the history of the Ainu and the sika deer, it becomes clear that there are many problems with this way of thinking. If, upon hearing the words "Ainu people" and "deer", one immediately envisions the Ainu people versed in traditional knowledge, their bows nocked, this is to ignore contemporary history as represented by the unilateral prohibition on poison arrows. The current situation in Hokkaido should not be linked uncritically to the earlier Ainu traditions. Takashi Kinase (1997) points to the existence of a constantly erased "present" as something that research writing on the Ainu has concealed. This idea applies to the relationship between the Ainu people and the sika deer, and in particular to hunting with guns.

To consider sustainability and the indigenous people in Hokkaido, one needs to dredge up the "present", which has been erased from history, particularly in the modern era. Or, to put it another way, one needs to open one's ears to the history of people living today without getting caught up in previous images of the Ainu people or in the ethnographic present. This task needs to be carried out together with the Ainu people, and the practice should be rooted locally. Through this process, cultural transmission from past to present becomes possible, thereby opening the way for an ongoing debate regarding what the situation should be on the island of Hokkaido.

Acknowledgements

Thanks are due to the following people for their assistance in writing this section: Hiroyuki Uno (Hokkaido Institute of Environmental Sciences), Shiro Tatsuzawa (Hokkaido University), Jeff Gayman (Kumamoto Gakuen University) and Masashi Kawakami (Hokkaido University).

Notes

1. "Hunter–gatherer people", as the phrase suggests, refers to people who subsist by gathering wild plants and by hunting or fishing. With the exception of the Inuit, at present most hunter–gatherer groups manage or cultivate plants to a greater or lesser degree. The term "hunter–gatherers" has come to be a convenient term for those whose way of life is supported solely by the gathering of plants and by hunting.
2. According to the 535th *Hokkaido Bulletin* dated 31 March 2000, the use of lead rifle bullets for hunting sika deer was prohibited from 1 January 2001. Furthermore, according to the 537th *Hokkaido Bulletin* dated 30 March 2001, the use of lead bullets in shotguns for hunting sika deer was prohibited as of 1 November 2001. From 2004, the use of lead bullets was restricted in the hunting of all large animals.
3. With regard to the pursuit and hunting of sika deer on Wakoto peninsula, there are data to indicate that they pursued the deer to the tip of the peninsula (Inukai, 1952). But the hunt in fact took place at the corridor-like point where the peninsula and the mainland are connected (Fujiwara, 1985).
4. The situation at Hakodate, Oshima, was slightly different. It can be interpreted that poison arrows were prohibited on weapons other than the spring bow. However, because of the issue of the "danger" of spring bows, the use of poison arrows was not proactively permitted (Yamada, 2001).
5. The author once accompanied Ainu people on a deer hunt. The hunters hunted with guns, but they had been beaters when they were children and had knowledge about deer

behaviour and traditional cooking methods. However, there are nowadays very few people like this among the Ainu.
6. What little information there is at the moment about deer hunting comes from interview surveys in Biratori (Committee for the Survey on Countermeasures to Preserve Ainu Culture and Environment, et al., 2006) and Shiranuka (Fujimura, 2009). The former survey was connected with construction plans for a dam, and the fact that detailed data from the land were collected and that the local Ainu people were mainly responsible for the survey makes it worthy of attention as a model for future reference.

REFERENCES

Akino, Shigeki (2004) "Considering Hokkaido Ainu Animal Spirit-Sending Ceremonies", in Committee for the Publication of Dr. Hiroshi Utagawa Memorial Essay Collection (ed.), *The Formation of Aynu Culture*. Sapporo: Hokkaido Publication Center, pp. 511–526.
Akino, Shigeki (2006) "Spirit-sending Ceremonies for Deer – A Reconsideration", *Report of the Obihiro Centennial City Museum* 24: 1–10.
Committee for the Survey on Countermeasures to Preserve Ainu Culture and Environment, Laboratory for the Survey on Countermeasures to Preserve Ainu Culture and Environment, and Biratori Board of Education (2006) *Report of Survey on Countermeasures to Preserve Ainu Culture and Environment – Saru River Region Cultural Assessment Activities*. Hokkaido: Biratori-cho.
Fujimura, Hisakazu, ed. (2009) *Report on Survey of Ainu Peoples' Culture 2008. Survey of Folk Technology 1 (Hunting Technology)*. Sapporo: Hokkaido Board of Education.
Fujiwara, Eiji (1985) *Story of Hokkai Ezo sika – History of the Destruction of the Environment in Hokkaido*. Tokyo: Asahi Shimbun Company.
Hezutsu, Tosaku ([1784] 1972) "Toyuki". In Kisaku Otomo (ed.), *Northern Gateway Series*, issue 2. Tokyo: Kokushokankoukai, 1972.
Higashimura, Takeshi (2006) *Introduction to the Relationship between Wajin and Ainu People in the Post-war Period: Late 1940s to Late 1960s*. Tokyo: Sangensha Publishers.
Hokkaido (2008) *Conservation and Management Plan for Sika Deer (3rd Period)*. Sapporo: Department of Environment and Lifestyle.
Honda, Yuko (2007) "Narratives of Partial Bark-stripping: A Study Based on Sarashina Genzo's Records", *Bulletin of the Hokkaido Ainu Culture Research Center* 13: 15–30.
Inukai, Tetsuo (1952) "On the Deer in Hokkaido, Its Decrease and Coming back", *Studies from the Research Institute for Northern Culture* 8: 1–68.
Kaji, Koichi (2006) "Hokkaido Natural Environment and the Sika Deer", in Koichi Kaji, Masami Miyaki and Uno Hiroyuki (eds), *Conservation and Management of Sika Deer*. Sapporo: Hokkaido University Press, pp. 3–9.
Kayano, Shigeru (1978) *Ainu Folk Crafts*. Tokyo: "Ainu Folk Crafts" Publication Support Committee.
Kinase, Takashi (1997) "Representation and Politics: A Sketch of Anthropological Discourses Surrounding the Ainu", *Japanese Journal of Ethnology* 62(1): 1–21.

Kinase, Takashi (1998) "The Heterophony of Otherness: A Prospect on Contemporary Ainu Images", *Japanese Journal of Ethnology* 63(2): 182–191.
Kojima, Kyoko (2003) *A Study of Ainu History – From the Viewpoint of the Transition of Images of Ezo and the Ainu*. Tokyo: Yoshikawakobunkan.
Kreiner, Josef (1993) "European Images of the Ainu and Ainu Studies", in Josef Kreiner (ed.), *European Studies on Ainu Language and Culture*. Munich: Iudicium, pp. 13–60.
Matsuura, Takeshiro ([1863] 1984) *Eastern Ezo Writings*. In Tsunekichi Yoshida (ed.), *Eastern Ezo Writings (New Edition)*. Tokyo: Jijitsusha, 1984.
Momose, Hibiki (2003) "Hunting Administration during the Development Period – 'Hokkaido Deer Hunting Regulations' Logic of the System and Hunting Limitations", in Koichi Inoue (ed.), *The North Eurasia from a Social Anthropological Perspective*. Sapporo: Hokkaido University Slavic Research Center, pp. 101–122.
Momose, Hibiki and Akihiro Yanaka (2005) "The Idea to Ainu by Japanese Seen in the Bear's Damage Report of News Paper at Early Meiji Era", *Iwamizawa Annual Report: Primary Education, Teachers and Education Research* 26: 45–56.
Nishikawa, Osamu, ed. (1995) *Atlas – Environmental Change in Japan*. Tokyo: Asakura Shoten.
Okada, Michiaki, ed. (2008) *To the Future – Messages from Young Ainu People*. Sapporo: Sapporo Television Broadcasting Company.
Saito, Reiko (2000) "Transition of Images of Ainu Culture in Hokkaido Tourist Guide Material", *Showa Women's University, Institute of International Culture* 6: 29–42.
Sasaki, Toshikazu (2004) *A Study on the Ainu-e Paintings*. Tokyo: Sofukan.
Sato, Yoshikazu (2005) "Food Habits of Brown Bear – Regional Difference and Annual Variation", *Mammalian Science* 45(1): 79–84.
Sato, Yoshikazu (2006) "Looking at the Relationship between Brown Bears and Humans – Background to Increased Damage by Bears and Conservation Management", *Biostory* 5: 32–39.
Sato, Yoshikazu, Toshiki Aoi, Koichi Kaji and Saeki Takatuki (2004) "Temporal Changes in the Population Density and Diet of Brown Bears in Eastern Hokkaido, Japan", *Mammal Study* 29: 47–53.
Sekiguchi, Yoshihiko (2007) *Ainu People Living in the Metropolitan Area – From the Daily Dialogue*. Tokyo: Sofukan.
Stewart, Henry (1996) "Foreword", in Henry Stewart (ed.), *Contemporary Hunters and Gatherers: Change and Regenesis of Subsistence Cultures*. Tokyo: Gensosha, pp. 3–10.
Takahashi, Osamu (2004) "A Look at Ainu Deer Spirit-Sending Ceremonies", in Committee for the Publication of Dr. Hiroshi Utagawa Memorial Essay Collection (ed.), *The Formation of Aynu Culture*. Sapporo: Hokkaido Publication Centre, pp. 493–510.
Takahashi, Osamu (2008) "Ceremonies and History: Interaction of Natural Causes", in Committee for the Publication of Dr. Chosuke Serizawa Memorial Essay Collection (ed.), *Dr. Chosuke Serizawa Memorial Essay Collection: Archeology, Anthropology and History*. Tokyo: Rokuichi Shobo, pp. 651–664.

Takarabe, Kae (2001) "American Ainu Image at the Beginning of the Last Century on the Basis of Frederick Starr's Newspaper Clippings", in Yoshinobu Kotani (ed.), *Reconstructing Ainu Culture Based on Overseas Ainu Collections*. Nagoya: Nanzan University Anthropological Institute, pp. 1–40.

Tawara, Hiromi (1979) *Hokkaido Nature Conservation: The History and Ideology*. Sapporo: Hokkaido University Press.

United Nations (2007) *United Nations Declaration on the Rights of Indigenous Peoples*. UN Doc. A/RES/61/295 Annex, adopted by General Assembly Resolution 61/295, 13 September 2007.

Uno, Hiroyuki (2006) "Current Status of the Conservation and Management for Sika Deer – Hokkaido Initiatives", in Yezo Deer Association (ed.), *2006 Yezo Sika Forum Report*. Sapporo: Yezo Deer Association, pp. 8–13.

Walker, Brett L. (2001) *The Conquest of Ainu Lands: Ecology and Culture in Japanese Expansion 1590–1800*. California: University of California Press.

Watanabe, Hitoshi (1952) "Utilization of Natural Resources by the Saru Ainu", *Japanese Journal of Ethnology* 16(3–4): 225–266.

Yamada, Shinichi (2001) "Hunting Regulation by the Colonial Department and the Ainu People: Especially about the Prohibition of Poisoned Arrow Hunting", *Bulletin of the Historical Museum of Hokkaido* 29: 207–228.

Yamada, Shinichi (2006) "Ainu Deer Hunting after the Enactment of Hokkaido Deer Hunting Regulations", *Bulletin of the Historical Museum of Hokkaido* 34: 129–156.

Yoneda, Yuko (1996) "Teaching 'Ainu Culture' at School: Problems and Controversies", *Bulletin of the Hokkaido Ainu Culture Research Center* 2: 123–148.

7-4
Sustainable rural/regional development by attracting value-added components into rural areas

Kiyoto Kurokawa

7-4-1 Introduction

In developing countries, there is an urgent need to reduce poverty and wage gaps by revitalizing the regional economy. This section describes the process of transferring Japanese initiatives for the local value-adding process, rehabilitating people's cultural practices, and initiating self-reliance strategies such as the "One Village, One Product" (OVOP) movement, *michinoeki* (roadside service stations), and *shokuiku* (dietary education). These unique approaches provide an important model of success in regional development policy.

The Yokohama Action Plan was designed to act as a road map for the support of African growth and development through the TICAD (Tokyo International Conference on African Development) process in 2008. The Government of Japan has taken the initiative to show its strong commitment by announcing a doubling of official development assistance to Africa by 2012. Community development and empowerment are vital elements of enhanced human security. Cultural considerations are also important to ensure sustainable community development (Ministry of Foreign Affairs, 2008).

The Japanese government has promised to expand the OVOP movement to Africa. This means a paradigm shift in international cooperation from a hard to a soft approach. In developing countries, a shortage of infrastructure is obvious. OVOP is encouraging a shift from infrastructure investment to human resource investment. But, more precisely, a good

Designing our future: Local perspectives on bioproduction, ecosystems and humanity,
Osaki, Braimoh and Nakagami (eds),
United Nations University Press, 2011, ISBN 978-92-808-1183-4

combination between infrastructure and human development should be examined by local people. The new target issues are how to sustain social, cultural and human well-being through international cooperation.

7-4-2 One Village, One Product (OVOP) movement

The world OVOP movement as a regional development approach

The "One Village, One Product" (OVOP) movement was initiated in 1979 by the former governor of Oita Prefecture, Dr Morihiko Hiramatsu. This subsection argues that the recent development of the overseas OVOP movement, especially in Thailand and Africa, can be defined as an "endogenous regional development policy" based on strong government initiatives.

The Government of Japan supported the initiation of the "One Village, One Product" campaign in developing countries as part of the "Development Initiative for Trade" advocated by then-Prime Minister Koizumi and announced on the occasion of the WTO Hong Kong Ministerial Conference held in December 2005 (Government of Japan, 2006). OVOP has been introduced to many overseas countries, including Thailand, Malaysia, the Philippines and China. The Thai government has developed this project to suit the Thai economy, calling it One Tambon, One Product (OTOP). Rural people are encouraged to identify their local resources and cultures as a way of adding value to their products. Thanks to the strong initiative by the Japanese government, more than 12 African countries, including Malawi and Kenya, are currently attempting to introduce this movement.

The original OVOP movement

The original OVOP movement in Oita is somewhat different from the overseas OVOP movement. It requires people to identify a product or an industry distinctive to their region and to develop it into a nationally, or even globally, accepted one. One of the main purposes is to eradicate citizens' heavy dependence upon government, and to promote autonomy and cooperation among regional peoples.

The local government of Oita played a catalytic role in disseminating the concept of the OVOP movement across the prefecture, providing technical support and facilitating market access to OVOP by organizing OVOP fairs and introducing OVOP antenna shops. OVOP antenna shops were established to promote OVOP marketing and to monitor consumers' evaluations of OVOP products. Financial support was provided by private banks and agricultural cooperative associations.

SUSTAINABLE RURAL/REGIONAL DEVELOPMENT

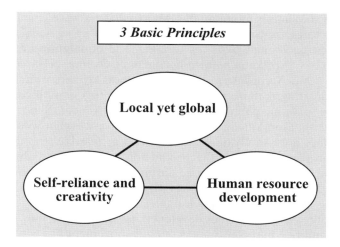

Figure 7.4.1 Three basic principles of the OVOP movement.
Source: Oita OVOP International Exchange Promotion Committee, <http://www.ovop.jp/jp/ison_p/haikei.html> (accessed 29 April 2010).

According to Kurokawa et al. (2008: 23–24), the impetus for the OVOP movement in Oita was due to:

i) The population shift from rural areas to major cities and the loss of vitality in various regions of the prefecture;
ii) The need to create new industries in regional areas;
iii) The need to reduce over-dependence of business on local government.

To overcome the above problems, the governor established three basic principles of the OVOP movement (Figure 7.4.1). These are as follows:

1) **Local yet Global:** being local and global simultaneously, called "Glocal". The idea is to make products that represent local areas/regions but which could also be competitive in global markets;
2) **Self-reliance / Creativity** or independence and new ideas. Villagers themselves [are] encouraged to decide which product(s) should be chosen as OVOP products; local governments [are] intended only to provide technical assistance;
3) **Human resource development:** OVOP would promote innovation and creativity and also encourage people to improve or harness their skills [Hiramatsu, 2005].

Oyama Town and the OVOP movement

The origin of the OVOP movement can be traced back to the New Plum and Chestnut (NPC) movement in Oyama Town, Oita, Japan, where 80 per cent of the land is mountainous. Many villagers were engaged in cutting down trees from the mountains or worked as seasonal labourers. They owned only small plots for farming. The population of Oyama in 1961 was 6,168. "Mr. Yahata, town headman of Oyama, declared the NPC strategy in 1961. According to the strategy, local farmers were advised to transform their agriculture from rice cultivation to a more diversified one by means of planting plum and chestnut trees" (Adachi, 2005).

Part of the reason Yahata pushed this initiative was that wild plum and chestnut trees grew in Oyama, and he saw it as feasible to bring agriculture to this region through the production of plums and chestnuts. In contrast to the promotion of rice farming, which was national policy, Yahata and others encouraged the growing of identical local products through the campaign "Let's plant plums and chestnuts to go to Hawaii!" After that, the NPC movement developed its concept into the "New Paradise Community". In this way, an economic movement has evolved into a social movement. In OVOP, "self-reliance and creativity" correspond to this concept.

Endogenous and exogenous approaches

Traditionally, there are two approaches to national development: exogenous and endogenous. The exogenous approach is based on supply-driven economic theory, which advocates using external resources. This strategy may boost the economy through large-scale investment and the creation of employment. But it is not applicable to all African countries. Although there has been increased foreign investment in Africa since the turn of the century, it must be noted that the major destinations of new investment are the oil- and mineral-rich countries. Moreover, such large-scale investment will not always benefit local communities. Accordingly, a new procurement procedure was under development to increase local procurements and community involvement. However, the sustainability of the local benefits from the project is beyond the communities' control.

The endogenous approach, in contrast, encourages local communities to undertake development by fully utilizing indigenous skills, know-how, local resources, cultural elements and the local market. As an endogenous approach, the OVOP movement is applicable to any country because it is predicated on utilizing locally available resources and services. As a community-based small business activity, the OVOP movement not only stimulates the local economy but also facilitates the empowerment of rural communities, including women, and thus develops community leaders. This is an important aspect of OVOP in the African context, where more than 60 per cent of the population is rural.

In light of the differences between these approaches, the Japan International Cooperation Agency (JICA) and major donors are trying to establish a new development strategy – capacity development. The basic idea of capacity development is very close to the OVOP concept in terms of endogenous development and a strong emphasis on self-reliance. The United Nations Development Programme (UNDP) defines capacity as the ability to perform functions, solve problems and set and achieve objectives (Fukuda-Parr et al., 2002). Through OVOP activities, many community leaders have been trained and supported. The essence of the OVOP movement, as Morihiko Hiramatsu puts it in his book on OVOP, is not to develop products but to encourage community leaders (Hiramatsu, 2005).

Originality of the Thai OTOP movement

The Thai government started its OVOP movement in 2001. The One Tambon, One Product (OTOP) movement has three characteristics. First, it has been part of a strong government initiative, so that it is different from the community-led Japanese type of OVOP. The Thai movement is in fact an element of government policy. The government plays an important complementary role in the field of human resource development. Second, the OTOP is widely accepted by the product rating system with a five-star rating, which is a new branding strategy for local products. Real efforts in product development have resulted in quality products. Third, the OTOP movement is strongly assisted by information and communication technology, including web-based marketing. The Thai government has developed many websites to promote the OTOP movement.

Within this OTOP movement, food-related industries are very important. The government has identified six targeted categories: Food, Beverages, Textile and Fashion, Accessories and Home Decoration, Gifts and Arts, and Herbs and Spa. Furthermore, about half of the five-star products were food-related products in 2003 and 2004. By using the five-star OTOP logo, the government has produced new OTOP brand markets (see Plate 7.4.1).

From observations and interviews conducted by Kurokawa et al. in 2008, the OTOP movement has been found to:
- motivate communities to develop confidence and a sense of community,
- create community awareness to examine, explore and use local resources,
- enhance a community's ability to think about shared problems and possible solutions,
- develop visionary leaders through community empowerment,

Plate 7.4.1 The OTOP star logo on Thai products (Kiyoto Kurokawa, November 2008, in northeast Thailand).

- build community confidence and avoid dependence on external supports.

The OTOP movement was started through strong government initiatives, but gradually it has brought the concept of "self-reliance" to village people. The community accumulates technical skills, know-how and practical knowledge learned by inference through experience while developing the human resources that are essential for sustained or continued innovation of their unique local products and a management system.

7-4-3 *Michinoeki* (roadside service stations)

Motorization and local communities along main roads

Motorization in developing countries is moving in step with rapid growth in people's incomes. The number of vehicles in China has been growing at an annual rate of almost 13 per cent for 30 years, nearly doubling every 5 years. India's fleet has been expanding at more than 7 per cent per year (World Resources Institute, 2009). And the progress of motorization has increased the number of long-distance travellers as well. Private roadside shops provide rest facilities, including toilets and restaurants. However, these establishments have no network service, such as the provision of traffic and sight-seeing information, and sometimes there is no service at all in rural areas.

In Japan, local communities along main roads provide retail goods and dining services to automobile drivers passing through their area. Growing demand for roadside rest facilities from both drivers and local

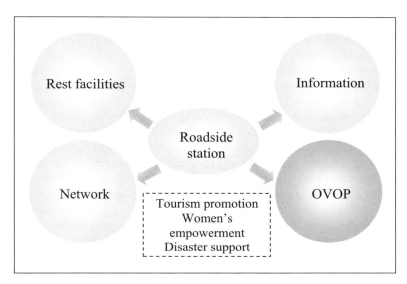

Figure 7.4.2 Roadside station and its additional functions.
Source: Prepared by the author.

communities resulted in the provision of *michinoeki* (roadside service stations). Since 1993, cities and towns that represent regions have been cooperating with highway administration bodies to develop *michinoeki* that function as rest areas, disseminate information and link regions. At these shops, community history and local cultural attractions are presented and locally produced products (including those developed by the OVOP movement) are sold to visitors and members of the local community (see Figure 7.4.2).

Michinoeki are also based on a unique concept to establish strong connections between travellers and local communities so that local residents can enjoy business opportunities through commercial, cultural, educational and healthcare activities.

Thai community centres as the first overseas michinoeki

The first overseas *michinoeki* was set up in Thailand. The Department of Industrial Promotion built a rural industrial village as a tourism promotion venture. This project was funded with a yen loan from the Japan Bank for International Cooperation (JBIC) for the construction of community centre buildings in 20 target villages, which would be used as co-ordination centres and venues for displaying and selling local products. JBIC has introduced the Japanese *michinoeki* as a model for these community centres. Additionally, JICA has invited local officers and leaders of the Thai community to Japan to extend OVOP training and the *michinoeki* concept in Japan.

The World Bank recognized the effectiveness of *michinoeki* and published a detailed guide in 2004. Pilot studies have already been carried out in East Asia and Africa (World Bank, 2004). Moreover, the possibilities of the *michinoeki* approach as a resource for community development have been widely recognized. According to Masahisa Fujita (2006: 1), "[b]oth OVOP and Michino Eki have attracted widespread attention in many developing countries as potential tools for bridging the gap between cities and rural areas through community-driven development, and are being implemented in many countries." In Japan, these service stations are usually connected online to provide useful local information for visitors. Sightseeing information on beauty spots, hotels and local food is provided by municipal government marketing activities. Accordingly, public–private partnerships have been organized at *michinoeki*.

Regional collaboration via michinoeki

According to the World Bank manual for roadside stations, in addition to the basic three functions of *michinoeki* – information point, rest area and community centre – tourism promotion, women's empowerment and disaster support should be included. The Bank has added tourism to fully utilize local resources such as cultural heritage and beautiful local scenery. And for women's empowerment, a significant positive impact has been observed in Japanese *michinoeki*. The disaster support function was quite a new issue even in Japan, but during the Niigata earthquake in 2004 *michinoeki* functioned as a disaster support station. Regional collaboration via the network of *michinoeki* is the most important feature of this system. Furthermore, the interconnections among *michinoeki* of different types and sizes generate synergy effects. "The resulting effects are revitalization of city centres, community redevelopment in suburbia, and the increase of regional collaboration and exchanges among different cities" (Hirotaka et al., 2005).

When scale effects are external to the firm, average costs at a factory decline as local industry output increases. This agglomeration effect brings people and firms together and sometimes it is a cause of over-urbanization. Well-balanced development between urban and rural areas is badly needed. In large cities, congestion effects have taken precedence over agglomeration effects, turning people away from the city. Accordingly, over-urbanization has prompted a re-evaluation of rural life. "Not only villagers but also some urban residents are interested in rural resources and are trying to make good use of them. This trend presents opportunities to utilize rural resources for community-based socioeconomic activities" (Sakurai, 2006: 105). There is some interest in networking among *michinoeki*-related people, but the main interest of the OVOP movement was the production of local products and OVOP had limited interest in networking. OVOP and *michinoeki* can be linked together by a common interest in marketing.

7-4-4 Dietary education

Food security and dietary education for all

Today's dietary education is somewhat different from traditional dietary education. In Japan, food safety and traceability have attracted a great deal of public attention. This is because of a series of controversies in the food industry, such as poisoned frozen Chinese *jiao-zi* dumplings and faked free-range chickens in Miyazaki prefecture in 2008. The Japanese government has been striving to conduct strict inspections of imported food products, to establish bilateral talks with other countries to request compliance with Japanese food sanitation regulations, to enforce inspection at quarantine stations, and to conduct field surveys. However, these measures cannot cover all imported food products.

The Ministry of Health, Labour and Welfare has published an overview of the actual situation as well as the results of inspections of imported foods. The total number of declarations, inspections and violations was about 1.87 million in 2005, and the volume was 31.8 million tons, a rough estimate on a gross declaration basis. Inspections were made of 190,000 declared products, accounting for 10.2 per cent of all declared products (Ministry of Health, Labour and Welfare, 2006). These food security issues reveal a strong need for dietary education for all citizens.

The Basic Law on Shokuiku

In Japan, the school meals system was introduced to protect children's nutrition after World War II. In addition, *shokuiku*, or dietary education, has been introduced recently. The Basic Law on *Shokuiku* (Law No. 63) was enacted in June 2005. It defines *shokuiku* as the acquisition of knowledge about food as well as the ability to make appropriate food choices. Japan's education system is built on three pillars: *chiiku* (intellectual training), *tokuiku* (moral education) and *taiiku* (physical education). *Shokuiku* is an additional concept that is part of school education, and it can be an innovative process for regional development. *Shokuiku* education is targeted not only at schoolchildren but also at all the people in a region. Healthcare and economic development are strongly linked to this concept. In almost every rural community, healthcare will play an important role in economic development. Healthy and competent people are an important part of an active rural economy.

> The objective of [the Basic] Law is to stabilize and improve people's lifestyle and to develop the national economy through comprehensively and systematically implementing policies on food, agriculture, and rural areas by means of establishing basic principles and basic matters for realizing them and clarifying the responsibilities of the state and local governments. (Government of Japan, 2005: Article 1)

In Japan, many reports on the eating habits of youth indicate that they have an unbalanced diet; they prefer Western food and dislike vegetables. In developing countries, one can observe the same tendency. Previous studies have indicated that the demand for many foods – especially meat, poultry, fish and dairy products – is responsive to income growth. After the implementation of the Basic Law on Dietary Education in 2005, specific items for "instruction in nutrition" were incorporated into school education; this step marked the beginning of health and nutrition education in schools.

Local community development by adding value into food

In Japan, a unique approach called *chisan-chishou* (grow locally, consume locally) has been quite popular recently. This approach has revealed the value of the local cultural aspects of the food industry and, accordingly, this perspective could provide new ideas to the value-adding process. To reduce one's carbon footprint and live a more environmentally friendly lifestyle, one should consume locally. This practice can reduce the transportation costs of the daily intake of energy. This is a new way of adding value to local products, and there are further possibilities. In terms of rural development, adding value to agricultural products is a very important step. Global awareness of healthy food means that new value can be added to local food products. It is said that the adoption of a Westernized diet and lifestyle results in heart disease. This is why Japanese dietary education can be a good solution both for a healthy life and for rural development.

7-4-5 Conclusion and discussion

This section has introduced a new approach to development by combining three concepts: OVOP, *michinoeki* and *shokuiku*. The main interests of these three approaches are different: OVOP encourages the production of local products; *michinoeki* encourages networking and shared marketing; *shokuiku* has revealed new values in local products. The recent financial crisis has exposed the economic burden of healthcare costs. These economic problems have also attracted public attention regarding preventive measures for chronic diseases and dietary education for everyone. Finally, the most important common interest is to believe in the power within the community. These uniquely Japanese approaches to development can be bound together by common interests and people's self-reliance and can be applied to sustainable social development.

REFERENCES

Adachi, Fumihiko (2005) "Marketing Perspectives of Community Enterprises in Oita Prefecture: Evaluation of Marketing Activities under OVOP Movement", Kinjo Gakuin University.

Fujita, Masahisa (2006) "Economic Development Capitalizing on Brand Agriculture: Turning Development Strategy on Its Head", IDE Discussion Paper No. 76, Institute of Developing Economies, Japan External Trade Organization (JETRO).

Fukuda-Parr, Sakiko, Carlos Lopes and Khalid Malik, eds (2002) *Capacity for Development: New Solutions to Old Problems*. London: Earthscan/UNDP.

Government of Japan (2005), Basic Law on Dietary Education, Ministry of Agriculture, Forestry and Fisheries. Available at: <http://www.maff.go.jp/soshiki/kambou/kikaku/NewBLaw/BasicLaw.html> (accessed 26 March 2010).

Government of Japan (2006) "'One Village, One Product' Campaign", Ministry of Economy, Trade and Industry, March. Available at: <http://www.meti.go.jp/english/information/downloadfiles/060314OVOPsymposium_EN.pdf> (accessed 26 March 2010).

Hiramatsu, Morihiko (2005) *For the Regional Leaders of the 21st Century*. Tokyo: Toyo Keizai Shinpo Co. (in Japanese).

Hirotaka, Koike, Teshima Takayuki, Morimoto Akinori, Yoshida Keiko and Tanaka Eiji (2005) "Machinoeki – A Challenge for Revitalizing Communities in Japan", Utsunomiya University.

Kurokawa, Kiyoto, Fletcher Tembo and D. W. te Velde (2008) "Donor Support to Private Sector Development in Sub-Saharan Africa: Understanding the Japanese OVOP Programme", JICA-ODI Working Paper 290. Tokyo: Japan International Cooperation Agency; London: Overseas Development Institute. Available at: <http://www.odi.org.uk/resources/download/1116.pdf> (accessed 26 March 2010).

Ministry of Foreign Affairs, Japan (2008) "TICAD IV: Yokohama Action Plan". The Fourth Tokyo International Conference on African Development, 30 May. Available at: <http://www.mofa.go.jp/region/africa/ticad/ticad4/doc/actoin.pdf> (accessed 2 April 2010).

Ministry of Health, Labour and Welfare, Japan (2006) "Results of Monitoring and Guidance Based on the Imported Foods Monitoring and Guidance Plan for FY2005". Available at: <http://www.mhlw.go.jp/english/topics/importedfoods/dl/6-4.pdf> (accessed 26 March 2010).

Sakurai, Seiichi (2006) "Role of Social Capital in Rural Diversification: A Case of Mountainous Villages in Japan", in Shigeki Yokoyama and Takeshi Sakura (eds), *Potential of Social Capital for Community Development: Report of the APO Survey and Symposium on Redesigning Integrated Community Development 2003–2005*. Tokyo: Asian Productivity Organization.

World Bank (2004) *Guidelines for Roadside Stations: "Michinoeki"*. Available at: <http://www.worldbank.org/transport/roads/rdside%20station%20docs/01_Intro-Note6.pdf> (accessed 26 March 2010).

World Resources Institute (2009) "Proceed with Caution: Growth in the Global Motor Vehicle Fleet", SDIS Global Trends, <http://www.mcrit.com/scenarios/files/webs/Global%20Motor%20Vehicle%20Fleet_archivos/autos2.htm> (accessed 29 April 2010).

7-5
Sustainable agricultural development in Asia and study partnerships

Takumi Kondo and Hong Park

7-5-1 Problematic areas in agricultural development in Asia

Rapid economic globalization has been observed since the end of the twentieth century and has been marked by a variety of economic indices: increasing opportunities for international trade, personal exchanges and foreign direct investment alongside the emergence of multinational enterprises. The level of interdependence in the global economy has risen rapidly since the 1980s.

East Asia's economic growth, commonly known as "the East Asian miracle", is understood as export-oriented growth. It resulted from the adoption of development strategies that aim at economic growth, focusing on exports and replacing the traditional strategies that relied upon imports.

In contrast to the economic growth achieved by East Asian nations, there are still a number of developing nations missing out on the wave of global economic expansion resulting from globalization. For nearly half a century, an effort has been directed towards low-income nations to eliminate the economic disparity. Nevertheless, increasingly serious problems remain, including poverty, food shortages, debt accumulation and the deterioration of natural resources and the environment. Over 1 billion people in low-income nations currently suffer from malnutrition and frequently face starvation. While confronting environmental disruption triggered by the notable economic growth, Asian countries also encounter serious poverty issues, with large numbers of poor people living in the

Designing our future: Local perspectives on bioproduction, ecosystems and humanity,
Osaki, Braimoh and Nakagami (eds),
United Nations University Press, 2011, ISBN 978-92-808-1183-4

region. It must hence be highlighted that poverty triggers further degradation of the environment. The Asian region, containing more than half of the world's population and having demonstrated rapid economic growth in recent years, has now gained significant influence regarding global environmental issues.

The expansion of food production is essential in order to feed growing populations. Improvements in farming productivity will continue to be critical as the world population increases. Meanwhile, it has recently become recognized that it is necessary to produce food without negatively affecting the environment, for example through deforestation, wetland destruction and soil degradation. It is therefore obvious that global food problems are inextricably linked with issues related to population, resources and the global environment. Furthermore, increasingly globalized food distribution has revealed the importance of the safety of food products, because they may be directly detrimental to health – causing diseases such as BSE (mad-cow disease), avian flu and illnesses caused by pesticide-laden food.

In order to create agricultural development strategies, it is extremely important to gain a clear understanding of how the regional environment, food production and the traditional sustainable farming community have become affected by population growth and rapid economic expansion as a result of globalization. For the development of sustainable agricultural communities, it is vital to uncover the relationship between economic growth and poverty reduction as well as that between food production and resources and environmental issues.

This section highlights China, South Korea, Taiwan, Mongolia and Nepal, and introduces sustainable agricultural development plans that take into consideration problems related to population, food, resources and the environment. This section also suggests redevelopment measures that enable farming communities to respond to increasingly penetrating market economies.

7-5-2 The development of sustainable agriculture targeting a mountainous area in China

Background and significance of environmentally friendly agricultural development

In south-western China, a number of limestone regions are widely spread across the Guangxi Zhuang Autonomous Region, Guizhou and Yunnan. This survey targets Qibainongxiang village, Dahuaxian prefecture and Guangxi Zhuang Autonomous Region, which contain dolines (basin-shaped valleys) formed as a result of erosion in karst topography. An ethnic minority called the Yaos has been living in these dolines for over 500 years within exclusively natural ecosystems. The Yaos have established a traditional society underpinned by a self-supporting food production system that is in harmony with the natural environment. However, China has been promoting economic growth by introducing a reform and opening-up policy, and this initiative has simultaneously triggered degradation of the natural environment and an increase in economic disparity between citizens. This trend is also observed in Dahuaxian prefecture. A local farming community development programme is under way in Qibainongxiang village in an effort to eliminate poverty, and such development is unavoidable in order to shift the semi-exclusive environment to an open society that is able to respond to market economies. Nonetheless, this programme has had seriously harmful effects on the village's natural environment.

This study, which is a result of a joint Japanese–Chinese research project, takes into account the local circumstances described above and has the following study goals:
(1) to reveal the material recycling system in both natural ecosystems (forest ecosystems and farmlands) and human activities (human livelihood activities and economic activities);
(2) to clarify the measures needed to rebuild the damaged forest ecosystems in order to identify the principles for coexistence between humans and nature in the near future that will enable human activities and natural ecosystems to coexist.

This study is based on the study project "Ecosystem Rebuilding and Comprehensive Development of Sustainable Biological Productivity in Southwestern China" (joint Japanese–Chinese research, 1998–2002), which highlights environmental issues observed in agricultural communities in China. This project was implemented after it had been adopted by another study project, "Environmental Conservation in the Asian Region", carried out through the Research for the Future Program (combined fields) undertaken by the Japan Society for the Promotion of Science.

The tasks of this study are divided into the following four categories: (1) human society, (2) material recycling, (3) forest evaluation/rebuilding, and (4) sustainable biological production. Because of limitations of space, this section introduces only the study tasks in categories 2 and 3.

In general, a variety of natural materials are procured and utilized for the purpose of production and human livelihoods. As regards material recycling and human activities, the study examined the mechanisms of circulation of basic natural materials (water, nutrients and energy) as well as the circulation of nitrogen, which has a negative environmental impact. It also revealed how these materials affect traditional farming society.

Material recycling in a populated doline in Qibainongxiang village: Challenges and solutions

With the main focus on surveying the settlement (*Nongshi-tun*), one of the populated dolines in Qibainongxiang village, the aim was to ascertain the principles for coexistence between humans and nature by examining the material recycling system in the doline as a way of obtaining reference information for the creation of a recycling-oriented society. The inhabitants were interviewed to estimate the doline's material recycling system and then the nutrient circulation system was analysed. It was discovered that the soil used for agroforestry and livestock-raising is calcareous, meaning that it is rich in calcium (Ca) and magnesium (Mg) but lacks a number of nutrients, such as nitrogen (N), phosphate (P), potassium (K) and sulphur (S). It was also revealed that biological production is sustained by an established material recycling system in which inadequate amounts of nutrients such as P and K are supplied from forest land, and that these nutrients are recycled within the doline. However, a large amount of N is supplied as a result of manuring, triggering the risk of groundwater contamination. Because the settlement has been able to sustain itself over a long period of time owing to low productivity and small amounts of resources being supplied externally, it might be highly likely to collapse if the amount of external resources were increased in order to improve the living standards of residents. A methodology was therefore sought to improve local biological productivity, the living standards of residents and environmental quality while maintaining the settlement's sustainability.

Nitrogen circulation in agriculture and food consumption and sustainable production systems

Traditionally, all human and animal excrement produced within the settlement is used as fertilizer, and it was suggested that this measure has been responsible for maintaining optimum nitrogen circulation. In Qibainongxiang village, however, modern farming technologies have been introduced in recent years in the form of chemical fertilizers and commercial animal feedstuff, and agricultural production has therefore been improving. Both compost and chemical fertilizers are applied, resulting in excessive levels of nitrogen in farmlands as well as the risk of groundwater contamination. Furthermore, the amount of animal excrement waste has grown with the increase in the amount of commercial feedstuff used. In this manner, modernized farming's use of chemical fertilizers and commercial feedstuff breaks the relationship between crop production and livestock raising, and the village very much lacks the capacity and structure to maintain sound environmental quality.

Use of bioenergy and its effects

Energy that generates no negative environmental impact on forests is needed, and biogas (a resources-recycling technology) is believed to be the most useful. Biogas production equipment was accordingly installed on local farms, and the effectiveness of biogas was examined with a particular focus on the efficient utilization of fertilizer ingredients. Laboratory experiments were also carried out to investigate the effectiveness of biogas. Interviews with local farmers revealed that the production of biogas and the accompanying digested slurry barely generates any negative effect on the material recycling system and is very suitable for the local circumstances. Furthermore, this technology is believed to contribute to the sustainability of future human lives. Laboratory experiments also proved that the concentration levels of the digested slurry's fertilizer ingredients change significantly according to the type of fermentation materials it contains.

The use of land and forests and the preservation of water and soil in Qibainongxiang village dolines

In relation to the category "forest evaluation/rebuilding", the study ascertained forest ecosystems' various environmental capacities and also identified the factors that have been able to sustain inhabitants' lives over a long period of time in such forest ecosystems. It also pointed to the negative effects of logging and presented measures to restore and rebuild forest ecosystems and the effective usage of such ecosystems.

Targeting the dolines located in Qibainongxiang village, the aim was to understand the dynamic behaviour of water, the status of water use and the environmental effects that result from the use of forests and land. The purpose of this study is to provide basic information concerning inhabitants' fundamental attitudes towards agricultural production and environmental preservation and to review new agricultural development measures that focus on protecting the environment. In the dolines located in Qibainongxiang village, rainwater soaks easily into the ground and therefore only a small amount of rainwater runs off. Furthermore, forests are able to slightly increase the percentage of both rainwater and surface water that soaks into the ground. This dynamic behaviour of water is believed to result in only small amounts of topsoil movement on the doline's slopes. Nonetheless, forests' capacity to increase the amount of water that soaks into the ground has negative effects if rainwater and surface water are wanted to be stored in a reservoir. It can be assumed that, based on these characteristic movements of water and soil observed in the past, the water and land usage system in the limestone mountainous areas in south-western China has been efficient because the intention has been to achieve coexistence with the natural environment.

Qibainongxiang village's geological conditions and features, and the restoration of forest soil and the ecology

In the Qibainongxiang region, trees are under strain as a result of long-term overuse of forests. In order to establish appropriate methods for handling forest soil, which are critical in restoring the ecology and producing sustainable wood products, the study investigated the mutual relationship between the properties of the forest soil, wall rocks (which generate soil), the village's geographical features and the status of vegetation. As a result, the physical quality of the forest soil in the following locations was discovered to be generally poor: at the top of the rock peak, on the ridge, in the upper and middle areas of the doline's slopes and on talus slopes. This soil consists of conglomerates or rock debris or it is extremely fine grained, mirroring the properties of the wall rocks. Its structure is not well developed and it is generally solid and firm. In order to restore a sound tree ecosystem with a variety of environmental capacities, the soil properties need to be improved through ecological and natural measures.

Change in the vegetation landscape in the Karst region ecology as a result of artificial disruption and the rebuilding of the ecology

Targeting four settlements with different levels of deterioration in forest vegetation, the study evaluated the change in the landscape triggered by artificial disturbance. It then proposed ecology-rebuilding measures that focus on the preservation and revitalization of forests to mitigate the negative effects of artificial disturbance. Common to all of the target settlements, woodland (ranging from evergreen bushes to trees) tends to localize in a conglomerate area on the ridge located between the upper and middle areas of slopes where the amount of above-ground biomass is insignificant (average of 69.3 tons/hectare). Meanwhile, sedimentary areas located between the middle and bottom areas of slopes consist mostly of cornfields. Grassy perennial herbs form the vegetation girdle in areas that were formerly used as cornfields. Composite annual herbs grow in clumps in the meadows designed to be used for goat grazing, and cogon grass grows gregariously in the meadows where goats had grazed. Deciduous bushes growing in the sedimentary areas are therefore likely ultimately to become deciduous trees. In this manner, the trend in vegetation transformation is believed to be different depending upon the soil properties (conglomerate or sedimentary).

New principles for coexistence between natural ecosystems and human activities

In order to understand the relationship between the natural ecosystem of severely damaged forests and human activities by using comprehensive indices, ecological footprint analysis was performed, based on the concepts of environmental carrying capacity, to obtain the maximum supportable population. The study compared the natural conditions of the four target settlements and analysed the natural ecosystem (the amount of natural resources) and human activities. As a result of this ecological footprint analysis, all of the target settlements were discovered to be facing environmental degradation owing to population growth beyond the environmental carrying capacity.

7-5-3 The identification of East Asian Corridor agriculture and a comparison of sustainable agricultural development platforms

The East Asian Corridor agricultural study

The Co-operative Associations Research Laboratory, Graduate School of Agriculture, Hokkaido University, began a study project designed to examine agricultural communities in East Asia at the beginning of the 1990s. This was the period when economic globalization was gradually becoming evident. To begin with, a comparative study was implemented to compare the status of domestic agriculture and local agricultural cooperative associations in Hokkaido and the homelands of the overseas students engaged in the study (South Korea, China and Taiwan). From 1997, a full-scale comparative study was initiated targeting Japan, China, South Korea and Taiwan under Grants-in-Aid for Scientific Research ("East Asian Agricultural Trends under the WTO System and the Reorganization of Agricultural Cooperative Associations"). The express purpose of this comparative study was to reveal: (1) the rapid agricultural structural changes observed in the East Asian region (China, South Korea and Taiwan) during the transition period of the shift to the World Trade Organization (WTO) system; (2) the operational reorganization of general agricultural cooperative associations, which are unique to Asia; and (3) the future direction of new Chinese agricultural cooperative associations.

Accordingly, a fixed-point observation monographic study was undertaken by referring to the framework designed to understand the structures of farming communities in Hokkaido (i.e. trends in land ownership, the differentiation of settlements and peasant classes, and the theory of the development of agricultural cooperative associations). This framework, underpinning the viability of Hokkaido's agricultural cooperative associations, is generally understood as the fundamental information source that is referred to when studying these associations (Sakashita, 1992).

The target survey locations are as follows: two locations in South Korea (a farming community located in a flat area and a hilly and mountainous area), two locations in Taiwan (an exclusively vegetable-producing area and an area carrying out both wet-rice farming and vegetable production), and two locations in China (a government-owned farm located in the north-eastern region and a coastal export-oriented, vegetable-producing district). The survey provided a range of information: the level of stability of the farmers (i.e. inflow and outflow of farmers as well as farming work structures), land ownership structures (i.e. inheritance system), the change in the usage status of farmlands, and the settlement's functions. As for the operational trends observed in general agricultural cooperative associations, South Korea and Taiwan displayed signs of shifting to credit cooperative businesses, as is the case with Japan, owing to the growing elderly population among farmers as well as the reduced income generated by rice production. On the other hand, in the case of some advanced farming communities in these two countries, the production of fruit and vegetables has been promoted to generate income and an effort has been made to establish the region as a fruit- and vegetable-producing district. In China, it was revealed that the development of both general and exclusive agricultural cooperative associations has been increasingly evident in regions with the potential to undertake commercial agriculture.

Based on the outcomes of the joint survey project, the number of fixed-point observation locations has increased since 2000, and individual surveys have been carried out on an ongoing basis. A new framework has been established called "East Asian Corridor agriculture" and a theoretical approach has been adopted to compare the agricultural structures observed within the region.

The purpose of this study is to identify both the common ground and the differences between agricultural communities in Hokkaido, Okinawa and the East Asian nations for the purpose of the relativization of the agriculture of Japan's inland/mainland. Conventionally, agricultural studies focusing on Japan mainly divide the nation into three islands – mainland, Shikoku and Kyushu – with Hokkaido and Okinawa occasionally treated as special regions. However, according to the outcomes of the research and study reported in this subsection, it has become increasingly obvious that Hokkaido is not necessarily unique within the East Asian region. Owing to the globalization of agriculture, an understanding of the diversity of this region is now required, and the significance of this diversity is becoming increasingly important. The study excludes Japan's mainland and classifies the area from Hokkaido to Okinawa via the East Asian countries as the East Asian Corridor. Having divided this particular region into continents, peninsulas and islands, fixed-point observation points were nominated in each survey location and individual characteristics seen in the survey location were examined through fieldwork for the purpose of making comparisons. Within the East Asian Corridor, the target survey locations selected in order to examine local agriculture are as follows: Hokkaido (domestic colony), north-eastern China (continental domestic colony), South Korea (continental peninsula), Shandong in China (continental coastal area I), Jiangnan in China (continental coastal area II), Taiwan (island) and Okinawa (island).[1] The base fields are Hokkaido, Okinawa, three locations on China's mainland (Heilongjiang: a government-owned paddy farm; Shandong: a Japan-bound vegetable-producing district in Tsingtao; Jiangsu: an industrialized farming village called Suzhou), three locations in South Korea (a paddy in a flat area, agriculture in a hilly and mountainous area, and horticulture in an urban area) and two locations in Taiwan (a vegetable-producing district and a farm alternating between paddy and fields) (see Figure 7.5.1).

Subsection summary

Unlike other Asian nations, farmland reform was implemented after World War II in China, South Korea and Taiwan, located within the East Asian Corridor, and the development of small-scale, family-run farms was widely observed within these nations (this type of farm management was restored in the early 1980s in the case of China). In contrast to inland and mainland Japan, which contain autonomous villages, the major factors in family-run farming management in these three countries are land, farmers and settlements. Therefore, in order to support the small-scale farms that hold the key to the development of sustainable agriculture, it is important to examine the common characteristics found in these countries and to identify the current functions of the agricultural cooperative associations that underpin family-run farms. The study will be working with researchers engaged in agricultural studies focused on Okinawa to deepen understanding of the seven base fields that are the subject of this study and to generalize the agricultural situation of these fields.

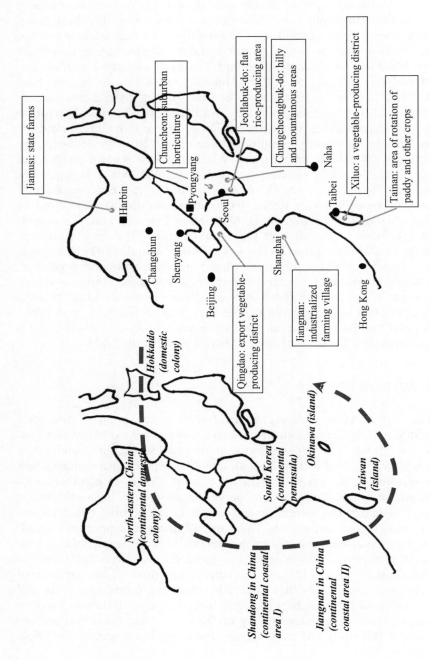

Figure 7.5.1 Agriculture of the Japanese islands and the East Asian corridor.

7-5-4 Irrigation management and agricultural productivity in Nepal

Significance of the study: Agricultural development and poverty in South Asia

When it comes to major South Asian nations, the level of national income per capita in 2006 was quite low, at USD 242 in Nepal, USD 419 in Bangladesh, USD 634 in India and USD 635 in Pakistan (based on USD as of the year 2000). In the case of Japan and South Korea, income per capita was USD 39,824 and USD 13,865, respectively. On the basis of gross domestic product (GDP), the agricultural sector of the major South Asian nations accounts for 34.4 per cent in Nepal, 20.1 per cent in Bangladesh, 18.3 per cent in India and 21.5 per cent in Pakistan. Although these figures may appear to be insignificant, they remain high when compared with those of Japan and South Korea, whose agricultural sector accounts for 1.5 per cent and 3.4 per cent, respectively, of GDP. The population growth rate in 2006 in the major South Asian nations was quite high, reaching 2.0 per cent in Nepal, 1.8 per cent in Bangladesh, 1.4 per cent in India and 2.1 per cent in Pakistan, with the number of farms per capita consequently declining.

In an effort to reduce poverty in these countries, it is essential to implement agricultural development, such as introducing new technologies and building infrastructure to ensure that agricultural productivity improves, because the majority of the poor engage in farming. In other words, the fundamental agricultural development goals in these developing nations are food security and rural poverty reduction.

The South Asian monsoon climate needs to be highlighted when promoting agricultural development in this region. It is extremely important to secure not only farmlands (the factors of production) but also water resources, because of the significant differences in the amount of solar radiation and precipitation between the monsoon and the dry seasons.

The primary benefit derived from irrigation is an extension of the crop planted area. The secondary benefit is an improvement in crop yields. The tertiary benefit is a reduction in the risks that decrease crop yields. Farmers who are able to use irrigation water demonstrate higher farming productivity, and the level of poverty in regions supplied with irrigation water is generally low. The supply of irrigation water during the dry season is likely to contribute to the reduction in poverty, with farmers earning an increased income and with an improved nutrition supply as a result of enhanced food production.

Although there are a number of irrigation methods, including gravity and well irrigation, this subsection deals exclusively with small-scale gravity irrigation systems and explains the actual irrigation situation under two different irrigation systems adopted in Kathmandu valley, Nepal. The small-scale gravity irrigation systems employed in this valley have been continuously surveyed to examine their effectiveness and maintenance status as well as the current usage of irrigation water.

This study refers to the following studies: Grants-in-Aid for Scientific Research "The Mechanism of Regional Public Property Supply and the Formation of Agricultural Irrigation Capital: Farmers' Irrigation Maintenance Activities in South Asia" (FY 2005–2007), "The Ecological Economics Evaluation of the Traditional Farming Technologies Adopted in Developing Asian Nations" (FY 2002–2004) and "The Comparative Study of Irrigation Efficiency in South Asia: From the Point of View of the Agricultural Technology Transfer to LLDCs" (FY 1998–2000).

The Shali Nadi irrigation system

The Shali Nadi irrigation district is located in Sankhu village, 17 km north-east of Kathmandu, and has an area of 176 hectares (Kondo et al., 2002). The focus was 98 farm fields, selected at random, in order to examine the usage and agricultural production profitability of these fields.

The most common cropping pattern observed in the upstream and midstream areas of the irrigation channel is "rice – summer potatoes – winter potatoes", with an area share of 72.3 per cent and 55.1 per cent, respectively. The farm fields along the upstream and midstream are used three times per year for cultivation purposes, which is quite frequent because the cultivation of summer and winter potatoes requires a large amount of water. This frequency is possible because farmers in these areas have a geographical advantage when it comes to gravity irrigation, and they are able to use irrigation water almost exclusively during the dry season. These farmers also benefit from rapid access to rainwater stored in the main channel during the monsoon season. They are therefore able to initiate rice cultivation approximately one month earlier than those working along the downstream, and can be relatively flexible in selecting crops for cultivation. In this way, the closer a farm is located to the upstream of the irrigation channel, the greater the advantage the farm has when it comes to the use of the irrigation water.

In contrast, the general cropping pattern observed in the downstream area is "rice–wheat", accounting for 81 per cent of the total area. Although rainwater is utilized for rice production during the monsoon season, wheat is a crop that does not require a large amount of water for cultivation. Because the irrigation water is used up by those working along the upstream, farmers in the downstream area are unable to cultivate potatoes, which provide high profitability, and therefore wheat is grown as an aftercrop of rice. In the vicinity of Sarumuturu village, located at the end of the irrigation channel, water does not run in a number of feeders during the dry season. It is therefore impossible to cultivate winter potatoes in the downstream area owing to the complete lack of water.

Summer and winter potatoes are very important cash crops. With the change in the cropping pattern from "rice–wheat" to "rice–potatoes", the demand for water also rises. This shift is causing a shortage of irrigation water supply during the dry season.

The differing incomes of farmers in the upstream, midstream and downstream areas are believed to occur because of those working along the upstream, who dominate the irrigation water supply. At present, there is no organization that manages the irrigation system in the Shali Nadi irrigation district. If farmers in this district are asked to whom the water belongs, they say it belongs to everyone (de jure). However, the use of irrigation water is not regulated and, as a matter of fact (de facto), no irrigation right is acknowledged for farmers in the downstream area. Moreover, because of the lack of a formal economic body that manages the use of water resources, some farmers working along the downstream are forced to take the initiative and undertake maintenance of the water channel as required, because they are the most vulnerable in the event of a water shortage.

The Khokana irrigation system

The farmlands subject to the Khokana irrigation system are located in Khokana village in the Lalitpur region in the Kathmandu valley (Kondo et al., 2005). The total area of these farmlands is approximately 250 hectares, and there are roughly 1,100 farm households. Although the predecessor of the Khokana Irrigation Association was informally developed approximately 200 years ago, the Association was formally established in 1996 when it was registered as a district irrigation office.

Under the Khokana irrigation system, farmers are required to pay irrigation fees. A different rate of fees applies to farmers belonging to the Khokana Village Development Committee (VDC) from those belonging to a non-VDC organization. Each farm household in Khokana village is obliged to provide 1 *pathi* rice (2.5 kg) and 4 *manas* wheat (1.6 kg) annually in the form of irrigation fees. The village's customary practice is that the irrigation fee is not determined according to the area of applicable farmland. In the case of farmers in the Champi, Bungamati and Sainbu districts located along the upstream of the channel, the requirement is to provide rice worth 1 *pathi* per *ropani* to the Association in exchange for being exempted from water channel cleaning duties.[2]

These traditional irrigation fee payment methods were instituted with the informal establishment of the predecessor of the Association. In 1999 and 2000, almost all the farmers provided the relevant crops in the form of irrigation fees.

Until recently, the village's customary practice in relation to maintenance was that all farmers belonging to the Khokana VDC were obliged to clean the water channel biannually: prior to rice planting (from mid-April to early May) and prior to or immediately after wheat sowing (at the beginning of November). In this manner, these farmers have worked towards the maintenance of the water channel since ancient times.

In 2000, the Khokana Irrigation Association outsourced maintenance work to a local cultural club and, consequently, farmers were no longer required to participate in the maintenance. The members of a local *thalackhi dapa khala* club are currently working as water channel supervisors (a *dapa khala* is a volunteer drumming group that performs at festivals and the weddings of local farmers). The club's water channel supervisory roles are as follows: (1) the maintenance of areas within the channel that are not lined (with a particular focus on the upstream and midstream areas) and (2) the segmentation of the irrigation area into a number of blocks, the establishment of a rough plan for rotational water supply and the implementation of the plan.

Since upgrade work carried out on the water channel, however, the very traditional cropping pattern "rice–wheat" has shifted to "rice–beans" or "rice–vegetables". According to the results of a survey concerning the usage status of 109 farm fields, of 52 fields with the traditional "rice–wheat" pattern, 22 fields (42 per cent) have shifted from wheat to vegetable cropping in one way or another. The vegetables that are cultivated include potatoes, cauliflowers, coriander, onions, garlic, fava beans, cabbages and daikon radish. Thus the upgraded water channel has been effective in replacing wheat, the traditional aftercrop of rice, with vegetables, which have an increasing market demand. This trend is particularly evident in the upstream area.

However, the water channel upgrade has not significantly affected the village's cropping intensity, which remains almost unchanged – 192 per cent before the upgrade and 195 per cent after. Thus, the level of cropping intensity under the Khokana irrigation system remains just below 200 per cent, which is the traditional standard level seen in the Kathmandu valley. In the case of the upstream and midstream areas of the Shali Nadi irrigation district in Sankhu village, the farmlands are able to be used three times per year for cultivation purposes, with a cropping intensity of approximately 300 per cent, which demonstrates a significant edge over the Khokana irrigation system.

Subsection summary

The primary finding of the comparison between the Shali Nadi and the Khokana irrigation districts is that, in general, an irrigation association is established where there are high water-carrying costs, which have a significant impact on the creation of systems for water channel maintenance and the distribution of irrigation water. In a region such as the Khokana irrigation district, with an extremely long distance between the diversion weir and the applicable farmlands, joint social action must be undertaken by the farmers in order to manage the water channel. An irrigation association's mission is to maintain the water channel, which is a regional public property. The predecessor of the Khokana Irrigation Association was established informally in ancient times and has continued to commit itself to the maintenance of the water channel as well as to the distribution of irrigation water by collecting irrigation fees. Based on these fees collected from farmers and the additional irrigation profits provided by a brick factory, the Association has recently started outsourcing the maintenance and supervision of the water channel to a local cultural club.

In the case of the Shali Nadi irrigation district, the distance between the diversion weir and the applicable farm fields located in the upstream area is small. Consequently, it is not necessary for the farmers in the upstream area to set up an irrigation association. It therefore becomes impossible to form an irrigation association that involves all of the farmers within the district. If one accepts that it is unfair in terms of the amount of irrigation water supplied that the further the farm field is located from the upstream, the higher the irrigation water-carrying costs become, it can then be suggested that a greater level of unfairness in the distribution of water ironically creates greater difficulty in facilitating cooperation between farmers. In other words, unfairness in the amount of water that is supplied explains the absence of rules for both irrigation water distribution and water channel maintenance.

7-5-5 The introduction of a "market economy" and the challenges for the development of sustainable agriculture in Mongolia

The introduction of a market economy and the domestic market system in Mongolia

Mongolia's socialist system came to an end at the same time as the Berlin Wall collapsed and socialist governments fell in several other countries. A market economy has since been rapidly promoted in Mongolia. In fact,

Mongolia had never experienced a full-scale market economy because its socialist planned economy system had lasted for more than 60 years.

Historically, it has never been possible to establish an orderly trade market (a market system in a broad sense) using flexible policies. Such a market can be established only by means of trial and error involving various business activities, with an effort to tailor it to the country's circumstances. A review of Mongolia's process of forming a market system provides useful information not only to other former socialist states but also to developing nations that are constructing a new domestic market system. In addition, Mongolia is a useful case study for identifying a market system that is effective for the purpose of developing sustainable agriculture.

Based on the viewpoint explained above and with reference to an actual condition survey conducted between 2005 and 2006, this study targets the distribution of meat, which is Mongolia's staple food, and focuses on the Khuchit Shonhor Food Market in Ulaanbaatar. During the course of this study, the Mongolian Ministry of Food and Agriculture and the Mongolian Office of the Japan International Cooperation Agency provided significant cooperation.

Development outcomes and future prospects

As a result of the major transition towards a market economy, a remarkable number of both international and domestic studies have been conducted to examine the Mongolian economy. These studies, however, were mostly based on macro data provided by the Mongolian government, and in most cases there was no reference to actual condition surveys. The outcomes of these studies may not have been accurate because of the ongoing socioeconomic turmoil and the significant lack of credibility in the fundamental data designed to be incorporated into macro data. As a result, it has become necessary to conduct a study based on the results of a detailed condition survey in order to examine and reveal the actual state of affairs even if such a survey may be able to provide only limited information concerning the national economy. As mentioned previously, this study targets the Khuchit Shonhor Food Market, which is located in Ulaanbaatar and deals in meat, the national staple food, and is increasingly becoming specialized in the wholesale market.

During the transition towards a market economy, the national food supply system collapsed and the agro-livestock cooperative system, which underpinned the national food supply system through agro-livestock production, also collapsed. In addition, national control over product distribution and pricing was lifted, triggering deregulation in these areas. In these circumstances, nomads engaged in independent, family-run farming are raising more goats than before and have been carrying out deregulated livestock sales for the purpose of greater economic reproduction. As a result, the meat supply has become increasingly unstable, prices have risen and meat consumption among city dwellers has started to fall. Moreover, the level of deterioration and destruction of meadows has intensified owing to goat overgrazing. In this way, the development of sustainable agriculture in Mongolia is about to come to a halt.

A major change has also been observed with the distribution of meat and the meat market. The number of products sold through meat-processing factories, those used to support meat distribution during the socialist era, gradually declined. In turn, livestock/meat brokers, who can be deemed to be "pedlars", have spontaneously come into the picture, and nearly half of livestock/meat products are now sold through these brokers. In line with the deregulated livestock sales promoted by nomads, these brokers may cause meat supply and demand to become increasingly unstable. As mentioned earlier, the Khuchit Shonhor Food Market is gradually establishing itself as a wholesale market that is able to provide these meat brokers with a main location for trading. Although this market was originally used as a retail market dealing in general food items, greater amounts of meat products have tended to be traded through this market, generating increased demand for the wholesale trade. Although there are a number of retail markets dealing in food items in Ulaanbaatar, the Khuchit Shonhor Food Market has obvious geographical advantages (i.e. a large lot area of 63,000 m^2 and a convenient location roughly 20 minutes away from the city centre) that have helped it to develop as a major meat market.

The brokerage methods adopted in Khuchit Shonhor Food Market are not fully organized to the extent that a bidding system is in use. Transactions are accomplished through negotiations in most cases, and this is also true of trading between retailers and buyers. According to the condition survey's results, trading prices vary significantly, and this strongly suggests that transactions are achieved as a result of arbitrary/incidental pricing. In other words, the trading of multi-valued goods is carried out in this market. An improved level of transparency is therefore required for pricing, and the equalization of prices also needs to be facilitated, for example by introducing a bidding system. Sanitary standards too must be upgraded, and the matching of supply and demand needs to be improved by the provision of freezers.

Meanwhile, the roles of brokers have rapidly become differentiated, and in the case of the Khuchit Shonhor Food Market there are three types of broker who directly connect with the market. Those who carry in animal carcasses are termed "local brokers" for convenience; they numbered 250–280 in 2005. They sell to brokers who exclusively engage in sales within the market ("market brokers"), who numbered 20–30 in the same period. There is, however, no clear distinction between these two types of broker – they happened to become differentiated in this manner owing to their roles at the time. It must therefore be noted that it is possible for these two categories of broker to switch roles or to carry out both roles simultaneously. Interviews revealed that these two types of broker, ranging from 270 to 310 in total, do not necessarily come to the market daily for the purpose of trading. It is reported that a maximum of only 90 engage in trading on a daily basis. The third type of broker ("wholesalers and retailers") purchases products from market brokers and sells such products at a designated booth within the market site to local retailers, caterers and general consumers. These brokers are required to register with Khuchit Shonhor Co., Ltd. (market proprietary company). They are also obliged to pay monthly tenant fees of 120,000 *tugrugs* for the use of the booth. The number of these brokers was approximately 360 in 2005.

Furthermore, according to the condition survey results, there are additional types of brokers, such as those who pick up livestock products from nomads and those who transfer products from a transit location (i.e. a wholesale-oriented local market) to slaughterhouses located on the outskirts of Ulaanbaatar. It can therefore be stated that brokers have been differentiated into multiple types since the collapse of the socialist system. Although it can be assumed that the broker differentiation process occurred spontaneously, and although it appears to be unproductive to have so many different types of broker, this process actually corresponds to rational commercial functionality given that meat products are distributed throughout the country and that the distribution infrastructure has not been fully developed. The differentiation of brokers seen in Mongolia provides an indicative example for former socialist nations with a high level of social division of labour and an established nationwide distribution system of how to create a new distribution structure. It may, however, be inappropriate to apply Mongolia's new economic development process unconditionally to developing nations that are currently establishing a social division of labour and a nationwide distribution system.

In addition, from the standpoint of the development of sustainable agriculture, a total reliance upon the mechanism of a market economy may trigger serious risks represented by increased goat-rearing in Mongolia. It is therefore necessary to undertake appropriate control measures.

Subsection summary

After the transition to a market economy, citizens were told that the rapid introduction of a market economy would bring about wholesale freedoms. This message may have resulted from the government's intention to avoid the re-emergence of a socialist system. Mongolia now seems to be facing major socioeconomic chaos. Typical examples that highlight this disruption are goat overgrazing, designed to produce cashmere by sacrificing food (meat) production, and arbitrary pricing, as seen in wholesale markets. Goat overgrazing in particular exhibits a strong likelihood of causing the deterioration and destruction of meadows that function as the basis of Mongolian agricultural production, and hence it is a grave problem for the development of sustainable Mongolian agriculture. From the standpoint of individual farming management, the current farming trend of nomads being at the centre of the shift to cashmere production (through goat-rearing) can be deemed to be a positive move, enabling nomads to obtain a better farming income. Nevertheless, from a general point of view and from the standpoint of Mongolia's sustainable agricultural development, this trend is not positive, because it risks destroying Mongolia's natural agricultural production base.

What drives nomads to initiate goat-rearing is the large gap in the incomes of family-run farms, with high relative prices of cashmere products and low relative prices of livestock products. This is mostly triggered by the chaotic and disorganized trade market in the wake of the breakdown of the national food supply system that resulted from the collapse of the socialist system and the introduction of a market economy.

7-5-6 Conclusion

A basic agricultural development strategy applicable to developing nations may be the effective utilization of market economy opportunities. This is because farm produce must win international price wars in this increasingly globalizing world. The wave of globalization demands that developing nations undergo rapid socioeconomic change. Moreover, the agriculture of developing countries and the traditional structures of domestic agricultural communities need to be reformed in an effort to undertake more systematic agricultural activities to respond to new economic opportunities.

A long-term adjustment period is required for domestic agricultural communities to adapt to new economic opportunities while easing the negative effects caused by globalization. Because many farmers in developing nations are still engaged in conventional farming, a variety of problems will occur if these nations rush to introduce new systems and technologies. In the case of low-income nations, it is necessary to execute economic change and socioeconomic infrastructure building (such as irrigation systems and wholesale markets) in ways that harmonize with traditional agriculture and the farming community's customary social system.

An effort must also be made to promote environmentally friendly agricultural production that accords with regional characteristics. Agriculture that is sustainable over a long period of time must value the diversity of ecosystems. In the quest to maintain a good balance between food production and the environment, it is valuable to share the knowledge and technologies acquired by advanced nations in their efforts to ensure environmental preservation. Asian nations must utilize advanced nations' experience by taking into account each country's factor endowments and climate requirements.

It must be highlighted that the individuals directly involved in the agricultural technology transfer described above are regional farmers, who are experts in the situation of regional agriculture and its problems. It is therefore important to make full and effective use of their potential abilities in order to enable a variety of market economic opportunities to become accessible.

Notes

1. Hokkaido, Okinawa and north-eastern China functioned like domestic colonies in the process of economic development.
2. 1 *ropani* = 0.051 hectares.

REFERENCES

Kondo, Takumi, et al. (2002) "Irrigation Efficiency in Small-Scale Irrigation System: The GIS Approach", in S. S. Acharya, Surjit Singh and Vidya Sagar (eds), *Sustainable Agriculture, Poverty and Food Security, Volume 1*, Jaipur: Rawat Publications, pp. 203–213.

Kondo, Takumi, et al. (2005) "Effects of Rehabilitation in the Performance of the Canal and Farmland: A Case of Khokana Irrigation System in the Kathmandu Valley, Nepal", in Jamaluddin Sulaiman, Fatimah Mohammed Arshad and Mad Nasir Shamsudin (eds), *New Challenges Facing Asian Agriculture under Globalisation*. Serdang: Universiti Putra Malaysia Press, pp. 709–726.

Sakashita, Akihiko (1992) *The Theory and Status of Middle-Class Peasantry Formation*. Tokyo: Ochanomizu Shobo (in Japanese).

7-6
Nature therapy

Yoshifumi Miyazaki, Bum-Jin Park and Juyoung Lee

7-6-1 Harmonization of humans and nature

Human beings feel comfortable when they are close to nature, but this feeling is difficult to describe in words, and science has yet not provided an adequate explanation for it. However, based on the recent development of assessment methods that can measure the physiological effects of relaxation, a considerable amount of scientific data has now been accumulated on the subject.

Five million years have passed since humans evolved into modern humans as they are today. The men and women living in modern times have therefore spent more than 99.99 per cent of their evolutionary history in natural environments. We have become the humans we are today, living in a modern civilization, through a process of evolution that took place in a natural environment. The human body is thus made to adapt to nature. However, as conveyed through terms such as "techno-stress", artificialization is taking place at such a rapid rate that we now find ourselves in stressful situations in our daily lives and are forced to deal with the resultant pressures.

If in such circumstances we receive a stimulus rooted in nature itself, such as nature therapy, we become aware of our own true nature, we can relax and can perceive this feeling to be comfortable and natural to us. All of this is accomplished without logical thought. Instead, we perceive it through *kansei* (a Japanese word meaning intuition) in an intuitive and illogical manner. Because this process cannot be described in words,

Designing our future: Local perspectives on bioproduction, ecosystems and humanity, Osaki, Braimoh and Nakagami (eds), United Nations University Press, 2011, ISBN 978-92-808-1183-4

physiological indicators play an important role. A great deal of attention is now being paid to the use of nature therapy to produce a condition of harmony between humans and nature, which is being addressed in a scientific manner using data on the physiological effects of relaxation.

7-6-2 The physiological assessment method for nature therapy

To provide assessment endpoints, the present authors simultaneously measure stress hormone (cortisol) levels in saliva, amylase activity in saliva and autonomic nerve activities (sympathetic and parasympathetic) as monitored by heart rate, heart rate fluctuation and blood pressure (Tsunetsugu et al., 2009). In addition, they have pioneered the technology needed to monitor the prefrontal cortex activity of the brain using near-infrared spectroscopy (Park et al., 2007). With Li (Li et al., 2008a, 2008b, 2007) as the principal investigator, the present authors are conducting a study on natural killer (NK) activity as a marker of immune functions and studying anti-cancer proteins. In addition, they have measured the levels of forest phytoncides (wood essential oils), urban exhaust fumes, temperature and humidity, illuminance, wind velocity and negative (minus) and positive (plus) ions. As a result, they have attempted to confirm the physiological effects of relaxation on 420 volunteers undergoing various kinds of nature therapy at 35 different forests throughout Japan.

Using amylase activity in saliva as a measure of stress (with each on-the-spot measurement being completed in about 1 minute), the present authors have been developing a system by which bus tourists can, on request, directly experience the relaxing effects of nature therapy.

7-6-3 The physiological effects of relaxation through nature therapy in forests

From 2005 to 2007, the present authors conducted studies, each lasting about a week, in forests at 35 different sites throughout Japan, ranging from the Yanbaru Kuina forest in Okinawa to the Kushiro Swamp in the large northern island of Hokkaido (Lee et al., 2009; Park et al., 2009a, 2009b, 2008; Tsunetsugu et al., 2007). Stress hormone (cortisol) levels in saliva, sympathetic and parasympathetic nerve activity (as monitored by heart rate fluctuation), blood pressure and heart rate were adopted as the measured variables (endpoints). It is known that cortisol levels, sympathetic nerve activity, blood pressure and heart rate become elevated when a person experiences stress, and that parasympathetic nerve activity is enhanced in relaxing situations. When parasympathetic nerve activity is enhanced, we become hungry because our digestive systems are activated. This explains why lunch tastes better when eaten outdoors, in a natural setting.

As a result of studies involving 420 volunteers at 35 different sites, the group of volunteers looking at natural surroundings while sitting down showed the following endpoint decreases compared with the urban control group: a 12.4 per cent decrease in cortisol level, a 7.0 per cent decrease in sympathetic nerve activity, a 1.4 per cent decrease in systolic blood pressure and a 5.8 per cent decrease in heart rate. This proves that stressful states can be relieved by nature therapy. It should also be pointed out that parasympathetic nerve activity was enhanced by 55.0 per cent, indicating a relaxed biological system.

7-6-4 The blueprint for "*shinrin* (forest) therapy" stations

The blueprint for *shinrin* therapy stations is based on an approval system. Prefectures, cities, towns, villages and corporations can submit approval requests for candidate forests. The final approval is made after reviewing (1) the actual validation study carried out on the physiological effects of relaxation, (2) intangibles such as accommodation plans and (3) tangibles such as the forest's environment and facilities. This blueprint is being promoted by Japan's Forestry Agency and has already been approved as the "Total Project for *Shinrin* Therapy".

This approval system represents a major goal in terms of generating economic benefits for prefectures, cities, towns, villages and corporations and for reviving community forests in Japan. Between 2005 and 2008, 38 sites were approved. The present plan envisages approvals for 100 *shinrin* therapy stations within the next 10 years. It is hoped that creating 50–100 *shinrin* therapy stations throughout Japan will contribute to better forest management and help revive forests all over Japan. Since urban residents living in stressful environments find such therapies an effective means of relaxation, from a preventive medicine perspective a reduction in national health care expenditure can also be expected. Furthermore, as pioneers in this kind of therapy, the present authors will be accumulating a vast amount of physiological data and scientific information.

Activities at shinrin therapy stations in Okutama Town

Okutama Town was officially designated as a *shinrin* therapy station in April 2008. Located a 90-minute train ride away from Shinjuku, the urban core of the Tokyo metropolitan area, the town is known for having the largest number of giant trees in Japan (about 1,000 such trees were confirmed by a nationwide forest survey by the Ministry of the Environment). The *shinrin* therapy station effectively went into operation in April 2009. Serving the Tokyo metropolitan area, its facilities will contribute to revitalizing Okutama Town and helping Tokyoites relieve stress.

During the five months between April and August 2009, events were held in Okutama Town to offer 11 daily tours and 9 overnight trips to the therapy station to 388 participants. The Welfare Department of the Education Bureau of the Tokyo Metropolitan Government has started using the *shinrin* therapy station facilities for government staff as well. The facilities around Lake Okutama help government employees relax and relieve stress. In addition to five trekking routes through the woods around the lake, the station provides individuals with facilities for activities such as bathing in natural hot springs, soba noodle making, pottery, stargazing, waterfall viewing and stretching. Professional health consultations and medical checkups are also available.

7-6-5 Nature therapy news around the world

Miyazaki attended the kick-off meeting of the Task Force on Forests and Human Health (ForHealth) held as part of the IUFRO (International Union of Forest Research Organizations) Division VI Symposium "Integrative Science for Integrative Management", on forest recreation, environmental policy and sustainable forest management, held in Finland from 14 August to 20 August 2007. The head of the Finnish Forest Research Institute, Professor Hannu Raitio, coordinated the ForHealth Task Force. According to his report, there has been increasing interest in Europe (especially in England) in the topic of forests and human health. In 2004, a project entitled "Forests, Trees and Human Health and Wellbeing" had been launched as part of a COST (European Cooperation in Science and Technology) programme, with 22 countries participating. However, this project ended in 2008. Professor Raitio mentioned that one reason he held the meeting was because he wanted to continue this project so that the existing movement in Europe could spread worldwide through the ForHealth Task Force.

The current situation in Japan and Korea, and future activities

Japan's Forestry Agency announced a blueprint for *shinrin-yoku* (forest bathing) in 1982. The present authors are currently accumulating scientific data on forests and human health, mainly through cooperation between the Forestry and Forest Products Research Institute and Chiba University, based on the "blueprint for *shinrin* therapy stations". This movement, which started in Japan, is now spreading to Korea, where a Korean Forest and Health Forum was established in 2005.

At the ForHealth Task Force meeting in Finland in August 2007, the present authors presented a lecture entitled "Forests and Human Health" in order to explain the approval system for *shinrin* therapy stations in accordance with the "blueprint for *shinrin* therapy stations" in which Japan's Forestry Agency plays a central role (Miyazaki et al., 2007). They also described the methods of the physiological studies required to gain approval, along with the results. Their data indicated that, compared with the urban control group, the forest group exhibited enhanced parasympathetic nerve activity (which is known to become elevated when relaxed) and suppressed sympathetic nerve activity (which is known to become elevated in response to stress), while other endpoints such as stress hormone level, pulse rate and blood pressure were all reduced. These results received much attention because no real physiological studies on the effect of forests on human health had previously been conducted outside Japan.

The Steering Committee of the ForHealth Task Force, which has 11 members, was established on 16 August 2007, with Miyazaki from Japan in attendance. Initially, it was surprising that so many countries shared a common interest in this subject. On second thoughts, however, it is a fact that most of the participating countries are experiencing many of the same problems of a so-called "stressful society" and are striving to find a way to solve these problems through forests and nature. As the present authors continue their collaborative studies with other countries, they will actively distribute further information from Japan.

REFERENCES

Lee J., B. J. Park, Y. Tsunetsugu, T. Kagawa and Y. Miyazaki (2009) "The Restorative Effects of Viewing Real Forest Landscapes: Based on a Comparison with Urban Landscapes", *Scandinavian Journal of Forest Research* 24(3): 227–234.

Li, Q., K. Morimoto, A. Nakadai, H. Inagaki, M. Katsumata, T. Shimizu, Y. Hirata, K. Hirata, H. Suzuki, Y. Miyazaki, T. Kagawa, Y. Koyama, T. Ohira, N. Takayama, A. M. Krensky and T. Kawada (2007) "Forest Bathing Enhances Human Natural Killer Activity and Expression of Anti-cancer Proteins", *International Journal of Immunopathology and Pharmacology* 20(S2): 3–8.

Li, Q., K. Morimoto, M. Kobayashi, H. Inagaki, M. Katsumata, Y. Hirata, K. Hirata, T. Shimizu, Y. J. Li, Y. Wakayama, T. Kawada, T. Ohira, N. Takayama, T. Kagawa and Y. Miyazaki (2008a) "A Forest Bathing Trip Increases Human

Natural Killer Activity and Expression of Anti-cancer Proteins in Female Subjects", *Journal of Biological Regulators & Homeostatic Agents* 22: 45–55.

Li, Q., K. Morimoto, M. Kobayashi, H. Inagaki, M. Katsumata, Y. Hirata, K. Hirata, H. Suzuki, Y. J. Li, Y. Wakayama, T. Kawada, B. J. Park, T. Ohira, N. Matsui, T. Kagawa, Y. Miyazaki and A. M. Krensky (2008b) "Visiting a Forest, But Not a City, Increases Human Natural Killer Activity and Expression of Anti-cancer Proteins", *International Journal of Immunopathology and Pharmacology* 21(1): 117–127.

Miyazaki, Y., B. J. Park and Y. Tsunetsugu (2007) "Forests and Human Health", August, <http://www.metla.fi/tapahtumat/2007/iufro-d6/presentations/15-08-wed/auditorio/Miyazaki_(Kick-off_18.30).pdf> (accessed 9 April 2010).

Park, B. J., Y. Tsunetsugu, T. Kasetani, H. Hirano, T. Kagawa, M. Sato and Y. Miyazaki (2007) "Physiological Effects of *Shinrin-yoku* (Taking in the Forest Atmosphere or Forest Bathing), Using Salivary Cortisol and Cerebral Activity as Indicators", *Journal of Physiological Anthropology* 26(2): 123–128.

Park, B. J., Y. Tsunetsugu, H. Ishii, S. Furuhashi, H. Hirano, T. Kagawa and Y. Miyazaki (2008) "Physiological Effects of *Shinrin-yoku* (Taking in the Forest Atmosphere or Forest Bathing) in a Mixed Forest in Sinano Town, Japan", *Scandinavian Journal of Forest Research* 23(3): 278–283.

Park, B. J., T. Kasetani, T. Morikawa, Y. Tsunetsugu, T. Kagawa and Y. Miyazaki (2009a) "Trends in Research Related to *Shinrin-yoku* (Taking in the Forest Atmosphere or Forest Bathing) in Japan", *Silva Fennica* 43(2): 291–301.

Park, B. J., T. Kasetani, T. Morikawa, Y. Tsunetsugu, T. Kagawa and Y. Miyazaki (2009b) "The Physiological Effects of *Shinrin-yoku* (Taking in the Forest Atmosphere or Forest Bathing): Evidence from Field Experiments in 24 Forests across Japan", *Environmental Health and Preventive Medicine* 15(1): 18–26.

Tsunetsugu, Y., B. J. Park and Y. Miyazaki (2009) "Trends in Research Related to *Shinrin-yoku* (Taking in the Forest Atmosphere or Forest Bathing) in Japan", *Environmental Health and Preventive Medicine* 15(1): 27–37.

Tsunetsugu, Y., B. J. Park, H. Ishii, H. Hirano, T. Kagawa and Y. Miyazaki (2007) "Physiological Effects of '*Shinrin-yoku*' (Taking in the Forest Atmosphere or Forest Bathing) in an Old-growth Broadleaf Forest in Yamagata Prefecture, Japan", *Journal of Physiological Anthropology* 26(2): 135–142.

7-7
Health risks and regional sustainability

Tatsuo Omura

7-7-1 A new concept of regional sustainability from the perspective of health risks

In the twenty-first century, the concept of regional sustainability has been introduced to promote sustainable development. The factors relating to it are diverse and complicated, even if limited, in terms of health risks. Therefore, it is necessary to know how to position regional sustainability at an academic level.

Today, almost 20 years after the Brundtland Commission, the conventional concept of global sustainability is discussed as a trade-off between sustainable development (in terms of gross domestic product) and the environment. To date, a growing number of research initiatives have been undertaken on sustainable development. These studies have made significant progress on global sustainability; however, their applicability is limited because they are fragmented. Hence, establishing a new concept of regional sustainability from the perspective of health risks will help to provide a focused pathway to achieve solutions that could lead to successful sustainable development in the future.

Taking into consideration the achievement of regional sustainability through health risk management, which aims to resolve the regional issues of health risks inhibiting sustainable development, it is interesting to take a different view from the conventional perspective of GDP, by developing a new framework in the social system that is able to realize regional sustainability. In this regard, the present author proposes a new

---

Designing our future: Local perspectives on bioproduction, ecosystems and humanity,
Osaki, Braimoh and Nakagami (eds),
United Nations University Press, 2011, ISBN 978-92-808-1183-4

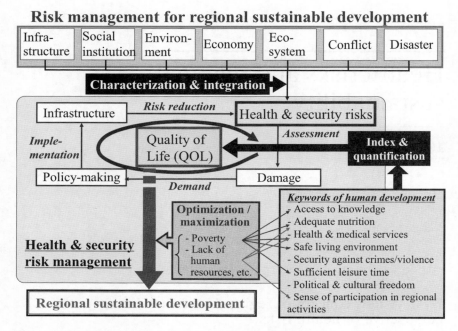

Figure 7.7.1 A new framework for regional sustainability from the perspective of health risks.

framework of regional sustainability (Figure 7.7.1) that could contribute to the development of a social framework for regional sustainable development from the perspective of health risks. In this framework, a "Quality of Life" (QOL) index is proposed instead of the conventional GDP. QOL is indexed and quantified at the society level and comprises four elements: health risks, damage, policy-making and infrastructure. In indexing and quantifying QOL, important aspects of human development should be considered, such as access to knowledge, adequate nutrition, health and medical services, a safe living environment, security against violence, sufficient leisure time, political and cultural freedom, and a sense of participation in regional activities.

In the first stage, a methodology will be developed for characterizing the relationships between health risks and infrastructure, social institution, environment, economy, ecosystem, conflict and disaster, and the integration of these relationships. Later, regional health risks will be evaluated. Based on the evaluated health risks, it will be possible to predict the social damage caused in the future by health risks, formulate a policy to overcome the damage and construct the infrastructure necessary to avoid health risks. Finally, regional sustainable development is realized by means of QOL optimization from the viewpoints of poverty, lack of human resources, and so on.

7-7-2 Health risks and regional sustainable development

In 2002, at the Environment and Development Summit in Johannesburg, South Africa, it was reported that about 6,000 children were dying every day in developing countries owing to unsafe drinking water. Moreover, the report predicted that, by 2025, 25 per cent of the world population would not be able to access safe and clean water. Hence, all countries participating in the Summit agreed to resolve these critical issues related to human health by committing technical and economic support to halve the predicted number of people who would not have access to safe and clean water by 2015.

Figure 7.7.2 shows the causes of death in developed and developing countries in 1998. In developing countries, the main causes of death were infectious and parasitic diseases, whereas in developed countries they were circulatory system diseases. This difference suggests that infectious diseases caused by the utilization of unsanitary water are the main cause of death in developing countries. Infant mortality rates (as deaths per 1,000 live births) in East and Southeast Asia in 2004 are shown in Figure 7.7.3. From this figure, we can see that there are higher mortality rates in developing countries in Southeast Asia. The prevalence of outbreaks of

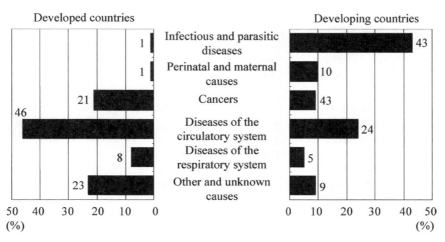

Figure 7.7.2 Causes of death in developed and developing countries, 1998.
Source: WHO (1999).

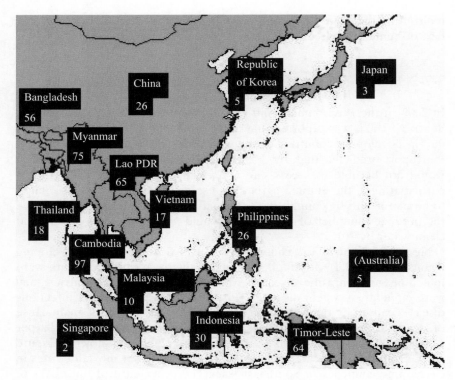

Figure 7.7.3 Infant mortality rates in East and Southeast Asia, 2004 (deaths per 1,000 live births).
Source: WHO (2006).

waterborne infectious diseases may pose a challenge to achieving sustainable development in the region.

Shigella, *Vibrio* and *Salmonella* are very common pathogenic bacteria. However, outbreaks of infectious diseases caused by other pathogens, such as protozoa (*Giardia* and *Cryptosporidium*) and enteric viruses (Norovirus, Rotavirus, Hepatitis A virus and adenoviruses), have now also been reported. Unlike bacteria, some of these pathogens are resistant to chlorine disinfection and so it is very difficult to remove or inactivate them by conventional water or wastewater treatment technologies. Therefore, innovative disinfection technologies are needed for the construction of new water utilization systems for sustainable development, not only in Southeast Asia but also in developed regions.

Figure 7.7.4 shows that sustainable development in developing countries is limited because of a lack of human resources and poverty caused

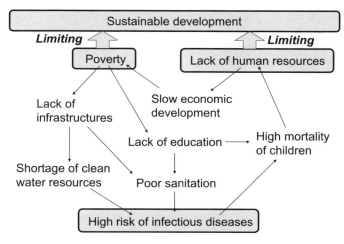

Figure 7.7.4 Relationships between infectious diseases and regional sustainable development.

by a high prevalence of infectious diseases. Many precious lives have been lost because of waterborne infectious diseases, resulting in a high infant mortality rate and, consequently, a lack of human resources, which is the cause of slow economic development. This leads to poverty in the society. A lack of infrastructure and education owing to poverty results in poor sanitation and a shortage of clean water. As a result, the society suffers from a high prevalence of infectious diseases. In order to resolve this situation, it is critical that infrastructures suitable to the culture and traditions of the region be constructed.

7-7-3 Transboundary movement of health risks associated with social issues

Internationalization, global warming and overpopulation are critical issues for the future. These issues will drive the current society to become a society with increased health risks through an increase in infectious diseases. This will interfere with the realization of the Johannesburg Summit agreement. The interaction of these issues and health risks has not been clearly formulated because of a lack of knowledge about the transboundary movement of health risks in the region.

Taking these issues into account, the transboundary movement of pathogenic micro-organisms is critical for understanding the prevalence of infectious disease outbreaks in the region. If an infectious disease outbreak occurs, it can easily spread to neighbouring regions via tourists or contaminated foods. Moreover, this outbreak would be expected to expand to other regions owing to internationalization and global warming. In this sense, it is important to protect the region from infectious disease outbreaks by understanding the transboundary movement of health risks stimulated by social issues, and it is necessary to find effective countermeasures for avoiding health risks.

7-7-4 Health risks related to infectious diseases in the Mekong watershed

The lower Mekong watershed, including most of the Lao People's Democratic Republic (PDR), Cambodia and substantial portions of Vietnam and Thailand, is one of the regions that suffers from outbreaks of waterborne infectious diseases. Under-five mortality rates in Lao PDR and Cambodia are 12.6 per cent and 12.5 per cent, respectively, which are 10 times higher than the average in developed countries (UNFPA, 2007). The proportions of people who have access to safe drinking water in Lao PDR and Cambodia are still low (51 per cent and 41 per cent, respectively) compared with other Southeast Asian countries (WHO and UNICEF, 2006). Although Vietnam and Thailand have achieved the Millennium Development Goals (MDGs) of the United Nations for drinking water targets, and the proportion of people with access to safe drinking water in these countries is 85 per cent and 99 per cent, respectively, the proportion of people with piped water in their households is still less than 40 per cent, even in Thailand. Therefore, most people use rainwater or well water, which does not have high water-quality standards (WHO and UNICEF, 2006).

The lower Mekong watershed is home to over 70 distinct ethnic groups whose water utilization varies depending on social conditions such as custom, income and education and geographical conditions such as access to water resources. Therefore, even if the microbiological quality is acceptable, drinking water often becomes contaminated with pathogens of faecal origin during transport and storage because of unhygienic storage and handling practices (Sobsey, 2002).

Table 7.7.1 summarizes drinking water sources in the lower Mekong watershed, based on the present author's field investigations. In urban areas of Khon Kaen in Thailand, Vientiane in Lao PRD and Phnom Penh in Cambodia, provincial water supply (PWS) systems comprising water treatment plants and distribution systems to households have been widely installed. In rural areas, village water supply (VWS) systems in which groundwater is pumped to the top of the water treatment plants have been provided as water sources in some regions. In urban areas, tap water

Table 7.7.1 Drinking water sources in the lower Mekong watershed

Country	City or village	Area	Drinking water source					
			River water	Rainwater	Well water	Tap water (VWS)	Tap water (PWS)	Bottled water
Thailand	Khon Kaen	Urban	N	Y	N	N	Y	Y
		Rural	N	Y	N	N	N	Y
Lao PDR	Vientiane	Urban	N	N	N	N	Y	Y
	Paylom	Rural	N	N	Y	N	N	Y (Rich)
	Khong	Rural	Y (Dry)	N	Y	N	N	N
Cambodia	Phnom Penh	Urban	N	N	N	N	Y	Y
	Kratie	Rural	Y (Dry)	Y (Rainy)	Y	Y	N	Y (Rich)
	Kompong Cham	Rural	Y (Dry)	Y (Rainy)	Y	Y	N	N
	Prey Veng	Rural	N	Y	Y	Y	N	N
Vietnam	My Tho, Vinh Long, Can Tho	Urban	N	N	N	N	Y	Y (Rich)
	Tan Thanh, Cai Lay	Rural	Y (Dry)	Y (Rainy)	N	Y	N	N

Notes: PWS: provincial water supply system; VWS: village water supply system; Y: drinking water source in households where the interview survey was undertaken; N: not used for drinking; Dry: used in dry season; Rainy: used in rainy season; Rich: used in relatively rich households.

from PWS systems and bottled water are consumed, and there is no difference in the seasonal drinking water sources. In rural areas, except Lao PDR, rainwater is commonly consumed in the rainy season. In the dry season, rainwater stored in large water tanks is used as the source of drinking water. Moreover, only relatively rich people are able to buy bottled water. In contrast, in households that do not have large water tanks, river, canal and well water are consumed in the dry season even if they are more heavily contaminated than rainwater.

Average concentrations of total coliforms in various water sources in the lower Mekong watershed are shown in Figure 7.7.5. In urban areas, coliforms are sometimes detected in tap water. In PWS systems, tap water is usually disinfected with chlorine. Therefore, it is possible that the tap water was contaminated during its distribution from the water treatment

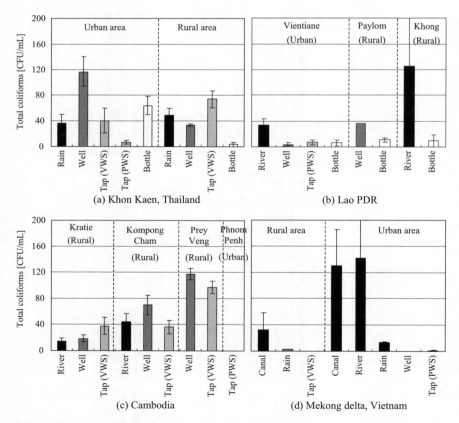

Figure 7.7.5 Average concentrations of total coliforms in various water sources in the lower Mekong watershed.

plant to households. In the urban areas, bottled water is the main source of drinking water, but the concentration of coliform groups in bottled water is as high as that in the rainwater in Khon Kaen in Thailand. There is a high possibility that bottled water is contaminated via unsanitary containers or the hands of consumers. Other water sources such as rainwater, well water and river water are contaminated with coliform groups at high concentrations. In VWS systems, the tap water in rural areas of Khon Kaen in Thailand and in Cambodia is heavily contaminated with total coliforms. However, tap water in the Mekong delta in Vietnam is not contaminated. This illustrates that the microbial quality of tap water varies between regions.

By analysing the data from these investigations, the annual risk of the infectious diarrhoea caused by *Escherichia coli*, attributable to drinking water, was evaluated in the lower Mekong watershed (Figure 7.7.6). The

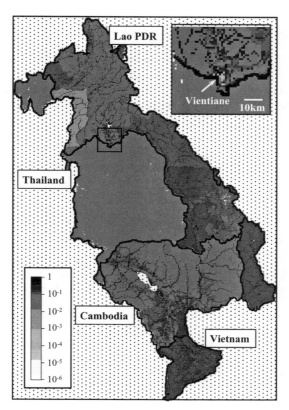

Figure 7.7.6 Estimated annual risk of infectious diarrhoea caused by *E. coli* in drinking water in the lower Mekong watershed.

risk in urban areas is as low as 10^{-4}, since people can drink tap water distributed by PWS systems, which is relatively less contaminated. On the other hand, the risk in rural areas surrounding urban areas is quite high (10^{-1}–10^{0}), since there is a high density of people living there without access to adequate sanitation. The risk of infection is also higher in remote rural areas than in urban areas, because river water is consumed and improved sanitation is not available.

From these results, it can be concluded that, in order to reduce the risk of waterborne infectious diseases and achieve sustainable development in the Mekong watershed, it is necessary to develop infrastructure, such as a water supply network and adequate sanitation, based on risk evaluation reflecting the characteristics of local water utilization.

7-7-5 Summary

Societies in developing regions, especially in Southeast Asia, have high health risks such as risks of infection by waterborne pathogens. These risks could be increased by internationalization, global warming and increasing population density. In order to achieve sustainable development in these regions, it is necessary to evaluate the risks and construct effective countermeasures that are applicable to these regions from the perspective of "Quality of Life" and are based on their own culture and traditions. For this, however, the present author needs to promote the research and engage in more collaboration. The belief is that it will be possible to achieve regional sustainable development in the near future.

REFERENCES

Sobsey, M. D. (2002) *Managing Water in the Home: Accelerated Health Gains from Improved Water Supply*. Geneva: World Health Organization.

UNFPA [United Nations Population Fund] (2007) *State of World Population 2007: Unleashing the Potential of Urban Growth*. New York: United Nations Population Fund.

WHO [World Health Organization] (1999) *The World Health Report 1999: Making a Difference*. Geneva: World Health Organization.

WHO (2006) *The World Health Report 2006: Working Together for Health*. Geneva: World Health Organization.

WHO and UNICEF (2006) *Meeting the MDG Drinking Water and Sanitation Target: The Urban and Rural Challenge of the Decade*. Geneva: World Health Organization; New York: United Nations Children's Fund.

7-8
An integrated solution to a dual crisis: An assessment of Green New Deal policy

Fumikazu Yoshida

7-8-1 Introduction – no time to lose

The world faces major economic and environmental crises. The economic difficulties, triggered by a glut of American subprime loans, nearly paralysed the United States and devastated manufacturing in Japan, a country largely dependent on exports. Corporate performance around the world recorded the biggest drop in the post-war period, but recovered through restructuring. However, the employment situation is still in crisis. To deal with these difficulties, the United States elected its first African American president, Barack Obama, with a mandate to resurrect the American economy. In Japan, the new Democratic Party of Japan has taken over government to cope with this crisis.

7-8-2 Global capitalism and its consequences

Since the demise of socialism in most of those countries that practised it, capitalism has reigned supreme. As the principles of neoliberalism and regulatory reform have taken over, there has been unprecedented participation in the world market by the developing countries in particular, and by those more advanced countries known by the acronym BRIC: Brazil, Russia, India and China. The subsequent deep competition has led to a lowering of prices and an increase in the demand for natural resources, with the consequent and complementary need to keep wages low and materials cheap.

The following are three adverse consequences of the current worldwide economic downturn:
1. The destruction of the global environment, increased global warming and, as a corollary, the destruction of nature and the loss of biodiversity.
2. The widening income gap between the developed and the developing countries, and its potential danger as a reason for terrorist attacks.
3. An unhealthy dependence on speculative money and surplus capital that has led to the bursting of the economic bubble, and may have doomed the world to recurrent worldwide financial crises.

As for wealth, the total income of the world's 500 wealthiest people is almost equal to the total income of 416 million of the poorest and also to the total income of the 2.5 billion people in the world who live on less than USD 2 a day, further widening the income gap. If the present mode of capitalism continues, the global population will not be able to sustain itself economically, socially or environmentally. Although global capitalism is subject to revision, the causes and effects of the problems brought on by the capitalist economy will have to be resolved, and, in most cases, it ought still to be possible to ameliorate the impacts of the global economic turmoil through public policy initiatives.

7-8-3 The first appearance of the Green New Deal

It is of critical importance that these two world crises – which have been isolated here as threatening the stability of the world – should not be dealt with separately but should instead be solved through the integration of policies. If it is hoped to sustain the world, there must be investment in public policies such as those that have been called a "Green New Deal" (also understood as green domestic demand). President Obama plans to invest JPY 150 trillion over the next 10 years to promote a clean energy economy and create 5 million new (green) jobs. China, Korea, France and Japan are also planning to implement similar policies.

The "Green New Deal" was originally proposed in July 2008 by the New Economics Foundation of the United Kingdom (New Economics Foundation, 2008), which offered suggestions for tackling what it identified as the three salient issues: the credit crunch, climate change and high-priced oil. The proposal sought the reconstruction of the financial and tax system and of energy policies, especially those in the United Kingdom. An essential component of the deal involved retraining people to work in the field of environmental regeneration; this would include definite prospects for renewable energy.

In line with President Obama's concrete policy objectives, the subsequent discussion has focused on how to provide facilities for renewable energy, on public transportation, on the introduction of insulation systems and alternative heat sources for millions of houses, and on local food production for local consumption. The economic recovery programme centres on the revamping of grand-scale public works projects, in particular. For example, space heating and lighting units in public facilities in the United States will be replaced with high-efficiency products and, as emergency countermeasures, public buildings such as schools and libraries will be renovated and roads and bridges will be repaired or newly constructed.

The Secretary-General of the United Nations, Ban Ki-moon, has spoken on behalf of the international community: "A financial crisis is a good opportunity to tackle climate change. Green growth does create new jobs" (UN News Centre, 2008). On 22 October 2008, the United Nations Environment Programme (UNEP, 2008) announced a Green Economy Initiative, which has three pillars: valuing and mainstreaming nature's services into national and international accounts; generating employment through green jobs and laying out the policies; instruments and market signals able to accelerate a transition to a Green Economy – all of which are indispensable for promoting real growth, combating climate change and bolstering employment.

In Japan, the Ministry of the Environment initiated its own Green New Deal in March 2008 and proposed that, within five years, the environmental market should be worth JPY 100 trillion, which, in comparison with 2006, would amount to a 40 per cent increase. At the same time, the Ministry's plan is also one of the growth strategies that the government will undertake. The details of the plan are that it will:
1. over the next 20 years, increase the amount of solar energy generation to 20 times its present level;
2. provide cash assistance when people purchase hybrid cars; the production of more than 200,000 cars for everyday use will herald a new generation of private vehicles;
3. promote the purchase of the latest energy-saving electrical appliances.

When public investment is implemented for the purpose of economic recovery, as in recovering demand and improving employment, the index of sustainability should be used to review the effects of CO_2 reduction and its impact on employment. Similarly, one must consider sustainability and employment indices when evaluating the infrastructure. The construction of public transportation networks and heat- and earthquake-proof structures should be a priority for road construction teams. As for the development of renewable energy to resolve global warming issues, the focus must be not only on research and development or investment in facilities for solar energy, wind power and biomass energy, but also on the institutionalizing of their widespread use, with incentives designed to maintain these trends. The development of renewable energy and the implementation of institutional reform can be seen as two wheels of one cart.

7-8-4 Strategies for a low-carbon and recycling-based society

Policies to deal with climate change and threats to the environment have become political issues internationally. In particular, the task of halving CO_2 emissions by 2050, in advanced countries even up to an 80 per cent reduction, is acknowledged as a long-term policy goal. Since the concept of creating a low-carbon society is based on an international consensus, it must be acknowledged that the concept of creating a recycling-based society is situated in the framework of the issues raised by climate change. If the concepts are reorganized from a social viewpoint, they will need to be ranked in order of descending priority: first, a sustainable society; second, a low-carbon society; and, third, a recycling-based society.

Currently, the Japanese government is proposing a 25 per cent CO_2 reduction by 2020 compared with 1990. According to an estimate by the National Institute for Environmental Studies (NIES, Kyoto University, and Mizuho Information and Research Institute, 2009), calculated on the basis of a model of a 70 per cent reduction by 2050 and in response to a suggestion by the Intergovernmental Panel on Climate Change, it will cost JPY 5–6 trillion to institute a 25 per cent reduction from 1990 levels. (The estimate was released in the 3rd medium-term objective review issued by the Committee on Global Warming on 23 January 2009.) The contents of the estimate deal with investments designed to create savings in the steel and chemical sectors, to install solar power for 1.77 million houses, to increase wind power by 10 times, to electrify 80 per cent of next-generation automobiles, to build energy-saving houses and to install high-efficiency water heaters. These costs also include expenditure for domestic demand expansion. This spending will help related industries, create new jobs and strengthen international competitiveness through the process of technological development. The burden of additional costs should not be borne by business owners alone, however, but should be covered by environmental taxes incorporated within the general revenue. The major topics to be discussed in Japan should concern this issue and the question of carbon emissions trading. The main points of contention will be how to set up the medium-term objectives for 2020, how to deal with carbon emissions trading and what actual measures will need to be taken in order to implement the objectives to formulate policies for the Green New Deal of Japan.

More specifically, Hokkaido, the present writer's home base, has the potential to become a low-carbon society, one that least depends upon fossil fuel. For instance, about 40 large-scale wind power generators are located in Tomamae Town (north-eastern Hokkaido), and policymakers have shown special interest in 11 wind turbines built in Hokkaido, Aomori and several other areas because they invest in renewable energy under the green electricity rates system (see below). Biomass residues from forest logging and methane gas generated from animal manure also offer considerable potential for the generation of power. According to research carried out by Hokkaido University (Yoshida and Ikeda, 2008), the area under cultivation required to supply enough food for 5.59 million people in Hokkaido is 280,000 hectares, and since 24 per cent of the present area will be sufficient to supply enough food, Hokkaido would be able to meet its energy needs from renewable energy sources alone if the remaining productive area is used for creating bioenergy. Hokkaido could offer Japan a nationwide model for renewable energy in the development of a low-carbon society.

With regard to Japan's Feed in Tariff policies, intended to expand biogas-generated power, the 2010 target of power purchase from new energies in accordance with the Special Measures Law Concerning the Use of New Energy by Electric Utilities – better known as the Japanese Renewable Portfolio Standard (RPS) Law – is only 1.35 per cent (12.2 billion kWh), which is extremely low; the targets for the European Union (EU) and Germany are, respectively, 22.0 per cent (if large-scale water power is excluded, the target ratio is 12.5 per cent) and 12.5 per cent (if large-scale water power is excluded, the target ratio is 10.3 per cent). As these numbers indicate, the EU countries are preparing to designate greater proportions of their budgets to purchase obligations, and because, on top of that, supplies are not adequate, although there will be more demand for new electrical power, a system of high-priced purchase is still required. If the Japanese target is upgraded to at least 5 per cent, the purchase price will increase rapidly. Electrical power companies in Japan are the main sources of most carbon emissions, and, in addition, they hold the monopoly in each region. The expansion of renewable energy purchase is therefore necessary as a global and social responsibility. In passing, it should be noted that the use of biomass energy achieves local production for local consumption, stimulates the local economy and promotes employment. Japan's neighbour, China, which legislated laws regarding renewable energy based on German legislation, has made it an obligation to purchase renewable energy using the national grid in accordance with a "Feed in Tariff", and, in turn, it has increased the usage of renewable energy. China is aiming at a target ratio of 10 per cent renewable energy in primary energy consumption by 2010.

7-8-5 Citizen windmills

The green electricity rates system organized by the Hokkaido Green Fund is noteworthy as a source of citizen-funded renewable energy power plants. Under this system, citizens pay a 5 per cent surcharge on their monthly electricity bill, which is invested in a "Green Fund" to construct citizen-funded power plants. This plan is similar to the current payments system, which usually requires payment by direct bank account withdrawal. The Hokkaido Green Fund deducts electricity charges and the Green Fund from an account held by each member of the fund and, on behalf of the members, it pays the bill to the Hokkaido Electric Power Company. The Hokkaido Green Fund recommends that 5 per cent of electricity should be "saved" rather than added as an extra 5 per cent. In September 2001, the first wind turbine, called "Hamakaze-chan" (sea wind), was built in north-eastern Hokkaido; a second wind turbine has since been constructed. Moreover, the Hokkaido Green Fund helped to construct the first citizen-funded wind turbine in Aomori Prefecture, across the Tsuguru Straits from Hokkaido. In 2005, a citizen-funded wind power plant was built in Ishikari Town. For some years, 11 wind power plants have been in operation and, by 2005, 1,960 tons of CO_2 emissions had been saved. So far, the total invested has been approximately JPY 2.3 billion, and the total number of investors has grown to around 4,000 local citizens.

The challenge for community wind turbines is how best to increase the local money flow. As a result of the current financial crisis, questions are now being asked about the structure of the financial regulations. It would be possible to direct financial organizations to loan money to projects that enhance social welfare. Policymakers could legislate to establish regulations. At least 5 per cent of the loan balance could be transferred to low-cost housing and another 5 per cent invested in environment-related projects, and environment-related investments are intended to convert existing houses into energy-saving homes or to utilize renewable sources of energy. If bank loans for low-cost housing or environment-related investments do not meet the 5 per cent ratio, a request would be made to save the corresponding sum as cash. This proposal was made by Professor Robert Pollin of the University of Massachusetts, United States, who worked out the concept and drafted the "Green New Deal" for President Obama (Pollin, 2003).

7-8-6 Environmental reclamation and the generation of employment

Although, in terms of their business expenses, the cost of land acquisition for public forestry projects that support forestry organizations is not a big proportion, the share of personnel costs is rather high and the number of directly employed people in relation to business costs is large. Consequently, the forestry organization is a fairly effective base for increasing employment and revitalizing the economy. To achieve the goals established by the Kyoto Protocol, the Japanese Forestry Agency has been working to manage 550,000 hectares of forest, which is twice the total area of Tokyo, with the aim of absorbing 13 million tons of carbon. However, there are challenges in terms of both quantity and quality (Kobayashi, 2008). Although the Forestry Agency had set a budget of some JPY 10 billion, the area of thinning in 2007 was 30–40 per cent less than the goal of 550,000 thousand hectares that had been proposed.

The primary issues are the economic feasibility of thinning wood and the high cost of doing so. Since the business of selling thinned wood is generally close to break-even, financial assistance of about 50 per cent is provided by the government in the form of a subsidy, and about 20 per cent comes from local prefectures. Nevertheless, local economies are under financial pressure. Another problem concerns employment. There are about 50,000 forestry workers in Japan, but their numbers tend to be decreasing, and they are ageing at an unprecedented, rapid pace – a quarter of them are over 65 years old. Although volunteer workers or low-skilled workers can carry out relatively simple tasks such as improvement cutting or the thinning of young trees, appropriate pre-commercial thinning or thin logging is difficult for them. Hence the urgent necessity of securing expert workers and offering short-term skill training. Some logging courses have been provided for construction workers in the form of skill retraining, and the experiment of the Shinshu Lumberjack Training Course, held during the term of the former Nagano governor, Yasuo Tanaka, may serve as a useful reference. One further issue is that forest boundaries are not always clear.

One biomass town project that is worthy of mention as a regionally specific effort is sponsored by Shimokawa Town, located in northern Hokkaido. The town has a population of 3,800 and 4,470 hectares of forest and, for the past 60 years, it has afforested 50 hectares of land annually. Hopes for sustainable forest management are high. While the managers circulate resources by continuously managing the forests, they are able to secure opportunities for employment and supplies of forest products, and, with log-processing producing zero CO_2 emissions, biomass energy is utilized in the region for such heating appliances as the pellet stove or the biomass wood boiler.

7-8-7 The transition from an "export-oriented" economy to expansion of domestic demand through cooperation in Asia

Japan, the Republic of Korea, China and other East Asian countries have, for some time, been "the world's factory", a base that exports goods to Western countries. Japan is still highly dependent on the export of digital products and automobiles, and it has permitted manufacturing industries to use temporary workers in order to enhance the country's export competitiveness. Yet this strategy has simply made Japan even more reliant on the United States, with the result that Japan has been deeply affected by the recent economic downturn.

When, just after the Lehman Brothers bankruptcy, the present author surveyed consumer electronics retailers in the United States, he found that there was competition among them to sell Japanese digital products cheaply and at prices even lower than those in Japan itself. This represents an attempt by an export-oriented economy to realize a profit by selling products, which were manufactured by cutting labour costs, cheaply overseas and expensively at home, but this arrangement is constantly exposed to the fluctuating changes in overseas consumer trends and currency exchange rates. Moreover, China and Japan hold large amounts of US government debt with the dollars accumulated from exports, which means they will be heavily affected by the US financial crisis. China, Japan and Korea must therefore expand domestic demand, improve their workers' "quality of life" and reduce their working hours. Here again, a Green New Deal policy would serve effectively to link public investment with environmental and social measures.

The countries of East Asia share an environmental destiny and, in order to preserve the environment of this region, Japan, China and Korea must cooperate. Since air pollution or coastal marine pollution caused by industrial activity in Japan and Korea are further exacerbated by pollution originating in mainland China, joint activities are urgently required to reduce air pollution, CO_2 emissions, water contamination and ocean pollution. In fact, circular economies, eco-town planning and pollution prevention have become obligatory. For example, Kitakyushu City and several cities in north-eastern China are planning to collaborate in such activities. Furthermore, it is necessary to take joint action for the reform of export-oriented economies and demand expansion.

Certainly, if no action is taken to reverse the current situation, worldwide CO_2 emissions produced by energy needs will increase by 57 per cent, and only if governments adopt a different policy scenario will CO_2 emissions level off. The participation and cooperation of China and India in this regard are therefore vital.

7-8-8 Conclusion: Two paths towards a Green New Deal

In the near future, every country will need to propose economic and fiscal policies categorized in terms of the Green New Deal, and it will be necessary, first of all and at a nationwide level, to discuss, investigate and review policies and practices that will create employment, protect the environment, stimulate the economy and lay the foundations for sustainability in the three fields of the nation's environmental, economic and social life.

From both mid- and long-term viewpoints, it is necessary to study concrete plans and measures directly connected with the mid-term target set for CO_2 emissions reductions for 2020 and the long-term target set for 2050. From a short-term viewpoint at least, it is more important to ensure employment for forest thinning, say, or other ways of supporting regional endogenous development or concrete policies and assistance for the revitalization of local industries than to secure market access for large companies.

If global financial and environmental crises caused by mechanisms associated with global capitalism remain unchanged and if actions are taken only for short-term results, the effects will be limited and similar situations will continue to recur. It is necessary to devise policies that will both eliminate the causes of these crises as well as encourage desirable social and environmental results. Measures are also needed that will protect vulnerable people who might suffer serious harm from environmental and financial damage and that will improve their situation.

It is high time to remind ourselves that what in English is referred to as "the economy" is, in Chinese characters, 経済, rendered as *jingji*, and that the term 経済 is derived originally from the phrase 経国済民 (*jingguo-jimin*), in Chinese characters, which represents the concept "governing the nation and improving the well-being of the people".

REFERENCES

Kobayashi, Noriyuki (2008) *Global Warming and Forests: The Benefits of Earth Protection*. Japan Foundation Information Center.

New Economics Foundation (2008) *A Green New Deal: Joined-up Policies to Solve the Triple Crunch of the Credit Crisis, Climate Change and High Oil Prices*. The first report of the Green New Deal Group. Available at: <http://www.neweconomics.org/publications/green-new-deal> (accessed 31 March 2010).

NIES [National Institute for Environmental Studies], Kyoto University, and Mizuho Information and Research Institute (2009) "Japan Roadmaps towards Low-Carbon Societies (LCSs)", Japan Low Carbon Society 2050 study, <http://2050.nies.go.jp/material/20090814_japanroadmap_e.pdf> (accessed 14 April 2010).

Pollin, Robert (2003) *Contours of Descent: U.S. Economic Fractures and the Landscape of Global Austerity*. London: Verso. Japanese translation by Yoshikazu Sato and Kenichi Haga, 2008, published by Nihon Keizai Hyouronsha Ltd.

UN News Centre (2008) "Secretary-General Calls for 'Green New Deal' at UN Climate Change Talks", 11 December, <http://www.un.org/apps/news/story.asp?NewsID=29264> (accessed 5 April 2010).

UNEP [United Nations Environment Programme] (2008) "'Global Green New Deal' – Environmentally-Focused Investment Historic Opportunity for 21st Century Prosperity and Job Generation", press release, 22 October. Available at: <http://www.unep.org/Documents.Multilingual/Default.Print.asp?documentid=548&articleid=5957&l=en> (accessed 31 March 2010).

Yoshida, Fumikazu, and Motoyoshi Ikeda (2008) *Sustainable Low-Carbon Society*. Sapporo: Hokkaido Publishing Company.

7-9
Towards a sustainable world through positive feedback loops between critical issues

Motoyoshi Ikeda

7-9-1 Recognizing the current status

Specialists at Hokkaido University have clearly stated that global warming may possibly increase through drastic change caused by positive feedback between ecosystems and climate systems through geochemical components. These drastic changes will have an impact on biological resources both on land and in the ocean. It is therefore necessary to keep a close watch on sensitive areas such as ice sheets as well as warning signals from particular indices, including sea ice cover in the Arctic. Information about these changes and their impact needs to be given appropriate publicity so that global citizens can make informed decisions.

In conjunction with global warming, other critical issues will become more serious through positive feedback loops. Once the loops of global warming, a deteriorating ecosystem, food and water shortages, physical and mental disease and poverty are activated, they cause more damage to the poor and the developing countries. Increased competition associated with economic globalization leads to the widening of the gap between rich and poor. Efforts at resolving one issue often have an opposite and adverse impact on other issues. Human beings cannot expect to have a rosy future without resolving these urgent issues. Specialists in various fields are thus requested to deepen their understanding of global change and its consequences, understand the sophisticated feedback loops that exist among them and ultimately find a possible route to a sustainable world.

Designing our future: Local perspectives on bioproduction, ecosystems and humanity,
Osaki, Braimoh and Nakagami (eds),
United Nations University Press, 2011, ISBN 978-92-808-1183-4

7-9-2 Interactions between six issues plus one primary cause

Human beings ought to live on the Earth in harmony with the whole ecosystem. Instead, they have an excessively negative influence on it. Since the industrial revolution, humans have brought about large-scale scientific and technological development, and have used huge amounts of energy and raw materials that cannot be replaced. Despite efforts to solve individual problems that have emerged, even more serious issues now confront us. This difficulty so far has been considered a natural consequence of having an enjoyable lifestyle.

Here, six critical issues are presented, along with their primary cause, in Figure 7.9.1. Then the interactions between them are described, proving that these issues need to be tackled systematically so that efforts may be directed towards correct and effective targets. The six issues are energy, global warming, biodiversity, food production, water resources and health. Their primary cause exists in the societal system, as a combination of industrial development and economic globalization.

People in developing countries usually suffer more from the damage caused by environmental change. In developed countries, the economically weak face serious harm. In addition, everyone potentially faces physical and mental harm from various environmental and resource problems, along with serious competition that grows with the ongoing economical globalization.

The critical issues interact with each other and worsen the others, as in the obvious example of deforestation directly enhancing global warming (e.g. Oechel et al., 2000; Rustad et al., 2001). One example of the interactions is depicted in Figure 7.9.1. An inefficient use of energy accelerates global warming, which leads to climate change and endangers biodiversity and water supply (IPCC, 2007a). Water shortages then

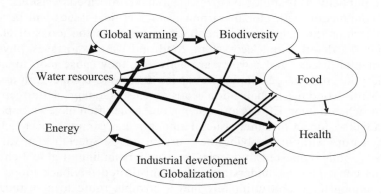

Figure 7.9.1 Interactions between critical issues.

adversely affect food production as well as the health of citizens living in areas with a pre-existing lack of clean water. Once these outcomes damage the social system, improvements in energy efficiency cannot be achieved (IPCC, 2007b).

The aim is to clarify these feedback loops and to achieve an adaptive world. The integrated system will be clarified by accumulating information and research through on-site data collection and an international network. A synthesizing method involves simulation of hindcasting and forecasting the natural social system of the world. The international network will function as a bridge between the approach implemented in Japan and that in other developed and developing countries.

A more complicated interaction occurs when efforts at resolving one issue have an opposite and adverse impact on the others (see Figure 7.9.2). A few immediate examples are shown in Figure 7.9.3. The first example is of an attempt at planting very efficient, rapidly growing trees to fix atmospheric carbon dioxide (CO_2). If this attempt is applied in an area with a high level of biodiversity, human efforts at reducing carbon emissions lead to a decline in biodiversity. Thus, the clean development mechanism of the Kyoto Protocol should also be evaluated from the viewpoint of biodiversity.

The second example is prevalent around the Aral Sea. A river system is utilized for agricultural water supply to facilitate secure crop production. If the area is dry under natural conditions, the additional water supply increases evaporation and leaves the ground salty. This change in turn reduces agricultural capacity. In addition, dust storms from the salty ground damage respiratory systems.

The third example has received a lot of attention in recent years. The high cost of petroleum naturally leads to the use of biofuel, which also helps to reduce CO_2 emissions, because the crops originally absorbed CO_2 from the atmosphere. As long as efficient methods are introduced to produce biofuel from material other than grains, the adverse impact will be relatively small. However, more investors are motivated towards felling trees in primeval forests, which could result in an increase in CO_2 emissions.

Human beings cannot expect a secure and promising future without resolving these critical issues. Numerous researchers have been working on individual issues by collecting data from various fields and analysing them on the basis of their respective specialties in the physical, chemical and biological sciences as well as social studies and the humanities. They should now realize the importance and necessity of close collaboration to overcome these issues. Since the issues are interrelated in a sophisticated manner, a collaborative initiative should be started as soon as possible.

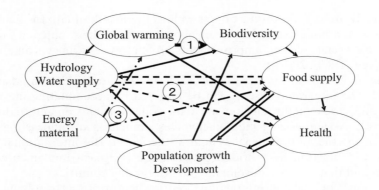

Figure 7.9.2 Throughout history, human beings have tried to solve one problem but often made others worse.
Note: The more complicated feedback loops are shown with three examples: (1) thick solid line; (2) dashed line; and (3) dot and dash line.

Figure 7.9.3 Examples of the short-sightedness with which we have tried to solve one problem but caused other problems.
Note: Refer to Figure 7.9.2 for the feedback loops that connect the critical issues.

7-9-3 The expected world

There is an urgent need to search for an optimal solution to these critical issues. This is the main agenda in the wake of the Kyoto Protocol. The representative method has been used to simulate the future under various scenarios in which the world proceeds through global warming and the other problems (e.g. Nakicenovic and Swart, 2000). This approach is useful for finding hypothetical routes towards a sustainable world. Searching for the different way suggested in this section is expected to offer the most feasible direction and effective actions to be taken by present-day humans to achieve a sustainable and pleasant Earth, as well as to save human lives. It is necessary to think about the grandchildren who will inherit the Earth in 2070.

Discussion is now extended to how global change can be mitigated. A typical example is the Green New Deal, which is intended to provide both a clean environment and economic success. A crucial aspect is usefulness from not only a short-term viewpoint but also a long-term one. Once one aims for a 50 per cent reduction in CO_2 emissions by 2050, for example, it is clear that the developed countries will have to take the initiative towards a low-carbon society by transferring technology to developing countries (Lefèvre, 2005). In order to achieve this goal, current lifestyles will definitely need to change.

In terms of time, the Earth is now almost at a mid-point between World War II and the year 2070. The developed countries, including Japan, rapidly increased their industrial production until 1990, followed by a majority of the developing countries. It is impossible for the developed countries to continue deriving trade benefits from the import of low-cost materials and the export of high-value products. Thus, the business model should also be revised. In summary, the measures taken to achieve sustainability at this stage should be evaluated from the viewpoint of how effectively they enable a shift from the current way of life to a future one.

Acknowledgement

The author acknowledges fruitful discussions with numerous specialists in the fields of climate change, water resources, food production, public health, energy systems and the social sciences.

REFERENCES

IPCC [Intergovernmental Panel on Climate Change] (2007a) *Climate Change 2007: Impacts, Adaptation and Vulnerability. Working Group II Contribution to the Fourth Assessment Report of the IPCC.* Cambridge: Cambridge University Press.

IPCC (2007b) *Climate Change 2007: Mitigation of Climate Change. Contribution of Working Group III to the Fourth Assessment Report of the Intergovernmental Panel on Climate Change, 2007.* Cambridge: Cambridge University Press.

Lefèvre, N. (2005) "Deploying Climate-friendly Technologies through Collaboration with Developing Countries", IEA Information Paper, International Energy Agency, Paris.

Nakicenovic, N. and R. Swart (2000) *Special Report on Emissions Scenarios.* Cambridge: Cambridge University Press.

Oechel, W. C., G. L. Vourtilis, S. J. Hastings, R. C. Zulueta, L. Hinzman, et al. (2000) "Acclimation of Ecosystem CO_2 Exchange in the Alaskan Arctic in Response to Decadal Climate Warming", *Nature* 406: 978–981.

Rustad, L. E., J. L. Campbell, G. M. Marion, R. J. Norby, J. Mitchell, et al. (2001) "A Meta-analysis of the Response of Soil Respiration, Net Nitrogen Mineralization, and Above Ground Plant Growth to Experimental Ecosystem Warming", *Oecologia* 126: 543–562.

Index

ABC. *See* allowable biological catch (ABC)
abiotic
   environment and diversity, 132, 329
   factors and jatropha cultivation, 210
   stresses, tree-borne oilseed crops tolerance of, 209
   stresses and energy crops, 93
acculturation, 256
Act on the Promotion of Producing Biofuels from Biomass of Agricultural, Forestry and Fisheries [Japan], 96
Ad Hoc Working Group on Further Commitments for Annex I Parties under the Kyoto Protocol (AWG-KP), 125
afforestation, 4, 115, 124, 214, 248
Agenda 21 document, 110, 184
agricultural
   chemicals, 4, 91–92
   communities, 388–89, 393–96, 405
   cooperative associations, 393–94
   land management, poor, 59
   mismanagement and soil degradation, 51
   production, 20, 101, 105, 165, 302, 390–91, 398, 405
   productivity, country-based, 102–3
agriculture
   of Japanese islands and East Asian corridor, 393–96

   natural habitats converted to, 38
   negative impacts of conventional, 39
   organic, 214
   sustainable, 39
   swidden, 52, 258–61
   urban organic, 215
Agriculture and Forestry Biomass Project of the National Long-Term Science and Technology Development Plan [China], 93
agro-ecosystems. *See also* ecosystem(s)
   ecological dynamics of, 38
   management, 41
   redesigning to maximize synergies, 39
   sustainable, 40
agroforestry, 106, 251–52, 390
agro-livestock cooperative system, 402
agro-livestock production, 402
Ainu people. *See also* sika deer
   deer hunting with guns, 370–71
   deer-hunting techniques, 365–66, 369
   ecological spirituality, 362
   hunter-gathers, 360
   "living in harmony with nature," 362
   perception of, 360–62
   poison arrows, 366–70
   sika deer and, 364–71
   spring bows, 365–67, 369
   traditional knowledge, 370

440 INDEX

Aioi-Nakajima district, 315–17
air pollution, 228, 430
Akita Prefecture [Japan], 96
albacore tuna *(Thunnus alalunga),* 137
albedo of earth, 8
Aleutian Low Pressure Index (ALPI), 131
allotment gardens, 215
allowable biological catch (ABC), 271–73
ALPI. *See* Aleutian Low Pressure Index (ALPI)
alternative
　energy resource, 228
　heat sources for homes, 425
　projects based on Choquet integral, 319
　river environment plan, 313, 324
　theory of sustainable fisheries yield, 228
amylase activity in saliva, 408
anthropogenic
　activities, 21, 51
　emissions, national inventory of, 114
　impacts, 50
　net primary production, 100
　trade in organic matter, 100
anti-cancer proteins, 408
Aomori Prefecture [Japan], 163, 428
Apau Kayan cultural ecosystem, 259–61
applied ecology-related journals, 335
aquaculture
　about, 20–21
　constraint mapping applied to, 139
　Japanese scallop, 140
　negative effects of, 132
　planning, and GIS spatial analysis, 139
　production, 130
　programmes, 130
arable land, 3–4, 93, 185, 188–89, 191
Aral Sea, 15, 435
artisanal fishers, 275, 277
ASEAN +3 Meeting of the Ministers of Agriculture and Forestry, 97
Association for the Promotion of Carbon Capture by Forest Biomass, 118
atomic power resources, 16. *See also* nuclear
Auroville Foundation Act [India], 214
Auroville (eco-village) [India], 214–15
avian flu, 388
AWG-KP. *See* Ad Hoc Working Group on Further Commitments for Annex I Parties under the Kyoto Protocol (AWG-KP)
axiology, 337

Bali Road Map, 124
Ban Ki-moon, Secretary-General of U.N., 425
bark charcoal, 78–80. *See also* charcoal
Basic Law on Dietary Education, 383–84
Basic Plan for Establishing a Resource-Circulating Society, 249
Basic Policy on the Promotion of Culture and the Arts [Japan], 353
"Becoming a Leading Environmental Nation Strategy in the 21st Century: Japan's Strategy for a Sustainable Society," 245
biochar
　as biomass-derived black carbon, 76
　experiments, 79
　in the future, 82–83
　long-term carbon sink in forest soil, 83
　as methods for sequestering carbon, 83
　micro-organisms and, 80–81
　organic farming and, 82
　reforestation and, 81–82
　as a soil conditioner, 81
　wastes carbonized as, 79
bio-control agents, 65
biodiesel
　in biomass town, 151, 153, 160
　from edible oil waste, sunflower and rapeseed oils, 96
　fuel, 90, 151, 153, 160
　fuel for Indian Railways, 209
　in India, 208–11, 215
　in Japanese test projects, 96
　jatropha as feedstock for, 209
　plantations on wasteland, 208
　sunflower and rapeseed for, 91
biodiversity. *See also* diversity
　carbon emissions and loss of, 435
　conservation of, 19, 276
　enhancement of, 59
　in the environment, 18
　loss of, 64, 330, 424
　in moso bamboo plantations, restoring, 66–72
　satoyama landscape, 246
　Second National Biodiversity Strategy [Japan], 244
　wild, 40
bioenergy
　facilities of Juehnde, 190
　forest based, 220–21, 232

organic waste, 17, 164–66, 171
tourism, 230
bioenergy village Juehnde [Germany]
  bioenergy projects in communities, guidelines for, 186–91
  bioenergy transfer to other villages, 192–93
  crop rotation, 187–88
  energy plants, 185, 187, 189–92
  social aspects and their realization, 191–92
  technical concept of, 185–86
bioethanol
  biomass energy fuel, 90–91
  in biomass towns, Japanese, 153
  from cellulose, 95–96
  in China, 92–93
  crops for, 91–92
  environmental load from, increasing, 91
  global warming, countermeasure against, 92
  in India, 94–95, 208, 211–12, 215
  in Japan, 95–97, 180–81
  liquid fuel, 90
  petroleum mix, 89
  production, increase in global, 91
  production, large-scale, 97
  production, leading countries in, 92
  production, raw material crops for, 92
  production and deforestation, 91–92
  production and greenhouse gases, 91
  production in Shinano Town, Nagano Prefecture, 96, 163
  production plants, 94
  production vs. food production, 91
  renewable energy, 92
  from sugar, starch or cellulose, 90
biofuel
  in Austria, 220, 222
  CO_2 emissions, reduction of, 435
  food vs., 91
  in India, 94, 207–9, 211
  in Japan, 95–97, 180
  from material other than grains, 435
  oil palm plantations and global, 261
  production in China, 93
  for transportation, 95
Biofuel Ethanol Plan [India], 94
Biofuel Steering Committee for Biofuel Development [India], 208

biogas (CH_4)
  as alternative fuel to oil, 174
  animal manure for, 189
  in Asia, 391
  in Austria, 222
  carbon neutral, 174
  CO_2 reduction and, 174, 178
  combined heat and power (CHP) generator, 185, 189, 192, 221
  desulphurized, 178
  generated power in Japan, 427
  in India, 94, 209
  organic waste, 164–65, 209, 216
  plant in Juehnde, 188
  plants, co-generation of, 179–80, 182
  plants, electricity sale from, 177–78
  plants, Green New Deal policies with, 181
  plants, wet-type and dry-type, 181
  plants in Hokkaido, 174–83
  plants on dairy farms, 175
  power generation and co-generation, 94
  production equipment, 391
  resources-recycling technology, 391
  sewage sludge or livestock manure, 90
biological
  allowable biological catch (ABC), 271–73
  cycles, 42
  diversity and ecosystems, 41
  diversity and forest management, 115
  diversity in soil, 43
  diversity in sustainable biosphere initiative, 333
  ecosystems in soil, 42
  energy flows, 21
  function of soil, 44–46
  interactions of wild and hatchery salmon, 133
  pest control, 40, 63–73
  process in biogas plants, 189
  production, sustainable, 389–90
  production functionality, 16
  resolution plane, space, 136
  resource recycling systems, 250
  resources on land in the ocean, 433
  sciences, 435
  species, 16
  system, relaxed, 409
  systems to support human needs, 50
  wastes produced in cities, 251
  water requirements, 30

biological control
  about, 63–64
  augmentative, 65–66, 68–72
  of cassava pests, 67–68
  classical, 64–65, 67–68
  conservation, 65–66
  in moso bamboo plantations, 66–72
  parasitoids, 63, 65
  polyphagous control agents, 67
  programmes, 73
  for two major cassava pests, 68
  using natural enemies, 66
biological diversity
  sustainability science, 333–34
biomass. *See also* Diagnostic Evaluation Model for Biomass Circulation (DEMBC)
  boiler, 227, 430
  CHP plants, 222, 228, 231
  from cropland and forest residues, 226
  electric capacity from, 230
  fired combined heat and power, 221
  forest, 226, 228, 230
  from forestry and sawmill residues, 221
  formed by plants, 9
  heat production, 222
  heat supply, 226, 228
  heating plants in Austria, district, 222–25
  lignocellulosic, 211–12
  manure, 187
  microbial, 44–45
  mitigation, 113
  power generation and cogeneration, 94
  power plant, municipal, 228, 230
  production, 16
  production areas, 16–17, 34
  production in Denmark, 194
  resources of *satoyama* landscapes, recycling, 249
  standing, 268–69
  utilization, 149, 151–52
  utilization in private households, 221
  utilization systems, 160–61
  of wild and hatchery populations of chum salmon in North Pacific, 131
  woody, 153, 157, 164, 187, 211, 222
biomass energy
  activities and political and socioeconomic impacts, 230
  carbon neutral, 90–91
  by gasification, 90
  by methane fermentation, 90
  to rural villages and towns, 221
  types and use of, 89–90
Biomass Nippon Strategy [Japan], 95–96
biomass town. *See also* Diagnostic Evaluation Model for Biomass Circulation (DEMBC)
  concepts in East Asia, 97
  economic efficiency, 157
  greenhouse gas (GHG), 156–57
  Japanese, 148, 152–53
  system, 95–96
biomes
  anthropogenic, 54
  data sets and soil resources, 52–54
  forest, 59
  groupings in Asia Pacific, 56
  ideal, pattern, 57–58
  by land class, 56–58
  spatial correspondence of, 54–59
  as units describing patterns of ecosystem forms and processes, 52
biophysical environment, 336
Bio-recycle Project in the National Institute for Rural Engineering, 149
Bio-Resources Centre [India], 215
bird's eye view approach, 255
black charcoal, 81. *See also* charcoal
blood pressure, 408, 411
blue whale, 271
bluefin tuna *(T. thynnus)*, 130
brainstorming (BS), 311, 313–14, 317
Brazil, Russia, India and China (BRIC), 6, 423
Brundtland Commission, 413
Brundtland Report, 310
BS. *See* brainstorming (BS)
BSE. *See* mad-cow disease (BSE)

calcium (Ca), 390
capitalism, global, 423–24, 431
carbon (C)
  biomass-derived black, 76
  compounds, 42
  content of soils and fungi growth, 81
  created by plants, microbes and other organisms, 42
  cycle and global warming, 92
  cycles, biogeochemical, 100
  cycles, forest contribution to global, 111
  deforestation of tropical forests, 113

dynamics in agro-ecosystems, 42, 44
efficiency, 152
emissions and loss of biodiversity, 435
emissions from electrical power plants, 427
emissions trading, 426
fixation as organic matter by photosynthesis, 99
footprint, 212, 238–39, 242, 384
greenhouse gas and, 149
monoxide, 90
net capture by land ecosystems, 113
offsetting credits system, J-VER, 109, 119–22, 125–26
organic, and human dependence on NPP, 100
organic, consumption as fuel for transportation, 100
organic, decomposition rate of, 101
plant and soil, 100, 102
pool, organic, 100
residence time of plant, 101–3
sequestration, 42, 46, 76, 83, 232
sink credits, J-VER, 119, 125, 127
soil organic, 101
stocks, farming practices to increase, 43
stocks in wood products, 113
terrestrial, 174
usage and biomass utilization, 155, 158
utilization, ratio of, 149
carbon dioxide (CO_2)
atmospheric, 90, 435
in Austria, 232
Bioenergy Village Project, 185–87
from biomass energy, 90
capture, economic and social value of, 118
capture and storage of, 108, 123
capture certification schemes for forest, 117, 119
Carbon Dioxide Capture and Storage, 112–13, 115
by China, 430
climate change and, 34
from combustion, 90
concentration, atmospheric, 22–23, 88
deforestation increases, 91
emissions, global, 184
emissions from humans, 23
forestation is essential for fixing, 83, 113
global warming and, 9, 22–23, 88
greenhouse gas, 8, 23
index of sustainability and, 425
neutral energy resources biomass, 186
nuclear power generation and, 23
reduction, 44, 113, 115, 118–19, 426
reduction and biofuels, 435
reduction and biogas, 174, 178
reduction certification system, 118–19
reduction targets, mid-term, 431
Samsø renewable energy project, 201
sea-level rise, 27
sink certification systems, 126
soil microbial flora, 80
world-wide, 431
carbon neutral
bioenergy, 231
biogas, 174
biomass, forest, 220, 232
biomass energy, 90–91, 230
biomass plant projects, 232
region, 230
carbon sinks
Action Plan on the Mitigation of Global Warming by Forest Carbon Sinks, 116
biochar application to forest soil, 83
CO_2 absorbed by, 110
CO_2 sink certification, 126
credits, 119, 127
crop rotation, 43
forest, and climate change, 112–19
forest, and *Marrakesh Accords*, 115–16
forest, countermeasures, 116–19
forest, economic and social added value of, 124–26
forest, evaluation of by IPCC, 112–13
forest, market mechanisms, 112–19
forest, outlook for post-Kyoto, 124–25
forest, post-Kyoto, 125, 127
forest, under the UNFCCC, 114–15
forests as, 108, 110–11, 114–17
global warming, as economic measure against, 108
initiatives after 2013, 127
IPCC's Special Report on *Carbon Dioxide Capture and Storage*, 112–13, 115
Japan Verified Emission Reduction (J-VER), 119–21, 126–27
Japanese domestic, 126–27
Japan's framework and Kyoto Protocol targets, 116–19

carbon sinks (cont.)
    Kochi Prefecture initiative, 118–19
    Kyoto mechanism general regulations, 110
    local government initiatives and, 117
    and reservoirs, 111
    Shimokawa-cho, Hokkaido initiative, 117–18
    for Tokyo area, 429
    tree thinning and, 116–17, 122, 126, 429
carbonization
    bacterial activity and acidity neutralizing, 81
    methane fermentation, 160–61
    of rice husk charcoal, 76
    technology, 79
    waste heat from, 152
    of wastes, 77, 79, 152–53, 160–61
    of wood of domestic broad-leaved trees, 79
    woody biomass, 153
carrying capacity
    climate change and, 131–32
    of earth, 11, 33–35, 131
    environmental, 393
    of human population on Apau Kayan lands, 260
    NEMURO model of, 133–37
    ocean ecosystem, 133
cassava, 67–68, 91, 93, 103
cation exchange capacity (CEC), 78
CBD. See Convention on Biological Diversity (CBD)
CDM. See clean development mechanism (CDM)
CEC. See cation exchange capacity (CEC)
cellulose
    about, 45, 90–91, 93, 95–96
    fermentation, 153
Central Pollution Control Board (CPCB) [India], 213
cereal(s)
    about, 99, 101–2, 104, 260
    cropland, 106
    crops, 91
    grain, 104
    production, 102–5
    productivity, 104–5
Certification Center on Climate Change, Japan (4CJ), 120
CH_4. See biogas (CH_4)

charcoal
    bark, 78–80
    black, 81
    compost, 77
    consumption in Japan, 79
    fertilizers, 80–81
    oak, 81
    pine, 81
    rice husk, 76–79, 82
    woody, 77–81
chemical fertilizers, 17, 19, 163, 178, 182, 390–91
Chiba Prefecture [Japan], 96
chisan-chishou (grow locally, consume locally), 384
chlorophyll, 138–39
Choquet integral, 319
CHP. See combined heat and power (CHP) generator
chub mackerel, 273
chum salmon *(O. keta)*, 131–32, 277–78
circulatory system diseases, 415
clean development mechanism (CDM), 108, 121, 210, 435
Climate Alliance (Klimabündnis), 228
climate change. *See also* global warming; Intergovernmental Panel on Climate Change (IPCC)
    bioenergy villages in Germany, 185
    carrying capacity of salmon and, 131–32
    Certification Center on Climate Change [Japan], 120
    CO_2 emissions, 34
    environmental challenges, global, 330
    financial crisis as opportunity to tackle, 425
    forest carbon sinks, 112–19
    global, 15
    Green Economy, transition to, 425
    "Green New Deal," 424
    inefficient use of energy accelerates, 434
    issues raised by, 426
    marine ecosystems, 140
    mitigating, 59, 220
    modern civilization, demands on, 59
    NEMURO.FISH, 134–36
    peak oil and, 235, 238–40, 242
    PICES CCCC model, 133
    policies to deal with, 426
    renewable energy to combat, 232

rising temperatures and extreme weather events, 42
sequestering carbon in standing forest biomass, 232
sustainability science and, 343
closed system, 8
Club of Rome, 33–34
coal-powered steam engine, 3
coastal marine pollution from China, 430
coastal water contamination, 33
Code of Conduct for the Import and Release of Exotic Biological Control Agents, 66–67
coexistence
 of all species, 276
 of cities and villages with natural environment, 30
 of human beings and the environment, 332
 of humans and nature, 4, 39, 389–90
 of natural ecosystems and human activities, 393
 with natural environment, 34, 392
 with nature, 2
 of nature and populations downstream, 32
 of oxic and anoxic conditions for nitrification and denitrification, 45
 of river channel ecosystems, 29
 of species, 267
co-generation
 of biogas plants, 179–80, 182
 system for electricity, 176
cognitive processing, 340
cognitive systems, 337
collaborative governance *(kyouchi)*, 254, 262–64
Collaborative Programme on Reducing Emissions from Deforestation and Degradation in Developing Countries (REDD), 108, 124–26
combined heat and power (CHP) generator, 185, 189, 192, 221
commercial feedstuff, 390–91
Commission for Certification Criteria for VER Using Carbon Offsetting [Japan], 120, 122
Commission for the Conservation of Southern Bluefin Tuna, 271
Committee on Global Warming, 426

Committee on Renovation in the City Centre (CRCC) [Furano], 299
common-pool resources (CPRs), 254–55, 263
community-driven development, 382
community-owned energy company, 190
compost
 of biodegradable wastes from households, 215
 biomass utilization system, 160–61
 charcoal, 77
 economic efficiency evaluation and, 151
 groundwater contamination and, 390
 livestock manure and kitchen waste, 152
 organic, 39, 81–82
 production process, 81
 residue as a fertilizer, 181
 rice husk, 77
 transformation technology, 153
 wastewater sludge, 32
conjoint analysis, 312–14, 321, 324
Conservation and Management Plan for Sika Deer in Eastern Hokkaido, 362
conservation area, 17–18, 34, 246
conservation farming, 39
contingent valuation method (CVM), 311, 314, 319–21, 324
Convention on Biological Diversity (CBD), 17, 125, 252, 270
Co-operative Associations Research Laboratory, Hokkaido University, 393
COP 7. *See* Seventh Conference of the Parties (COP 7)
COP 10. *See* Tenth Conference of the Parties to the Convention on Biological Diversity (COP 10)
COP 11. *See* Eleventh Conference of the Parties (COP 11)
COP 13. *See* Thirteenth Conference of the Parties (COP 13)
COP 15. *See* Fifteenth Conference of the Parties (COP 15)
COST. *See* European Cooperation in Science and Technology (COST)
Council on Sika Deer Management Policy, 362
cows and food sufficiency, 171
CPCB. *See* Central Pollution Control Board (CPCB) [India]
CPRs. *See* common-pool resources (CPRs)

CRCC. *See* Committee on Renovation in the City Centre (CRCC) [Furano]
crises
  biodiversity, 244
  economic and environmental, 423, 431
  energy, 241
  food and energy, 4
  food shortage, 188
  income gap, widening the, 424
  worldwide financial, 424
critical issues, interaction between, 434–35
crop microclimates manipulation, 65
crop production
  Aral Sea, around the, 435
  biochar application benefits, 76
  biomass for electricity and heat, 187
  chemical fertilizers and commercial feedstuff, 391
  cropland sustainability issues, 106
  forests provide land for, 106
  organic fertilizers and charcoal for, 77
  shifting cultivation and, 106
crop rotation
  bioenergy village Juehnde, 187–88
  carbon sinks, 43
croplands, 4, 42, 50, 54, 105–6
cropping pattern, 398–400
cultural
  adaptation of diverse people for sustainable future, 214
  amenability, 341–42
  ecosystems, 254, 257–58, 260
  sustainability of urban or rural areas, 214
currency exchange rates, 430
CVM. *See* contingent valuation method (CVM)

Dahuaxian Prefecture [China], 388–89
Danish Energy Authority, 194, 198–99
debt accumulation, 387
deep-well pumps, 19
deforestation
  bioethanol production and, 91–92
  in developing countries, 124
  food production and, 388
  global warming and, 108
  greenhouse gas (GHG) emissions, 124
  in Japan, 4
  Kyoto Protocol, 115

  soil degradation and, 51
  stopping, 108
  of tropical forests, 113
DEMBC. *See* Diagnostic Evaluation Model for Biomass Circulation (DEMBC)
democracy, deliberative, 264
Democratic Party of Japan, 423
denitrification (anoxic process), 45
Department of Industrial Promotion [Thailand], 381
desertification, 59, 93
desulphurization systems, 176
desulphurized biogas, 178
Diagnostic Evaluation Model for Biomass Circulation (DEMBC)
  about, 148–49
  biomass circulation, development of evaluation model, 153–54
  biomass in database, sources of, 151
  biomass towns by resource distribution, 159
  biomass utilization, 155–56, 160–61
  biomass utilization scenario, 158–59
  biomass utilized, quantity of, 155
  boundary conditions, 149, 151
  cash flow, basic, 150
  data collection, 152
  economic efficiency, 157
  economic efficiency evaluation, 151
  evaluation criteria and index, 152
  evaluation factors, 149
  fossil fuel energy, 155–56
  greenhouse gas emissions, 156–57
  Japanese biomass towns, 152–53
  management expenses of business and incentives, 155
  present conditions, evaluation standard for, 158
  results, 155–61
  scenario assessment, evaluation standard for, 158
  transformation technologies, 153
dietary education *(shokuiku),* 375, 383–85
digestate fertilizer, 189
discursive democracy, 264
diversity. *See also* biodiversity
  abiotic environment and, 132, 329
  biological, 41, 43, 115, 333
drinking water, 28, 30, 415, 418–21
Dublin Principles, 30–31
dust storms, 435

Earth (Earth's)
  albedo of, 8
  carrying capacity, 11, 33–34, 131
  energy flow through, 8
  fauna and flora's dependence on water, 24
  heat balance, 9
  human animal mass on, 18
  life support systems of, 51
  production areas of, 16
  protective areas of, 16
  raw materials and space, insufficient, 15
  solar energy input, 9
  surface temperature rise, 23
  urban-industrial areas of, 16, 34
  water, demand for, 24–27
  water resources per capita, 26–27
earthquake proof structures, 425
East Asian Corridor agriculture study, 393–96
ecocentric worldview, 338–40
eco-cities, 207, 212, 215
eco-city programme (ECP) [India], 213, 215
Eco-Industrial Estate projects [India], 213
ecological
  food chain, 21
  footprint, 4, 11, 34
  oriented cultivation system, 187
  responsibility, 341–42
  science, 333, 335, 342–44
  spirituality, 362
  systems, 328, 330, 333–35
Ecological Society of America (ESA), 333
Ecologically Noble Savage, 329, 343
ecology-rebuilding measures in karst landscape, 392
"Eco-Longevity Societies," 250
economic downturn, global, 207, 424, 430
economic globalization, 255, 387, 393, 433–34
economic growth
  in China, 92, 389
  domestic, in mature countries, 7
  East Asia's, 387–88
  in India, 92, 207
  Japan's rapid, 347–48, 350
  post-war recovery period, 347
  poverty reduction and, 388
  rapid, 14–15, 92
  rates of countries undergoing modernization, 7

  in rural Japan, 250
  sustainable policy, 285
  of twentieth century, 22
economic indicators, 311
economic indices, 387
economic sustainability, 285
ecosystem(s). See also agro-ecosystems
  based fisheries management, 133, 276
  based management, 133, 269
  conservation, 17–18, 28
    ocean, 131
  conservation area, 18
  deteriorating, 433
  extratropical, 99
  forest, 79, 261, 389, 391
  human activities and, 389, 393
  Integrated Marine Biogeochemistry and Ecosystem Research (IMBER), 140
  marine, 140, 142, 270, 275–76, 278
  maximum sustainable ecosystem services (MSES), 267–69, 278
  Millennium Ecosystem Assessment, 38, 330
  ocean, 130, 133–37
  productivity, country-level data on, 100–101
  river channel, 29
  services, 40–41, 59, 267–69, 278, 329–30
  Sustainable Green Ecosystem Council (SGEC), 110
  temperate, 8
ecosystem approach
  dynamic cultural, 257–59
  to fisheries, 140
  single species management, 270
  stock management methods based on, 138
  sustainable conservation based on, 130, 132–33
  sustainable fisheries of Pacific salmon based, 134
"Ecosystem Rebuilding and Comprehensive Development of Sustainable Biological Productivity in Southwestern China," 389
ecotourism, 202, 204. See also tourism
eco-villages [India], 207, 214–15
ECP. See eco-city programme (ECP) [India]
education system, Japan's, 13
egocentric worldview, 338–39
electric vehicles, 179

electrical power companies, 427
Eleventh Conference of the Parties (COP 11), 124
emissions trading, 108, 118–20, 125, 127, 426
empowerment of women, 213
energy
　"back-up system," 192, 205
　company, community-owned, 190
　conservation at public facilities, 179
　efficiency, 14, 138, 183, 186, 189, 241, 435
　farming in India, 209–11
　flow through Earth, 8
　saving electrical appliances, 425
　security, 90, 92, 209, 220, 232
energy consumption
　cities with high population density and, 30
　fossil, 155–58, 161
　fossil, reduction in, 149
　fossil fuels, based on, 9, 14
　in Furano city, 166, 295
　industrialization growth rate, 14
　municipal/industrial water, reuse and recirculation systems to secure, 28
　offshore wind turbines to compensate for, 200–201
　per capita, 14
　rate of commercial, 9
　renewable, 194–95, 197, 203, 428
　Samsø's, 204
　satoyama landscape, 249–50
　in twentieth century, 88
　water treatment processes and, 29
　world population growth rate and, 21
Energy Descent Plan, 241–42
energy plants
　bioenergy village Juehnde, 185, 187, 189–92
　in Denmark, 203
　in Furano City, 169, 171
energy production
　biomass crops for, 187
　community resources for, 187
　digestate as fertilizer, 189
　domestic livestock waste and, 174
　human population growth and, 50
　from manure and other organic residuals, 188
　raw materials for, high cost of, 180
　renewable, 82, 194, 199
　rice or wheat used for, 181
　sustainable, 185
　wood demand for, 225
environment (environmental). *See also* global environment
　conservation, 17, 25, 34, 286
　degradation, 39, 93
　destruction, 3
　deterioration, 387
　force, 256–58, 260
　lakes, 31–32
　loads, 30, 174, 311
　pollution, 64, 347
　structural capacity, 34
　sustainability in developing countries, 47
　taxes, 426
Environment and Development Summit in Johannesburg, 415, 417
"Environmental Conservation in the Asian Region" project, 389
ESA. *See* Ecological Society of America (ESA)
Escherichia coli, 421–22
An Essay on the Principle of Population (Malthus), 20
ethanol for gasoline, 211–12
ethical defensibility, 341–42
European Cooperation in Science and Technology (COST), 410
European Union (EU) countries, 222
Exclusive Economic Zone [Japan], 278
excrement
　biogas from livestock, 164–65
　biomass energy from livestock, 171
　cow, for fertilizer, 165
　human, in cities, 17
　human, returned to nature, 30
　human and animal, 390
　waste, animal, 390
exotic species, 211, 244
export-oriented
　economies, 430
　growth, 387
　vegetable-producing district, 394
Ezo wolf *(Canis lupus hattai),* 363, 369

FAC. *See* Furano Agricultural Cooperative (FAC)
Fallen Angel, 329, 343

Farmers Field School (FFS), 73
farming
  cover crop, 43
  crop rotation, 43
  management, family-run, 395
  manure application, 43
  no-till, 43, 45
  practices to increase carbon stocks, 43
  sustainable, 39
  sustainable, low-input, 40–41
  sustainable vs. modern, 41
farmland areas, 18, 296
farmlands, 389–90, 394, 397, 399–401
fast-breeder nuclear reactors, 24
FCAs. *See* fisheries cooperative associations (FCAs)
FCCI. *See* Furano Chamber of Commerce and Industry (FCCI)
FCEC. *See* Furano Citizens' Environment Council (FCEC)
FCPF. *See* Forest Carbon Partnership Facility (FCPF)
Feed-in Tariff [Japan], 175, 179, 182, 427
fermentation unit, biogas plant, 188
Fertile Crescent in Mesopotamia, 27
fertilizers
  charcoal, 80–81
  chemical, 17, 19, 163, 178, 182, 390–91
  consumption, 51
  cow excrement, 165
  digestate, 189
  as farm energy inputs, 4
FESLM. *See* framework for evaluating sustainable land management (FESLM)
FFS. *See* Farmers Field School (FFS)
Fifteenth Conference of the Parties (COP 15), 125
Fifth Comprehensive National Development Plan [Japan], 350
Finnish Forest Research Institute, 410
fish
  catches, 20–21, 130
  stock dynamics, 268
  yields, 20–21, 277
fisheries cooperative associations (FCAs), 275
fisheries management
  ecosystem-based, 133, 276
  limitations of, 132
  Marine Management Plan for the Shiretoko World Natural Heritage Site, 141–42, 276–77
  modern, 278
  MSY theory, classical, 270
  multi-species, target switching in, 277
  Shiretoko World Natural Heritage site, 131, 141–43
  transaction cost of, 277
fishing-ban, seasonal, 275, 277
flood
  control, 315–18, 320–21, 324
  damage, 27–28, 316–18
food
  imports, mass, 17
  pesticide-laden, 388
  products, safety of, 388
  sanitation regulations, Japanese, 383
  security, 91, 93, 271, 383, 397
  self-sufficiency and biogas plants, 181
  shortages, 20, 180–81, 387, 433
  stocks, worldwide, 20
  supply system, national, 402, 405
  web of Shiretoko World Natural Heritage area, 141–42
Food and Agriculture Organization (FAO), 101
  framework for evaluating sustainable land management (FESLM), 59
food self-sufficiency
  rate, 167, 171
  ratio in Japan, 20, 163
  with rice for livestock feed, 250
foreign direct investment, 387
Forest Agency [Japan], 116, 429
Forest and Forestry Basic Act, 111
Forest Carbon Partnership Facility (FCPF), 125–26
forest carbon sinks
  about, 108, 110–11, 114–17
  Action Plan on the Mitigation of Global Warming by Forest Carbon Sinks, 116
  climate change, 112–19
  economic and social added value of, 124–26
  evaluation of by IPCC, 112–13
  marketing mechanisms, 112–19
  Marrakesh Accords, 115–16
  outlook for post-Kyoto, 124–25
  post-Kyoto, 125, 127
  under UNFCCC, 114–15

450   INDEX

Forest Certification (PEFC), 110
Forest Certification System [SEGC],
  110–12, 117, 125–26, 232
forest CO_2 capture certification schemes,
  117
forest management
  activities, formula for total capture by
    domestic, 123
  carbon offsetting, 122
  defined, 115
  historical perspective, 109
  ISO 14001 Environmental Management
    System, 110
  J-VER system, 119–21, 126–27
  in Kochi Prefecture, 118–19
  Kyoto Protocol, Article 3, 115
  Marrakesh Accords, 116
  permanency of, 122–23
  profitability of, for Japan's forestry
    industry, 126
  projects in Japan, 122
  role of forests in tackling global warming,
    126
  in Shimokawa-cho, Hokkaido, 117–18
  Statement of Forest Principles of
    UNCED, 110–11, 123
  sustainable, 108, 110–11, 113–15, 117,
    122–23, 125–27
Forest Products Research Institute (Japan),
  411
Forest Stewardship Certification, 220
Forest Stewardship Council (FSC), 110
forest(s)
  based feedstocks, 225
  bathing *(shinrin-yoku)*, 410–11
  bioenergy, 220–21, 232
  biomass, 221, 226, 228, 230
  biomass and carbon neutral, 220, 232
  biomes, 59
  CO_2 sinks in Kochi Prefecture, 118–19
  CO_2 sinks in Shimokawa-cho, 117–18
  conservation, 60, 261
  ecosystems, 79, 261, 389, 391
  environment tax [Japan], 118
  evaluation/rebuilding, 389, 391
  fire, vulnerability to, 4
  global warming prevention, role in, 113,
    123
  rainfall capture and moderate runoff, 4
Forests, Trees and Human Health and
  Wellbeing project, 410

ForHealth. *See* Task Force on Forests and
  Human Health (ForHealth)
fossil and nuclear energy, 7, 23
fossil energy
  alternatives energies, shift to, 97
  biogas power plant replaces, 189, 192
  in biomass towns, 155–58, 161
  centralized energy system replacement
    for, 23
  CO_2 in atmosphere, 88
  consumed, 165
  demand for, by China and India, 92
  growing concerns about, 215
  mass consumption of, 21
  mineral fertilizer requires high input of,
    189
  modern agriculture fueled by, 3
  modern rice production depends on, 163
  reducing and substituting, 140, 207
  renewable energy replacement, 188
  Western-type society and, 23
  world depends on, 88
fossil fuel free movement, 234
fossil fuels
  availability of, 22
  cheap, 3, 234
  chemical fertilizers and insecticides made
    from, 19, 163
  CO_2 and global warming, 88–89
  CO_2 reduction by switching from, 118
  dependence on, 22, 228
  dependence on, reducing, 88
  electricity derived from, 189
  energy consumption based on, 9
  energy security and low consumption of,
    209
  finite resources of, 6
  imported, increase in, 79
  oil reserves, 22
  renewable energy *vs.*, 89
  resources, 3, 16
  substitute for, 210
4CJ. *See* Certification Center on Climate
  Change, Japan (4CJ)
framework for evaluating sustainable land
  management (FESLM), 59–60
freshwater, 15, 24–25, 28
FSC. *See* Forest Stewardship Council (FSC)
FSM. *See* fuzzy structural modelling (FSM)
FTA. *See* Furano Tourism Association
  (FTA)

fuel cells, 97
fuelwood, 100, 221–22, 225, 228, 232
Fukuoka city
  agriculture and dairy farming in, 164–65
  model, limitations of, 171
  potential energy and material flow in, 164
  thermal energy, mathematical analyses of, 165–69
Fukuoka Prefecture, 96
Fundamental Law for the Promotion of Culture and the Arts [Japan], 353
Furano. *See also* Hokkaido
Furano Agricultural Committee, 290, 293–94, 296
Furano Agricultural Cooperative (FAC), 295–96, 299
Furano Chamber of Commerce and Industry (FCCI), 298–99, 305
Furano Citizens' Environment Council (FCEC), 300
Furano City
  ageing society, 295, 301–2
  agriculture, 295–97
  case study interviews, 290–91
  case study methodology and background, 289–95
  case study problem solving figures, 292–94
  commerce, 298–300
  Committee on Renovation in the City Centre (CRACK), 299
  communications network, 286
  employment opportunities for young, 301
  energy consumption, 166, 295
  energy plants, 169, 171
  expectations, mutual, 304–6
  issues, 302–4
  neighbourhood associations, 290, 300–301
  social welfare, 300–302, 305
  sustainability governance, 288–89, 295, 306
  tourism, 297–98
  waste disposal and recycling, 300, 303, 305–6
Furano Hotel Association, 297–98
Furano Tourism Association (FTA), 297–99, 305
Future Forests [United Kingdom], 120
fuzzy integral method: multi-criteria analysis, 313–14, 324
fuzzy structural modelling (FSM), 311, 314

G7, 11, 14
genetic diversity, natural, 38
Geographical Information Systems (GIS), 52, 131, 137–39
geothermal energy/power, 8, 89
German Renewable Energy Act, 190–91
German Technical Cooperation, 213
GHG. *See* greenhouse gas (GHG)
GIS. *See* Geographical Information Systems (GIS)
glacier thaw, continental, 27
global ecosystem and economic subsystems, 7–9
global environment (environmental). *See also* environment
  challenges, 330
  constraints in the twenty-first century, 33
  destruction of, 424
  deterioration in, 4
  food production and soil fertility, 42
  forest reductions and, 109–10
  greenhouse gas emissions impact on, 3
  issues, 388
  issues for future generations, 309
  problems, 254–55, 360
  tropical rainforests and, 109–10
global population. *See* world population
global warming. *See also* climate change
  about, 22–24
  bioethanol as countermeasure against, 92
  carbon cycle and, 92
  carbon offsetting strategy to prevent, 119–20
  carbon sinks, 108, 116
  on chum salmon in the North Pacific Ocean, 132
  CO_2 and, 9, 88–89, 114
  Committee on Global Warming, 426
  conventional tillage and, 44
  countermeasures, 124
  deforestation, 108
  energy, inefficient use of, 434
  flooding and, 28
  forest management and, 126
  Framework Convention on Climate Change (UNFCCC), 114–15
  greenhouse gas (GHG) emissions, 23
  increase, 433
  Japan's countermeasures, 116
  marine animals, affect on, 131

global warming (cont.)
  positive feedback between ecosystems, climate systems and geochemical components, 433–36
  poverty, 433
  prevention, role of forests in, 113, 123
  prevention and sustainable forest management, 113
  sea-level rises, 27
  sustainable farming practices and, 47
  10-year Action Plan on the Mitigation of Global Warming by Forest Carbon Sinks [Japan], 116
  water, thermal expansion of, 27
Global Warming Countermeasures Law. See Law Concerning the Promotion of the Measures to Cope with Global Warming (Global Warming Countermeasures Law) [Japan]
global water consumption, 25
globalization, 6–7, 433–34
glocalization, 258, 262–63
goat overgrazing, 405
goat-rearing in Mongolia, 404–5
Government of Hokkaido, 362
Government of India, 208
Government of Indonesia, 262
Government of Japan, 142, 274–76, 278, 375–76
Government of Oita, 376
grain production
  bioenergy village Juehnde, 187
  world, 19–20, 26–27
Grants-in-Aid for Scientific Research, 398
Green Economy Initiative [U.N.], 425
green electricity rates system, 427–28
green jobs, 424–25
Green New Deal policy, 179, 181–82, 424–28, 430–31
Green Revolution, 17, 19, 27, 105
greenhouse gas (GHG) emissions
  global volume of, 124
greenhouse gas (GHG) emissions
  bioethanol production, 91
  of biomass town, 156–57
  capture and storage by ecosystems, 114
  deforestation and forest degradation, 124
  energy reflected from, 8
  forests, emissions from, 108
  fossil energy consumption, 149
  global environment, impact on, 3

  global warming and, 23
  precipitation distribution, impact on the, 3
  reduction and capture projects, 122
  reduction of, 115–16, 120, 124–26, 149
greenhouse materials, 8
gross domestic product (GDP)
  annual growth rate, 12–13
  of Bangladesh, 397
  as basis of current industrial structure, 9
  in China, 13
  of developed countries, 11–12
  of developing countries, 11–13
  human population, per capita rate of increase in, 101
  India, 207
  of India, 397
  industrialization and increase in, 14
  of Japan, 13, 397
  of major South Asian nations, 397
  in mature regions, 12–13
  of Nepal, 397
  of Pakistan, 397
  personal income and, 11–12
  of South Korea, 397
  in Switzerland, 13
groundwater
  basins, 31
  contamination, 390
  deep-well pumps to pump, 19
  depletion of, 15
  irrigation farming and, 27
  pollution, 188
  quality, 4
  from wells, 31
grow locally, consume locally (chisan-chishou), 384
Guangxi Zhuang Autonomous Region [China], 388
"Guidelines for Carbon Offsetting in Japan," 120
Guizhou [China], 388

habitat loss, 132
HANPP. See human appropriation of NPP (HANPP)
hatchery, salmon, 131–33
HDI. See Human Development Index (HDI)
health and nutrition education in schools, 384

health risks
  infectious diseases in Mekong watershed, 418–22
  regional sustainability and, 413–17
  transboundary movement of, 417–18
  from waterborne pathogens, 422
heart rate, 408
heat pump power, 222
heterotrophs, 99–100
High Pressure Gas Safety Act [Japan], 178
Hokkaido. *See also* Ainu people; Furano; sika deer
  agricultural cooperative associations, local, 393–94
  Aioi-Nakajima district, 315–17
  area under cultivation for food, 427
  bioethanol production, 180
  biogas plant construction, 174–83
  Conservation and Management Plan for Sika Deer, 362
  Council on Sika Deer Management Policy, 362
  dairy farm, 177
  Deer Hunting Regulations, 365, 367–69
  Development Bureau, 182
  Development Commission, 367–69
  East Asian Corridor agricultural study, 393–95
  Electric Power Company, 428
  farming systems, large-scale mechanized, 289
  farmland, abandoned, 296
  fisheries yields for 11 major exploited taxa, 278
  food self-sufficiency rates, 163
  forest CO_2 sinks in Kochi Prefecture, 118–19
  forest CO_2 sinks in Shimokawa-cho, 117–18
  Green Fund, 428
  Japanese scallop aquaculture in Funka Bay, 140
  methane-fermented digested slurry, 178
  Pacific saury fishing fleets trajectories, 139
  Prefecture, 96, 117, 163, 275–76, 279
  reclamation era, 290
  Regional Development Bureau, 177, 182
  renewable energy sources, 427
  research in Urahoro region of Eastern Hokkaido, 363
  Shimokawa Town biomass town project, 429–30
  Shiretoko World Natural Heritage Site, 131, 141–42, 273–79
  structures of farming communities in, 394
  wind power plants, 428
  wind turbines built in, 427
Hokkaido University, 393, 427, 433
holocentricity, 338–41
Hopkins, Rob, 234–35, 237
human
  ecological systems, 343
  excrement in cities, 17
  excreta, 76–77
  force, 256–58, 260
  health risks, 64
  nature interface, 257–59
  nature relationships, 245–46, 248, 250–52
  population density, pattern of country-based, 104–5
  population density and annual production of cereals, 104–5
  population growth, food as restricting factor of, 3
  relationships with nature, 4
  society, 105–6, 256, 309, 328, 389
  view approach, 255–56
human activities
  biomass production, 16–17
  carbon dioxide emissions from, 23
  consequences of, 99
  fishing, intensive, 140
  food production, 16
  hunting and trapping of deer, 364
  Kyoto Protocol allows limited, 115
  livelihood and economic activities, 389
  material recycling and, 389
  natural ecosystems and, 389, 393
  NEMURO and, 137
  urban-industrial area, 16
human appropriation of NPP (HANPP), 99
Human Development Index (HDI), 35
hybrid cars, 425
hydrocarbon-yielding plants, 209, 211
hydroelectric power, 89
hydrogen plants, 205
hydrological cycle, 30
Hyogo Prefecture [Japan], 96

ice sheets, 433
IMBER. *See* Integrated Marine Biogeochemistry and Ecosystem Research (IMBER)
"Implementation Rules for Offsetting and Carbon Credit (J-VER)", 120
independence, food and energy, 2
index of sustainability, 425
Indian medical systems, 214
Indian Railways, 209
Indonesian decentralization policy, 262
Indonesian Minister of Agriculture, 261
industrial pollution, 51
industrial production, 437
infant mortality rates, 415–18
infectious diarrhoea, 421–22
infectious diseases, waterborne, 416–18, 422
inorganic nutrients, 100
inorganic soil, reductive, 106
insecticides, 64, 163, 187
institutional culture, 256–58, 260–62
insulation systems, 425
Integrated Marine Biogeochemistry and Ecosystem Research (IMBER), 140
Intergovernmental Panel on Climate Change (IPCC)
  about, 23, 132, 330, 426
  Fourth Assessment Report (AR4), 23, 113
  "Good Practice Guidance for Land Use, Land-Use Change and Forestry," 116
  Special Report on *Carbon Dioxide Capture and Storage*, 112–13, 115
  Special Report on Emissions Scenarios (SRES), 132
  Third Assessment Report (AR3), 23, 113
Intergovernmental Panel on Forests (ITTO), 112
internal combustion engine, 3
International Biochar Initiative, 76
International Organization for Standardization (ISO), 110
international trade, 64, 387
International Tropical Timber Organization (ITTO), 109
International Union for Conservation of Nature (IUCN), 130, 142, 271, 274–75
International Union of Forest Research Organizations (IUFRO), 409
Intruding Wastrel, 329, 343
intuition *(kansei)*, 407

IPCC. *See* Intergovernmental Panel on Climate Change (IPCC)
irrigation
  channel, 398–99
  district, Shali Nadi, 398–99, 401
  drip, 33
  farming and groundwater, 27
  fee, 400–401
  gravity, 398
  local circulation-type, 29
  methods, 397–98
  system, Khokana, 399–401
  systems, 3, 398
  water, 397
  water supply, shortage of, 399
ISO. *See* International Organization for Standardization (ISO)
ISO 14001 Environmental Management System, 110
ITTO. *See* Intergovernmental Panel on Forests (ITTO); International Tropical Timber Organization (ITTO)
IUCN. *See* International Union for Conservation of Nature (IUCN)
IUFRO. *See* International Union of Forest Research Organizations (IUFRO)
Iwate Prefecture [Japan], 163

Japan
  arts and culture, 353
  bubble economy, 355–56
  economic development, 346, 353
  energy self-sufficiency rate, 163
  GDP, 13
  impoverishment, regional, 350
  Kyoto, modernization of, 351
  Kyoto, urban centrality of, 349
  living standards, 13
  middle class, 349, 358
  national organic body, 350
  order, deconstruction of traditional, 349
  Osaka, economic centrality of, 349
  regional formation, 350–51
  regional impoverishment and income disparity, 346
  regional structure, 349
  special industrial zone, 355–56
  tourism, 351–53
  upheavals in 1960s, 347–49
  urban spaces, 354, 356–57
  welfare state, 350

Japan Bank for International Cooperation (JBIC), 381
Japan Biochar Association, 76
Japan Certification Center on Climate Change (4CJ), 120
Japan International Cooperation Agency (JICA), 78, 379, 402
Japan Society for the Promotion of Science, 389
Japan Verified Emission Reduction (J-VER)
  about, 119–21, 126–27
  carbon sink credits, 119, 125, 127
  carbon sinks, 119–21, 126–27
  carbon-offsetting system, 119–20, 125–26
  forest management, 119–21, 126–27
Japanese common squid *(Todarodes pacificus)*, 137
Japanese domestic carbon sinks, 126–27
Japanese sardine, 271–72
Japanese scallop culture, 138–40
Japanese–Chinese research project, 389–90
Japan's Forestry Agency, 409–11, 429
jatropha *(Jatropha curcas)*, 209–10
JBIC. *See* Japan Bank for International Cooperation (JBIC)
JICA. *See* Japan International Cooperation Agency (JICA)
J-VER. *See* Japan Verified Emission Reduction (J-VER)

K. *See* potassium (K)
Kenyah people, 257–61
Khokana Irrigation Association, 399–401
Khokana irrigation system, 399–401
Khokana Village Development Committee (VDC), 399
Khuchit Shonhor Co., Ltd. [Mongolia], 404
Khuchit Shonhor Food Market [Mongolia], 402–3
K-J method: morphological analysis, 313–14, 317
knowing systems
  paradigm shift to holocentric, 341–42
  systematic development and sustainability, 336–37
  systems approaches, hard and soft, 339
  three dimensions of, 340–41
  worldviews and nature of, 337–42
Kochi Prefecture [Japan], 96 118–19
Kusasenri (grassland landscape), 248

Kyoto Protocol
  Ad Hoc Working Group on Further Commitments for Annex I Parties, 125
  Article 3, paragraph 4, 115–16, 123–24
  carbon sink initiatives in Japan, 124
  clean development mechanism (CDM), 121, 435
  CO_2 capture by forests, quantitative evaluation of, 110
  emission reduction projects, incentives for, 108
  emissions trading, 120, 125
  forest carbon sinks and global warming, 108
  forest carbon sinks and UNFCCC, 114–15
  forests role in prevention of global warming, 110
  Japanese Forestry Agency, 429–30
  Japan's global warming countermeasures, 116
  Special Report on *Carbon Dioxide Capture and Storage*, 115
  urgent need for solutions, 436

land
  management, 59–60, 115
  ownership structures, 394
  tenure system, 259–60
  use planning, 57
land class
  in Asia Pacific, 52, 54
  biomes by, 56, 58
  urban and agricultural land use on different, 56–57
  urban expansion and, 59
Land Use, Land-Use Change and Forestry (LULUCF), 113, 125
Law Concerning the Promotion of the Measures to Cope with Global Warming (Global Warming Countermeasures Law) [Japan], 116
learning process, 332, 338
Lighthouse Project "Bioenergy Village," 184–93
light-water nuclear reactors, 23–24
The Limits to Growth, 33, 36
livestock
  excrement and biogas, 164–65
  excreta, 76, 81, 89

livestock (cont.)
  farming, modern, 10
  feed and self-sufficiency, 250
  manure and composting, 152
  raising, 390
  wastes, 38, 174
living environments, 311
living standards, 3, 9, 13, 390
local agriculture, collapse of, 18
Long-Term Renewable Energy Development Plan [China], 93
low-carbon future, 207
low-carbon society, 182, 234, 245, 426–27, 437
LULUCF. *See* Land Use, Land-Use Change and Forestry (LULUCF)

mad-cow disease (BSE), 388
magnesium (Mg), 189, 390
magnesium phosphate, 78
maize, 78, 91–93, 187–88, 259
malnutrition, 387
Malthusian trap, 20
marine
  ecosystem, 140, 142, 270, 275–76, 278
  food webs, 130
  social–ecological systems, 140
Marine Management Plan for the Shiretoko World Natural Heritage Site, 141–42, 276–77
Marine Working Group [Japan], 274
market brokers, 404
market economy in Mongolia, 401–2, 405
Marrakesh Accords, 115–16
material recycling, 389–91
maximum sustainable ecosystem services (MSES), 267–69, 278
MDGs. *See* Millennium Development Goals (MDGs)
mean trophic level (MTL), 270–72, 278
meat-processing factories, 403
mental culture, 256–58, 260
mental disease, 433
methane
  from animal manure, 427
  fermentation, 90, 153, 155, 160–61, 174, 180, 183
  fermented digested slurry, 178–79
  gas, 32, 174, 427
  greenhouse gas, 8
Mg. *See* magnesium (Mg)

michinoeki (roadside service stations), 375, 380–85
microbiological quality, 418
micro-grid heating, 226
Millennium Development Goals (MDGs), 47, 418
Millennium Ecosystem Assessment, 38, 330
Ministry of Agriculture, Forestry and Fisheries [Japan], 95, 180
Ministry of Economy, Trade and Industry (METI) [Japan], 119, 179
Ministry of Health, Labour and Welfare [Japan], 383
Ministry of New & Renewable Energy [India], 208
Ministry of Petroleum and Natural Gas (MoPNG) [India], 94
Ministry of the Environment (MOE) [Japan], 274–76, 425
Miyazaki prefecture [Japan], 383
MOE. *See* Ministry of the Environment (MOE) [Japan]
Mongolian Ministry of Food and Agriculture, 402
Mongolian Office of the Japan International Cooperation Agency, 402
monoculture, 4, 69–70
monsoon climate, 397
monsoon season, 28
Montreal Process meeting in Santiago, Chile, 111
MoPNG. *See* Ministry of Petroleum and Natural Gas (MoPNG)
mortality
  causes of, in developed and developing countries, 415
  rates, infant, 415–18
  rates in developing countries, 415
  rates in Mekong watershed, 418
  of zooplankton, 134
moso bamboo plantations, 68–72
MSES. *See* maximum sustainable ecosystem services (MSES)
MSW. *See* municipal solid waste material (MSW)
MTL. *See* mean trophic level (MTL)
Multiple Use Integrated Marine Management Plan [Japan], 276
municipal solid waste material (MSW), 212

mycorrhizal fungus, 81
mystical dimension, 339

N. *See* nitrogen (N)
N₂O. *See* nitrous oxide (N₂O)
Nagano Prefecture [Japan], 96, 249–50
National Biofuel Coordination Committee [India], 208
National Development and Reform Commission (NDRC) [China]
  Agriculture and Forestry Biomass Project of the National Long-Term Science and Technology Development Plan, 93
  Long-Term Renewable Energy Development Plan, 93
National Development Planning Agency [Indonesia], 261
national grid
  electricity fed into, 189
  hot water, 189
  purchase renewable energy using, 427
  sales of electricity to, 190
national income per capita, South Asian, 397
National Institute for Environmental Studies (NIES), 426
National Mission on Biodiesel [India], 208
National Policy on Biofuels [India], 208
National Spatial Plan [Japan], 350
natural
  control of pests, 63
  disasters, 57
  eco-systems, 329
  environments, 244–45, 310–11, 313, 407
  killer (NK) activity, 408
  resources, degradation of, 38
  resources deterioration, 387
nature of nature, 337, 339
nature therapy, 407–11
NDRC. *See* National Development and Reform Commission (NDRC)
NEMURO ocean ecosystem model, 130, 133–37
neoliberalism, 423
net primary production (NPP), 99–101
  civilization maintenance and high, 105+106
  country-based, 101–2
  geographical pattern of, 101
networking, synergistic, 2
networking of diverse communities, 2

New Economics Foundation [United Kingdom], 424
New Plum and Chestnut (NPC) movement, 377–78
NIES. *See* National Institute for Environmental Studies (NIES)
Niigata Prefecture [Japan], 96
nitrification, 45
nitrogen (N)
  atmospheric, 19
  balance for energy crops, 187
  biomass production and, 187
  charcoal neutralizing soil pH and increases, 78
  circulation, 390
  concentration in organic waste, 166
  consumption in Furano, 166
  cows contribute, 171
  cycle, 17, 42, 389–90
  cyclic usage of, for fertilization, 149, 152, 158
  from digestate of energy crops and manure, 189
  dioxide, 8
  dynamics in agro-ecosystems, 42–44
  fertilization, 42
  as fertilizer component, 32
  fertilizer self-sufficiency rates, 165
  fixation in soil, 78
  fixing bacteria, 80
  greenhouse gas and, 149
  inorganic, 44
  leaching, 43
  as major water pollutant, 38
  paddy fields, applied to, 165
  plant nutrient, 44
  production in Furano, 166
  recycled, 169, 171
  rice contributes, 171
  self-sufficiency rate, 167
  soil used for agroforestry and livestock-raising, 390
nitrogen oxides (NO$_x$), 201
nitrous oxide (N₂O), 42–46
NK. *See* natural killer (NK) activity
Non-Legally Binding Authoritative Statement of Principles for a Global Consensus on the Management, Conservation and Sustainable Development of All Types of Forests, 110

non-profit organizations (NPOs), 248, 285, 301, 313, 358
non-renewable energy resources, 21
North Pacific Marine Science Organization, Climate Change and Carrying Capacity Program (PICES CCCC), 133
North–South divide, 110
NO_x. See nitrogen oxides (NO_x)
NPC. See New Plum and Chestnut (NPC) movement
NPOs. See non-profit organizations (NPOs)
NPP. See net primary production (NPP)
nuclear
  energy, 7, 23
  power generation, 7, 23
  reactors, light-water, 23–24

oak charcoal, 81. See also charcoal
Obama, President, 179, 423–25, 428
obesity, 351
Obihiro City, Hokkaido, 285, 316
ocean ecosystem
  model, NEMURO, 130, 133–37
  protection of, 130
  structure and function of, 133
oceanic fish catch, annual, 20
OECD. See Organisation for Economic Co-operation and Development (OECD)
Ogallala Aquifer (U.S.), 15
oil
  palm plantations, 261–62
  peak, 235, 237–40, 242
  reserves, 14, 22
Oita Prefecture [Japan], 376
One Tambon, One Product (OTOP), 376, 379–80
"One Village, One Product" (OVOP) movement
  michinoeki (roadside service stations), 375, 380–85
  shokuiku (dietary education), 375, 383–85
  Tokyo International Conference on African Development (TICAD), 375
  Yokohama Action Plan, 375
ontology, 337, 339
Open Space Technology, 240
organic
  agriculture, 214–15
  carbon, 100–101, 103

compost, 39, 81–82
  farming and biochar, 82
  matter, anthropogenic trade in, 100
  matter, microbial degradation of, 45
  matter fixation by photosynthesis, 99
  Rankine cycle system, 228
  recycling, 215
  residuals and manure for energy production, 188
  soil and carbon (C), 101
  soil organic carbon (soil-C), 101, 103
  soil organic matter (SOM), 42, 44, 106
  substances, 8
  waste and nitrogen concentration, 166
organic farming
  with compost and biochar, 82
  environmental damage, 39
  large-scale, 39–40
  soil fertility, 39
  systems, 39
  techniques, 296
organic waste
  from agriculture, 164
  bioenergy, 17, 164–66, 171
  biogas (CH_4), 164–65, 209, 216
  from citizens, 164
  renewable energy and, 222
  two-year cycle of, 17
Organisation for Economic Co-operation and Development (OECD), 6
OTOP. See One Tambon, One Product (OTOP)
overfishing, 130, 132, 269–71, 278
OVOP. See "One Village, One Product" (OVOP) movement

P. See phosphate (P); phosphorus (P)
Pacific salmon (*Oncorhynchus* spp.), 131, 133–34
Pacific saury *(Cololabis saira),* 137
paddy fields, 27, 29, 33, 244, 247
parasitic diseases, 415
parasitoids, 63, 65
parasympathetic nerve activity, 408–9, 411
pathogenic bacteria, 416
pathogenic micro-organisms, 417–18
pathogens, waterborne, 422
peak oil, 235, 237–40, 242
PEFC. See Forest Certification (PEFC)
pelagic fish species, 270–71
pellet home-heating systems, 222, 225

INDEX   459

pellets
  biomass, 161, 164
  wood, 90–91, 169, 201, 222, 225, 232, 250
Peruvian anchovy, 271
pest control
  biological, 40, 63–71
  chemical methods of, 64
pest management, 63–64
pesticide(s)
  annual application of, 63
  poison water and soil, 38
  usage, 38
petroleum dependence, 88–89
petroleum-based civilization, 3
phosphate (P)
  charcoal neutralizing soil pH and
    increases, 78
  soil used for agroforestry and livestock-
    raising, 390
  world resources of, 32
phosphorus (P), 38
  in domestic waste water, recovery from,
    32
  as fertilizer component, 32
  greenhouse gas and, 149
  as indispensable for modern civilization,
    14
  as major water pollutant, 38
photovoltaic
  energy, 222
  power production, 179
  solar panels, 228
phytoplankton, 21
PICES CCCC. See North Pacific Marine
    Science Organization, Climate
    Change and Carrying Capacity
    Program (PICES CCCC)
pine charcoal, 81
pine wilting disease, 82
plant nutrients removal, 51
plant pathogens, 63
plantation(s)
  forests, 17
  moso bamboo, 66–72
  oil palm, 261–62
  rubber, 262
  on wasteland and biodiesel, 208
plutonium-239, 23–24
plutonium-241, 23
policy, plan or programme (PPP), 310
political democratization, 255

polyphagous control agents, 67
population growth rate of major South
    Asian nations, 397
Post Carbon Institute [U.S.], 234
potassium (K)
  digestate from energy crops and manure,
    189
  greenhouse gas and, 149
  for liquid fertilizer, 189
  rice husk charcoal, 77
  soil used for agroforestry and livestock-
    raising, 390
poverty
  alleviation projects, 213
  in Asia, 387
  deforestation and, 59–60
  in developing countries, 416
  elimination, 389
  environmental degradation and, 387–88
  global warming and, 433
  human resources, lack of, 416–17
  infrastructure and education, lack of,
    417
  Millennium Development Goals, 47
  QOL optimization and, 415
  reduction, 375, 388, 397
  relative, 347
  in South Asia, 397
  of tropical regions, 106
PPP. See policy, plan or programme (PPP)
prefrontal cortex activity of brain, 408
preventive medicine, 409
principle of polluter (user) pays, 30
production areas of Earth, 16
Programme for the Endorsement of Forest
    Certification Schemes [Austria], 220
protective areas of Earth, 16
public health, 10, 13, 285
public transportation networks, 425
pumped-storage hydroelectricity, 205

Qibainongxiang village doline
  material recycling in populated, 390–91
  natural ecosystem and human activities,
    393
  preservation of water and soil, 391–92
  restoration of forest soil and ecology,
    392
  vegetation landscape in Karst region
    ecology, 392–93
Quality of Life (QOL) index, 414

rainforest destruction, 109
rangelands, 54
rate of energy consumption, 21
recycling
  agricultural products between humans and livestock, 164
  biological resource, 250
  biomass in *satoyama* landscapes, 250
  biomass resources, 249
  culture, 303
  facilities, 303
  organic, 215
  oriented legislation, 213
  policies, 291
  processes of water, 30
  rate, 290
  resources, 117
  re-use and, of wastes, 213
  society, 97, 285, 390, 426
  system, material, 390–91
  system for citizens, 291, 295
  systems, 304
  technology, resources, 391
  of urban wastes, 212
  waste disposal and, 290–91, 295, 300, 303, 305–6
  zones, regional, 251
REDD. *See* Collaborative Programme on Reducing Emissions from Deforestation and Degradation in Developing Countries (REDD)
reforestation, 81–82, 115, 117, 124, 126
regional self-sufficiency, 4
regional sovereignty, 17
regulatory reform, 423
relaxation, 407–9
remote sensing, 130, 137
renewable energy
  as alternative to fossil fuel energy, 89
  in Austria, 221–33
  bioethanol, 92
  biofuel, 97
  biofuels and biomass, 207
  from biomass, 82, 89, 220
  convincing people about, 192
  in Denmark, 194–205
  development of, 24, 89
  eco-city programme, 214
  economic incentive for, 203
  electrical power, converted into, 89
  energy self-sufficiency, 169
  environmental emission reduction, 201–2
  examples of, 89, 195
  fossil energy replacement with, 188
  fossil fuel depletion *vs.,* 164
  Green New Deal, 425–28
  heat and electricity produced by, 201
  to heat public buildings, 226
  heating systems, district, 197–98
  investment as part of Green New Deal policies, 179
  local acceptance of, 203
  natural conditions of location, 89
  people involvement with, 186
  self-sufficiency in, 201, 204
  sources, local, 186
  sources, natural, 4
  systems, alternative, 191
  systems, individual, 198–99, 201
  systems, local, 189
  technical and social implementation of, 185
  tourism, 202
  Transition Network movement, 235
  wind turbines, land-based, 195, 199–200
  wind turbines, off-shore, 200–201, 203
Renewable Portfolio Standard (RPS) law, 178, 182, 427
Research for the Future Program, 389
rice
  cultivation, 76–77, 95, 378, 398
  husk charcoal, 76–79, 82 (*See also* charcoal)
  production, 102, 104, 163, 289, 394, 398
  terrace in Bali, Indonesia, 251
risk management
  in fisheries, 130
  health, 413
river channel ecosystems, 29
River Structure Working Group [Japan], 274
roadside service stations *(michinoeki),* 375, 380–85
rooftop gardening, 215
root nodule bacteria, 78, 80
root nodules of soybean plants, 80
RPS. *See* Renewable Portfolio Standard (RPS) law
rubber plantations, 262

S. *See* sulphur (S)
Saitama Prefecture [Japan], 180

salmon
  chum *(O. keta),* 131–32
  hatchery, 131
  Pacific *(Oncorhynchus* spp.), 131, 133–34
  populations, 132
salmonids, wild populations of, 274
salt accumulation in fields, 27
salt damage, 27, 435
Samsø renewable energy island [Denmark]
  background, 194–95
  backup from conventional power plants, 205
  ecotourism, 202, 204
  energy status, changes in, 197
  environment, improvement in, 201–2
  fuel oil consumption, 201
  heating and electrical power, 204–5
  heating plants, district, 197–98
  job creation, 204
  local support and ownership, 203–4
  map of, 196
  renewable energy, 194–96
  renewable energy share, 201
  renewable systems, individual, 198–99
  subsidies, 203
  wind turbines, land-based, 195, 199–200
  wind turbines, offshore, 200–201, 203
sanitary standards, 403
sanitation, 417, 422
sanitation regulations, 383
Santiago Declaration, 111
satellite view approach, 255–56
satoyama
  coexistence of humans and nature, 4
  concept, 245–48
  initiative, 245
  landscapes, 244–46, 248–50, 252
  traditional Japanese forests, 71
  village-vicinity mountains, 17
  woodlands, 246
SBI. *See* Sustainable Biosphere Initiative (SBI)
school meals system [Japan], 383
Scientific Council (SC) [Japan], 274–75
scientific knowledge, 286, 330, 332
SCMEA. *See* sustainable conservation management based on the ecosystem approach (SCMEA)
SEA. *See* strategic environmental assessment (SEA)
sea ice cover in Arctic, 433

sea surface temperature (SST), 138–39
sea-level rises, 27
Sea-viewing Wide Field-of-view Sensor (SeaWiFS), 139
seawater, 24, 29
SeaWiFS. *See* Sea-viewing Wide Field-of-view Sensor (SeaWiFS)
second law of thermodynamics, 8
Second National Biodiversity Strategy [Japan], 244
Seventh Conference of the Parties (COP 7), 115
sewage treatment, 79
sewerage systems, 28
SGEC. *See* Sustainable Green Ecosystem Council (SGEC)
Shali Nadi irrigation system, 398–99, 401
Shimokawa-cho Prefecture, 117
Shinano Town, Nagano Prefecture, 96, 163
shinrin therapy stations in Okutama Town, 409–10
shinrin-yoku (forest bathing), 410–11
Shinshu Lumberjack Training Course, 429
Shiretoko World Natural Heritage Site, 131, 141–42, 273–79
shizen-nou farming system, 40
shokuiku (dietary education), 375, 383–85
sika deer. *See also* Ainu people
  Ainu people and, 364–71
  brown bear *(Ursus arctos),* 363–64, 366, 370
  Conservation and Management Plan for Sika Deer, 362
  Council on Sika Deer Management Policy, 362
  deer hunting with guns, 370–71
  deer spirit-sending ceremonies, 369
  Ezo wolf *(Canis lupus hattai),* 363, 369
  in Hokkaido, 362–64
  Hokkaido Deer Hunting Regulations, 365, 367–69
  Hokkaido Development Commission, 367–69
  hunting and trapping of, 364
  Meiji Restoration, 364–65
  population and cost of damages to agriculture and forests, 362–64, 369
  taxable licence system, 367
Sim-CYCLE, 100–101
skipjack tuna *(Katsuwonus pelamis),* 137
SO_2. *See* sulphur dioxide (SO_2)

social–ecological systems, 140, 330, 335
sociocultural environment, 336
soil
  acidity, 329
  biological function, 44–45
  degradation, 51, 59, 388
  fertility, 39
  land quality map for Asia Pacific, 52–53
  microbial flora, 80
  performance, 52
  quality, 4, 51
  resilience, 52
  resources, anthropogenic demands on, 59
  resources, human impact on, 51
  resources and biome data sets, 52–54
  resources and biomes, 54–59
  salinity, 57
  water content and N_2O and N_2, 46
soil organic carbon (soil-C), 101, 103
soil organic matter (SOM), 42, 44, 106
solar
  energy, 7, 21, 23, 426
  energy generation, 425
  energy input, 7
  panels, photovoltaic, 228
  power, 89, 426
  radiation, 8
  thermal power, 222
solid waste management programme, 215
SOM. *See* soil organic matter (SOM)
sorghum, 78, 93–94
South Asian nations, 397
southern bluefin tuna *(Thunnus maccoi)*, 271
soybean, 78–80
Special Measures Law Concerning the Use of New Energy by Electric Utilities [Japan], 427
Special Report on Emissions Scenarios (SRES), 132
species extinction, 18–19
spiritual
  beliefs and values, 340
  benefits, 267
  needs, 111
  sensitivity, 341–42
  sustainability, 214
SRES. *See* Special Report on Emissions Scenarios (SRES)
SST. *See* sea surface temperature (SST)
standing biomass, 268–69

starvation, 364, 387
stock dynamic models, 267
strategic environmental assessment (SEA), 309–12
stress hormone (cortisol), 408–9, 411
sulphate, 44
sulphur (S)
  desulphurization systems, 176
  desulphurized biogas, 178
  soil used for agroforestry and livestock-raising, 390
sulphur dioxide (SO_2), 201
supra-systems, 336
sustainability governance
  brainstorming (BS), 311, 313–14, 317
  case study: environmental improvement planning of river basin, 313–24
  Choquet integral, 319
  concept of, 284–86, 310–11
  conjoint analysis, 312–14, 321, 324
  contingent valuation method (CVM), 311, 314, 319–21, 324
  decision-making process, 309–10
  decison-making using analytical, 311–13
  fuzzy integral method: multi-criteria analysis, 313–14, 324
  fuzzy structural modelling (FSM), 311, 314
  K-J method: morphological analysis, 313–14, 317
  at local level, 288–89, 295, 306
  policy, plan or programme (PPP), 310
  process, 289
  strategic environmental assessment (SEA), 309–12
sustainability science
  about, 184–85, 285, 328
  biological diversity, 333–34
  characteristics of, 333
  definitions and core questions, 330–32
  ecological science *vs.*, 342–44
  ecology and, 329–30
  ecology and, from systems perspective, 336–37
  global change, 333–34
  journals, 335
  manuscripts, 335
  research priorities and common issues, 333
  Sustainable Biosphere Initiative (SBI), 333–34
  sustainable ecological systems, 333–35

sustainable (sustainability)
  agricultural development, 51, 77, 388, 393, 405
  agro-ecosystems, 40
  as balance between human society and environment, 309
  biological production, 389
  development and human survival, 34
  ecological systems, 333–35
  economic, 285
  farming, 40–41, 47
  farming community, traditional, 388
  fisheries, 134, 278
  forest management, 108, 110–11, 113–15, 117, 122–23, 125–27
  index of, 425
  indicators, 311
  land management, 60
  marine resources in marginal seas, 142
  multi-sectoral policies, requires, 288
  as a normative concept, 285
  regional development, 286
  regional society, 284
  society (societies), 14, 88–89, 204–5, 245, 251, 426
  society, building, 285–86
Sustainable Biosphere Initiative (SBI), 333–34
sustainable conservation management based on the ecosystem approach (SCMEA), 132–34
Sustainable Green Ecosystem Council (SGEC), 110
swidden agriculture, 52, 258–61
sympathetic nerve activity, 408
synergistic networking, 2
synthetic planning, 315

Taj Eco-City Project [India], 213
Task Force on Forests and Human Health (ForHealth), 410–11
taxation system for the maintenance of forests [Japan], 118
TBOs. *See* tree-borne oilseed crops (TBOs)
Technical Research Association for Multiuse of Carbonized Materials (TRA), 79
technocentric worldview, 337, 341
techno-stress, 407
temperate ecosystem, 8

10-year Action Plan on the Mitigation of Global Warming by Forest Carbon Sinks [Japan], 116
Tenth Conference of the Parties to the Convention on Biological Diversity (COP 10), 252
terrestrial biosphere, 52
TFAP. *See* Tropical Forestry Action Plan (TFAP)
Thai community centres, 381–82
"Think globally, act locally," 254–55
Thirteenth Conference of the Parties (COP 13), 124
TICAD. *See* Tokyo International Conference on African Development (TICAD)
tidal power, 89
Tokyo International Conference on African Development (TICAD), 375
Tokyo Metropolitan Government, 410
Tokyo Prefecture [Japan], 96
TOREDAS. *See* Traceable and Operational Resource and Environment Data Acquisition (TOREDAS)
total allowable catch (TAC), 271–73
"Total Project for *Shinrin* Therapy," 409
tourism. *See also* ecotourism
  bioenergy, 230
  Furano City, 299, 301, 303–6
  Japan, 351–53
  promotion venture, 351–53, 381–82
  renewable energy, 202
TRA. *See* Technical Research Association for Multiuse of Carbonized Materials (TRA)
Traceable and Operational Resource and Environment Data Acquisition (TOREDAS), 137–38
traditional knowledge, 352, 370
transesterification plant, 209
transformation technologies, 153
The Transition Handbook, 237, 240
Transition Initiatives, 234–36, 240, 242
Transition Initiatives Primer, 236–38
Transition Network movement
  Energy Descent Plan, 241–42
  manual with 12 key steps, 236–37
  sixteen criteria, 235–36
  towns, designated, 237
  twelve steps of Transition, 238–42
Transition Town (TT), 235–37

Transition Training, 235
tree planting, 4
tree thinning, 116–17, 122, 126, 429
tree-borne oilseed crops (TBOs), 209
tropic regions, land-use planning in, 106
tropical
  development strategy, 105–6
  and extratropical environments, 106
  forest timber trade, 109
  rainforests, 34, 101, 109
  and subtropical forests, 112
  zone, 17
Tropical Forestry Action Plan (TFAP), 109
TT. *See* Transition Town (TT)
tuna *(Thunnus* spp.)
  albacore *(Thunnus alalunga)*, 137
  bluefin *(T. thynnus)*, 130
  farming in Japan, 21
  industrial fishing of, 270
  overfished species, 130, 271, 273
  skipjack *(Katsuwonus pelamis)*, 137
  southern bluefin *(Thunnus maccoi)*, 271
  stock decline, 270–71
typhoons, 28

UN. *See* United Nations (UN)
UNCED. *See* United Nations Conference on Environment and Development (UNCED)
UNDP. *See* United Nations Development Programme (UNDP)
UNESCO. *See* United Nations Educational, Scientific and Cultural Organization (UNESCO)
United Nations (UN)
  city definition, 15
  Collaborative Programme on Reducing Emissions from Deforestation and Degradation in Developing Countries (REDD), 108, 124–26
  Declaration of Human Rights, 236
  Forum on Forests, 111, 125
  Green Economy Initiative, 425
  Human Development Index (HDI), 35
  Intergovernmental Panel on Forests (ITTO), 112
United Nations Conference on Environment and Development (UNCED), 109–10
  Agenda 21 document, 110, 184
  in Rio de Janeiro in 1992, 184
  Statement of Forest Principles of UNCED, 110–11, 123
United Nations Development Programme (UNDP), 109, 379
United Nations Educational, Scientific and Cultural Organization (UNESCO), 142, 274–75
United Nations Framework Convention on Climate Change (UNFCCC), 114–15, 124–25
United States, 14, 18
uranium resource depletion, 24
uranium-238, 24
urban
  expansion, 59
  industrial areas of Earth, 16, 34
  land use, 57
  organic agriculture, 215
  spaces, 354, 356–57
  waste recycling, 212–13
  water resources, conserving, 215
urbanization, 6, 50–51, 57
usufruct rights, 260

Village Development Committee (VDC) [Khokana], 399

walleye pollock fisheries, 273–78
waste management, 59, 213, 215
wastewater
  concentrated domestic, 32
  domestic, 32
  domestic, phosphorus in, 32
  reuse of, 25
  systems, 32
  treatment, 28
  treatment technologies, 416
water
  access to safe and clean, 415
  consumption, 14–15, 25, 32–33, 38
  contamination, 15, 33, 390, 430
  culture system, 76
  demand, 25, 31, 185
  demand for, 24–27, 31, 185
  districts, 30–32
  drinking, in lower Mekong watershed, 418–20
  erosion, 59
  global warming and thermal expansion of, 27
  heaters, high-efficiency, 426

international conflicts over, 32
management, 3, 45, 57, 214
metabolism, 30, 32–33
pollution control in river basins, 28
purification, 50, 79
recycling, 30
resource conservation, 28
resources, conserving urban, 215
resources, exhaustion of, 29
resources per capita, 26–27
shortages, 26–27, 399, 433–35
systems, modern centralized, 29
treatment, 29, 416, 418, 420–21
treatment technologies, 416
treatment technology, 32
unsafe drinking, 415
waterborne infectious diseases, 416–18, 422
waterborne pathogens, 422
waterlogging, 57
weather conditions, extreme, 28
Westernized diet and lifestyle, 384
wetland destruction, 388
wheat, 16, 91–93, 164–66, 178, 180–81, 187, 212
wildlands, 54
willingness to pay (WTP), 320–21, 324
wind
  energy, 195, 222
  generated power, 89
  power, 23, 182, 194, 426–27
wind turbines
  community, 428
  electricity production from, 201–3, 205
  Hokkaido, 428
  land-based, 195, 199–201
  offshore, 200–201, 203
  public support for, 203–4
  on Samsø, 197, 201, 203–4
winter potatoes, 398–99
wood
  biomass, 90–91
  chips, 185, 197–98, 222, 227–29, 232, 250
  demand for energy production, 225
  as a fuel, 51
  peat deposition, 106

pellets, 232
waste, 151
woody
  biomass, 153, 157, 164, 187, 211, 222
  charcoal, 77–81
World Bank, 109, 382
  Forest Carbon Partnership Facility (FCPF), 125–26
world food production, 63
world population
  in 2030, 38
  in 2080, 38
  in 2100, 50
  current, 18, 20, 38, 50, 99
  demographic decline, 10
  as destabilizing factor in future, 11
  of developing countries, 7, 11
  grain production and, 19
  growth in, 9–10
  of mature countries, 7
  patterns, 11–12
  in twentieth century, growth in, 14
  in year 2100, 9
World Trade Organization (WTO), 18, 393
World Water Forum, 28
worldview perspectives, 337–39, 343
WTO. *See* World Trade Organization (WTO)
WTO Hong Kong Ministerial Conference, 376
WTP. *See* willingness to pay (WTP)

Xinjiang-Uygur Autonomous District [China], 89

Yamagata Prefecture [Japan], 163
Yokohama Action Plan [Japan], 375
Yunnan [China], 388

Zero Garbage Town (ZGT) [India], 213
ZGT. *See* Zero Garbage Town (ZGT) [India]
zooplankton, 21
Zushi-Onoji Historical and Natural Environmental Conservation Area, 246

ISBN 978-92-808-1180-3 • paper •
488pp • US$37.00

Sustainability Science
A Multidisciplinary Approach

Edited by Hiroshi Komiyama, Kazuhiko Takeuchi, Hideaki Shiroyama and Takashi Mino

Sustainability Science

Contents overview:

Sustainability science: Building a new academic discipline
Hiroshi Komiyama and Kazuhiko Takeuchi

The connections between existing sciences and sustainability science
Contributors include: Yuya Kajikawa, Hiroshi Komiyama, Riichiro Mizoguchi, Kouji Kozaki, Osamu Saito, Terukazu Kumazawa, Takanori Matsui, Takeru Hirota and Kazuhiko Takeuchi

Concepts of "sustainability" and "sustainability science"
Contributors include: Motoharu Onuki, Takashi Mino, Masaru Yarime, Kensuke Fukushi and Kazuhiko Takeuchi

Tools and methods for sustainability science
Contributors include: Hideaki Shiroyama, Hironori Kato, Mitsutsugu Hamamoto, Masahiro Matsuura, Nobuo Kurata, Hideyuki Hirakawa, Makiko Matsuo and Hirotaka Matsuda

The redefinition of existing sciences in light of sustainability science
Contributors include: Akimasa Sumi, Hiroyuki Yohikawa, Mitsuru Osaki, Gakushi Ishimura, Magan Bailey, Takamitsu Sawa, Jin Sato, Makio Takemura and Kazuhiko Takeuchi

Education
Contributors include: Mitsuhiro Nakagawa, Michinori Uwasu, Noriyuki Tanaka, Makoto Tamura, Takahide Uegaki, Hisashi Otsuji, Harumoto Gunji, Motoharu Onuki, Takashi Mino, Akihisa Mori, Michinori Kimura, Keishiro Hara, Helmut Yabar, Yoshiyuki Shimoda, Nobuyuki Tsuji and Yasuhiko Kudo

Building a global meta-network for sustainability science
Kazuhiko Takeuchi

Climate Change and Global Sustainability
A Holistic Approach

Edited by Akimasa Sumi, Nobuo Mimura and Toshihiko Masui

Sustainability Science

Contents overview:

Introduction: From climate change to global sustainability
Akimasa Sumi and Nobuo Mimura

Structuring knowledge of climate change
Ai Hiramatsu

Communicating climate change risk
Seita Emori

Climate change impacts and adaptation
Contributors include: Nobuo Mimura, Yasuaki Hijioka, So Kazama, Hiroyuki Kawashima, Yasuhiro Yamanaka, Masahiko Fujii, Hisamichi Nobuoka, Satoshi Murakami and Makoto Tamura

Designing climate policy
Contributors include: Hiroshi Hamasaki, Tatsuyoshi Saijo, Seiji Ikkatai, Shell International BV and Shunji Matsuoka

Transformation of social systems and lifestyles
Contributors include: Keisuke Hanaki, Tokuhisa Yoshida, Hideo Kawamoto

Integration of a low-carbon society with a resource-circulating and nature-harmonious society
Toshihiko Masui

Future vision towards a sustainable society
Akimasa Sumi

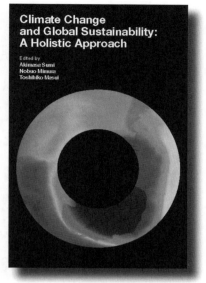

ISBN 978-92-808-1181-0 • paper •
325pp • US$35.00

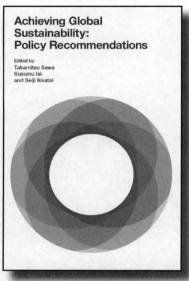

Achieving Global Sustainability
Policy Recommendations

Edited by Takamitsu Sawa, Susumu Iai and Seiji Ikkatai

Sustainability Science

Contents overview:

Introduction,
Takamitsu Sawa

Global sustainability
Contributors include: Michinori Uwasu, Kazuhiro Ueta and Takamitsu Sawa

Paradigm shift of socio-economic development
Contributors include: Kazuhiro Ueta, Takashi Ohshima and Masayuki Sato

ISBN 978-92-808-1184-1 • paper • 375pp • US$37.00

Strategies for sustainable society
Contributors include: Seiji Ikkatai, Satoshi Konishi, Shiro Saka, Akihisa Mori, Kosuke Mizuno, Haris Gunawan and Yukari Takamura

Adaptation for environmental change
Contributors include: Hans-Martin Füssel, Susumu Iai and Jiro Akahori

Policy recommendations towards global sustainability
Contributors include: Takamitsu Sawa, Kazuo Matsushita and Seiichiro Hasui

Establishing a Resource-Circulating Society in Asia
Challenges and Opportunities

Edited by Tohru Morioka, Keisuke Hanaki and Yuichi Moriguchi

Sustainability Science

Contents overview:

Introduction: Asian perspectives of resource-circulating society – Sound material metabolism, resource efficiency and lifestyle for sustainable consumption
Tohru Morioka

The Asian approach to a resource-circulating society – Research framework, prospects and networking
Contributors include: Yasushi Umeda, Yusuke Kishita, Tohru Morioka, Terukazu Kumazawa, Takanori Matsui, Riichiro Mizoguchi, Keishiro Hara, Hiroyuki Tada, Helmut Yabar and Haiyan Zhang

ISBN 978-92-808-1182-7 • paper • 375pp • US$37.00

Initiatives and practices for a resource-circulating society
Contributors include: Motoyuki Suzuki, Masao Takebayashi, Tsuyoshi Fujita, Rene van Berkel, Yong Geng, Xudong Chen, Kunishige Koizumi, Weisheng Zhou and Yuichi Moriguchi

Characterization and local practices of urban-rural symbiosis
Contributors include: Kazutoshi Tsuda, Toyohiko Nakakubo, Yasushi Umeda, Tohru Morioka, Mitsuru Osaki, Nobuyuki Tsuji, Toshiki Sato, Noriyuki Tanaka, Youji Nitta, Hiroyuki Ohta, Tasuku Kato, Ken'ichi Nakagami, Hironori Hamasaki, Myat Nwe Khin, Ai Hiramatsu, Yuji Hara and Keisuke Hanaki

Biotic resources utilization and technology development
Contributors include: Shinya Yokoyama, Kiyotaka Saga, Toshiaki Iida, Takashi Machimura, Akio Kobayashi, Yoshihisa Nakazawa, Keisuke Hanaki, Noboru Yoshida, Tohru Morioka and Yugo Yamamoto

Exploring opportunities for sustainable city-region design
Contributors include: Tohru Morioka, Shuji Kurimoto and Yugo Yamamoto

Conclusion: Challenges to a resource-circulating society in Asia
Tohru Morioka, Keisuke Hanaki and Yuichi Moriguchi

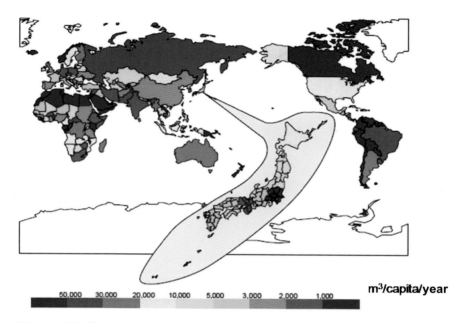

Figure 1.2.9 Water resources per capita.
Source: Ministry of Land, Infrastructure, Transport and Tourism, Water Resources Department (2003).
Note: Please see page 26 for this figure's placement in the text.

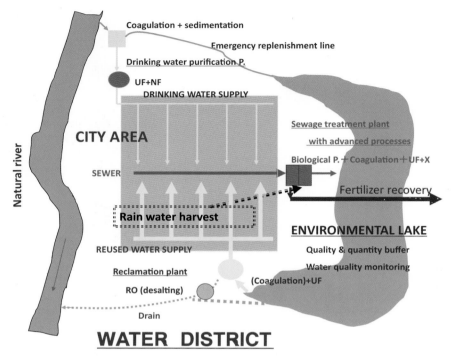

Figure 1.2.10 A new water system in a large city.
Sources: Tambo (1976, 2002, 2010).
Note: Please see page 31 for this figure's placement in the text.

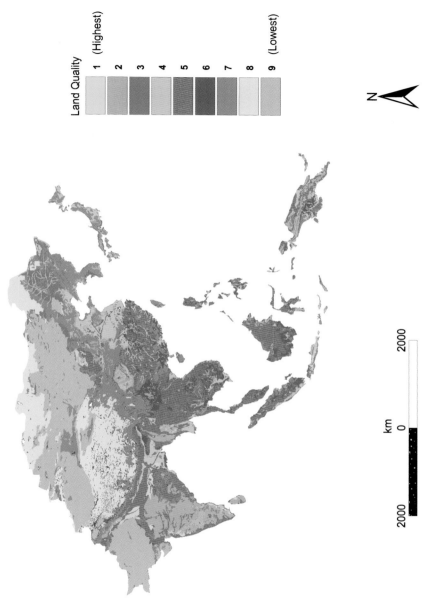

Figure 2.2.1 Inherent soil/land quality map for the Asia Pacific.
Source: Based on Beinroth et al. (2001).
Note: Please see page 53 for this figure's placement in the text.

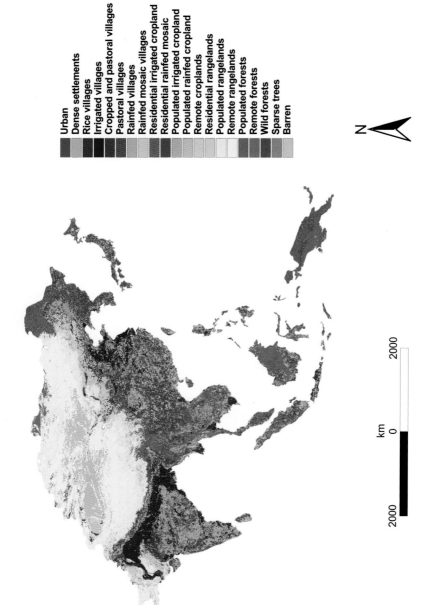

Figure 2.2.3 Biome map for the Asia Pacific.
Source: Based on Ellis and Ramankutty (2008).
Note: Please see page 55 for this figure's placement in the text.

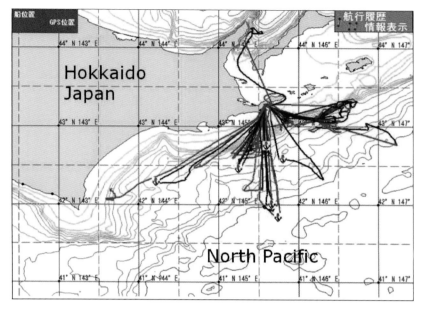

Figure 3.4.7 Trajectories of Pacific saury fishing fleets showing a direct approach to fishing grounds from the mother port off eastern Hokkaido, Japan, for the period 1 September to 30 October 2006.
Source: Saitoh et al. (2009).
Note: Please see page 139 for this figure's placement in the text.

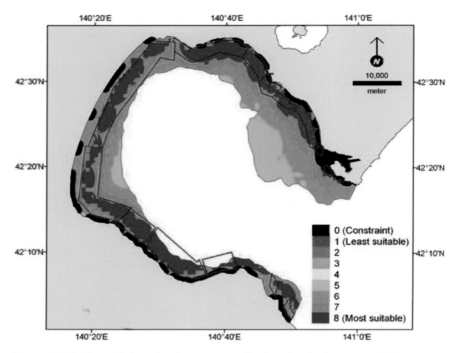

Figure 3.4.8 Overall site selection map, masked to depths in excess of 60 metres, for Japanese scallop aquaculture potential in Funka Bay, south-western Hokkaido.
Notes: 1. Red polygons represent existing Japanese scallop aquaculture area.
2. Please see page 140 for this figure's placement in the text.

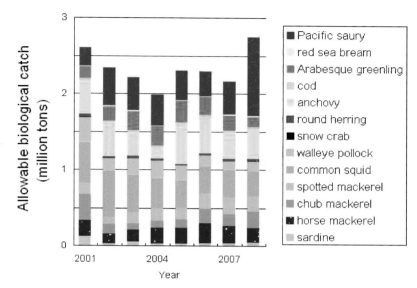

Figure 5.5.4 Total allowable biological catch for several major species.
Notes: 1. From top to bottom of each bar: Pacific saury, red sea bream, Arabesque greenling, cod, anchovy, round herring, snow crab, walleye pollock, common squid, spotted mackerel, chub mackerel, horse mackerel and sardine.
2. Please see page 273 for this figure's placement in the text.
Source: Fisheries Research Agency, Japan, unpublished.